Waste Treatment Processes in Environmental Engineering

The Editors

Dr. S.N. Kaul has obtained his bachelor and master degrees from IIT, Kanpur and was awarded Ph.D. degree from the University of New Castle Upon Tyne, UK. Dr. Kaul joined CSIR-National Environmental Engineering Research Institute [NEERI] Nagpur in 1970 and worked for over three decades. Dr. Kaul superannuated as Acting Director and Director Grade Scientist from NEERI and was also associated with MIT College of Engineering, Pune as its Principal.

Dr.Kaul has published over 350 Papers in National and International journals of repute beside authorship of over 30 Books on Environmental Engineering and Science, 10 Patents and 150 conference papers. Dr. Kaul has designed and operated several waste treatment plants in India. Dr. Kaul has supervised over a dozen of Ph.D. students and 40 M Tech students. In addition, Dr. Kaul was responsible for organizing several National and International Conferences and has traveled extensively in India and aboard concerning various R and D programs. Dr. Kaul has received several National and International awards and is expert member in importants National committees.

Dr. D.R. Saini obtained his B Tech and M Tech from HBTI,Kanpur and PhD [Chem.Engg.] from IIT-Delhi. Presently working as Principal Scientist at CSIR-NCL, Pune. Dr. Saini is co-author of 8 books, six chapters and published seventy papers in international journals. He has guided three PhD students and few M.Tech. students.

Dr. B.N. Rai is presently working as Professor in the Deapartment of Chemical Engineering and Technology, IIT (BHU), Varanasi. Dr. Rai has done B.E. (Chem. Eng.) from Bangalore University and M.Tech. (Chem) and Ph.D. from Banaras Hindu University. He has more than 20 years of teaching, research and administrative experiences in various capacities. He has more than 40 publications in international journals, book chapters, conferences, etc. and delivered invited key note lectures in various conferences. He has guided more than 20 M.Tech. and two Ph.D. students. Dr. Rai was awarded by the Ministry of Non-Conventional Energy, Govt of India for writing a book in Hindi on fundamental topics. His research interests include Environmental Biotechnology, Industrial Pollution and Control. He has also organized international conferences, refresher courses and workshops on waste management. Besides above, he has served the Universitiy in various capacities such as President, IT Gymkhana, Coordinator, National Youth Festival, etc.

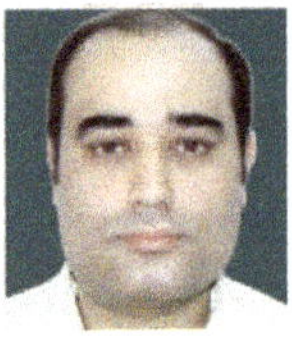

Er. Prateek Kaul has obtained M.Tech. from V.NIT, Nagpur. Er. Kaul travelled respectively in India and abroad. Er. Kaul has several research publications, books and patents to his credit. Er. Kaul has worked in several multinational industries and is presently associated with Symantec (USA), Pune.

Waste Treatment Processes in Environmental Engineering

Volume 1

Editors

Prof. (Dr.) S.N. Kaul
Principal
MIT College of Engineering, Pune
Former Acting Director and Director-Grade
Scientist, NEERI [CSIR], Nagpur, M.S.

Dr. D.R. Saini
Scientist
National Chemical Laboratory [CSIR],
Pune, M.S.

Dr. B.N. Rai
Professor, Department of Chemical Engineering & Technology,
Indian Institute of Technology (BHU),
Banaras Hindu University, Varanasi, U.P.

Er. Prateek Kaul
Sr. Manager,
Symantec Corporation (USA), Pune

2016

Daya Publishing House®
A Division of
Astral International Pvt. Ltd.
New Delhi – 110 002

Waste Treatment Processes in Environmental Engineering
(4 Volumes Set)
Volume 1: Page 0001-0306
Volume 2: Page 0307-0628
Volume 3: Page 0629-0908
Volume 4: Page 0909-1184

Cataloging in Publication Data—DK
Courtesy: D.K. Agencies (P) Ltd. <docinfo@dkagencies.com>

Waste treatment processes in environmental engineering / editors, Prof. (Dr.) S.N. Kaul, Dr. D.R. Saini, Dr. B.N. Rai, Er. Prateek Kaul.
4 volumes cm
Includes indexes.

ISBN 9789351309116 (Vol. 1)
ISBN 978-93-5124-372-4 (International Edition)

1. Sewage—Purification—India. 2. Sewage—Purification. 3. Environmental engineering—India. 4. Environmental engineering. I. Kaul, S. N., editor.

DDC 628.30954 23

Published by : **Daya Publishing House®**
A Division of
Astral International Pvt. Ltd.
– ISO 9001:2008 Certified Company –
4760-61/23, Ansari Road, Darya Ganj
New Delhi-110 002
Ph. 011-43549197, 23278134
E-mail: info@astralint.com
Website: www.astralint.com

Laser Typesetting : **Classic Computer Services**, Delhi - 110 035

Printed at : **Thomson Press India Limited**

Dedicated to the Sacred Memory of

Sri Sri Sarda Maa

&

Swami Vivekananda

Preface

The Editors of the present edition on Waste Treatment Processes in Environmental Engineering, have been associated with various educational institutes, research organizations, universities etc. It was felt that a comprehensive selective topics in Environmental Engineering concerning India with Indian Experiences from various part of the India with Indian data from various Indian experts must be collected to disseminate the authentic information to our readers.

Text books and reference books are already available in the market but most often they relate to information and experiences of the western world, which is not relevant to Indian conditions and economics. The editors have been contacting various experts throughout the country in order to present the Indian experiences on which the contributing authors have been working. Often the good and invaluable work gets published in foreign journals which are expansive and, therefore,beyond the reach of most of the readers to acquire. To reduce this gaps the editors have made some efforts in this direction, and the editors feel that readers will appreciate this and try to benefits from this compilation. It is also primarily with this view to make aware Indian readers about the work of Indian experts and their experiences.

Waste Treatment Processes in Environmental Engineering is a compendium of knowledge concerning Environmental Issues from several experts who are associated with prestigious institutes in India, *viz.*: IITs, CSIR, NITS, Central Universities, Pollution Control Board. This book endeavors to explore various important topics of immediate concern to all our stake holders, especially students, researches, practitioners, teachers, NGOs etc. This book also introduces the reader about areas of research of the experts in India.

Economic progress is directly connected with environment and ecology. Material progress goes linear while environmental degradation follows geometric progression. Therefore, environment and ecology is the rate determining step in the progress of the country. This aspect is extensively exhibited in this book.

Emphasis has been laid on environmental pollution and control and the editors feel that these this specific book will be immensely useful to all concerned with Environmental and Ecology encompassing all the three component of Environment, *viz.*: Air, Water and Solid Waste besides EIAs, cleaner technology, problem associated with implementation including environmental education etc.Finally, the editors would like to thanks all the contributing authors for accepting our invitation to contribute their valuable research and development work for our readers and editors would also like to thanks them for the time and efforts devoted while compiling their work.

The trust reposed by the Astral International (P) Ltd., New Delhi is thankfully acknowledged.

Prof. (Dr.) S.N. Kaul

Dr. D.R. Saini

Prof. (Dr.) B.N. Rai

Er. Prateek Kaul

Contents

Volume 2

Volume 3

Volume 4

Chapter 1

Water Quality Management in Reuse and Recycling and R&D

S.N. Kaul[1], J.K. Pandey[3], Vinay Prakash[2], D.R. Saini[3] and P. Kaul[4]

[1] *MIT, Pune – 411038*
[2]*HEMRL, Pune*
[4]*Symantec (USA), Pune*

ABSTRACT

Water is a prime natural resource and is a basic human need. The underlying assumption in traditional water resources planning process still continues to be: "freshwater is a gift of god which would continue to be available in a perpetuity and in abundance". This is not valid as both quantity and quality of water poses series problem.

It is estimated that out of the total precipitation of around 400 x 10^4 million cubic meter per year in India the surface water resources cover as 176.8 x 10^4 million cubic meter. Out of this only 50 per cent can be put to beneficial use because of topographical and other constraints. Utilization of water resources in India is classified depending on nature of use into three main categories; domestic, agriculture, and industry. An estimated total water use in the country is 4074 million meter cube per day for the year 2000-2001, the figures of distribution among various uses are 2.1 per cent, 88 per cent and 2.4 per cent respectively.

Wastewater is a misplaced resource and can be reused to advantage through proper management. A differentiation, however, needs to be made between water reuse and recycling. Reuse implies utilization of water that has been previously used for another purpose, and recycle implies reuse of water more than once for the same purpose. India can take advantage of using bright sunshine, high temperatures and possibility of water usage for irrigating agriculture land for wastewater management. The approaches in wastewater management should essentially entail: waste volume reduction, waste strength reduction; and by-product recovery.

In India, majority of water treatment plants adopt conventional methods of treatment leading to sludge generation of around 0.1-0.2 per cent.

The chapter presents the following issues, *viz.*

- Water and wastewater status in India
- Domestic category
  - Land application of sewage for irrigation
- Industrial category
  - Pulp and paper industry
  - Tanneries
  - Textile and dyeing industries
  - Energy from waste
- Agriculture category
- Problems of wastewater management in domestic, industrial and agriculture sectors
- Problems of environmental management in small scale industries
- Towards pragmatic solutions

Preamble

Water is a prime natural resource and is a basic human need. The underlying assumption in traditional water resources planning process still continues to be: "freshwater is a gift of god which would continue to be available in a perpetuity and in abundance." This is not valid as both quantity and quality of water poses series problems.

Wastewater is misplaced resource and can be reused to advantage through proper management. A differentiation, however, needs to be made between water reuse and recycling. Reuse implies utilization of water that has been previously used for another purpose, and recycle implies reuse of water more than once for the same purpose. India can take advantage of using bright sunshine, high temperatures and possibility of water usage for irrigating agriculture land for wastewater management. The approaches in wastewater management should essentially entail: waste volume reduction, waste strength reduction, and by-product recovery.

It is estimated that out of the total precipitation of around 400×10^4 million cubic meter per year in India the surface water resources cover as 176.8×10^4 million cubic meter. Out of this only 50 per cent can be put to beneficial use because of topographical and other constraints. Utilization of water resources in is classified depending on nature of use into three main categories, domestic, agriculture, and industry. An estimated total water use in the country is 4074 million meter cube per day for the year 2000-2001, the figures of distribution among various uses are 2.1 per cent, 88 per cent and 2.4 per cent respectively.

Water and Wastewater Status in India

Domestic Category

It is estimated that 31 per cent of rural population and 77 per cent of urban population in India has access to potable water supply. Surface water is the dominant source of organized urban water supply. In rural area groundwater is the basic source. Municipalities and state Water Supply and sewerage boards own and operate the urban water supply schemes, however in rural areas there is no such organized water supply. Table 1.1 gives the status of water supply, wastewater collection and treatment in Metropolitan cities of India.

The situation in sewerage collection treatment and disposal is discouraging as only 63 per cent of the metropolitan population is provided with sewage collection and treatment system. The data obtained from remaining municipalities shows less than five per cent of the total wastewater generated in Indian urban areas is collected and only about 2 per cent is receiving treatment leaving remaining 98 per cent discharged into the environment untreated.

The indiscriminate discharge of large quantity of untreated sewage into the surface water has deteriorated all the rivers in the country. India has 14 major rivers whose basin area is 83 per cent of the total land area of the country, they contribute 85 per cent of the surface water flow of the country and 80 per cent of the total population is living in the basins of these rivers.

The preservation and restoration of river water quality in India started with the launch of much published Ganga Action Plan (GAP) in 1984. One of the important part of GAP was collection and treatment of domestic wastewater discharged from urban centers located on the bank of Ganga, GAP phase I has created an infrastructure for diverting 730 million litres per day (MLD) and treating 540 million litres per day of municipal sewage. The second phase of GAP aims in collection and treatment of domestic wastewater from 95 towns. The Central Government has also initiated the National River Conservation Plan (NRCP) involving population abatement works in polluted stretches of 18 rivers covering 46 towns.

Often construction of conventional sewage treatment plan requires huge capital cost. The Indian municipalities and public health departments are short of funds, thereby leaving most of the domestic wastewater untreated. The viable option is to adopt cost effective/low cost sewage treatment scheme. Sewage farming is one such option, which utilizes partially treated sewage for irrigation to grow crops and vegetables.

Land Application of Sewage for Irrigation: A Case Study of Chandigarh

70 MLD of community wastewater having 210 mg/L of BOD, 423 mg/L of TSS and 32 mg/L of oil and grease were generated from urban area of Chandigarh. About 98 per cent of the entire volume was collected at the sewage treatment plant (STP) and rest was abstracted from the sewer trunk for land irrigation before it reaches the treatment plant. This sewage was partially treated in STP to reduce the pollutant levels to 190 mg/L of BOD, 405 mg/L of TSS and 29 mg/L of oil and grease. The

Table 1.1: Status of Water Supply, Wastewater Collection and Treatment in Metropolitan Cities

Name of the City	*Water Supply Population Covered*	*Sewerage System Population Covered*	*Treatment Plant*		*Mode of Disposal*
			Level	*Capacity (MLD)*	
Bombay	99	80	Primary	82.0	Creek, Sea
Delhi	96	75	Primary and Secondary	745.0	River Yamuna and Agriculture Land
Calcutta	95	–	Primary and Secondary	–	River Kulti and Hoogly
Hyderabad	100	75	Primary and Secondary	140.0	Land
Ahmedabad	90	75	Primary and Secondary	382.0	River Sabarmati
Bangalore	100	85	Primary	286.0	Vrushabhavathi Valley, Ballandur Tank
Kanpur	75	60	Secondary	160.0	Ganga, Agriculture Land
Lucknow	100	–	–	–	–
Madras	85	77	Secondary	–	Sea, Irrigation
Nagpur	75	66	Primary (Not in Operation)	45.4	Local Nallah, Agriculture
Pune	78	53	Primary	90.0	River and Agriculture Land
Jaipur	80	–	–	–	Agriculture Land
Total	93	63		1930.4	

Source: Status of Water Supply and Wastewater Collection Treatment and Disposal in Class I Cities, CPCB, Delhi, CUPSS/30/1989-90.

partially treated wastewater is applied on land for cultivation of paddy crop. The sewage farm has 1250 acres of land adjacent to STP and is being utilized for past 10-15 years.

However, this sewage farm is not operated properly. There is heavy application of sewage water for cultivation of paddy crop, which has raised the water table of the sewage farm upto the ground level. The quality of the grain and yield of paddy is not satisfactory. The accumulation of heavy metals in sewage irrigated soils were high compared to the soils receiving tube well irrigation. The groundwater near the sewage farm was showing high bacteriological contamination.

A study was conducted to improve the quality of the sewage farm. This study has recommended following important modification and suggestions:

- ☆ Land preparation and drainage to avoid wastewater logging
- ☆ Arrangement for periodic desludging
- ☆ Provision of subsurface drains for water table control
- ☆ Pretreatment of wastewater before application
- ☆ Optimum loading of wastewater and scientific irrigation schedule for crop
- ☆ Optimum fertilizer dosing
- ☆ Soil Management practices
- ☆ Maintenance of soil pH and organic matter
- ☆ Crop rotation

Industrial Category

Industrial growth in India took place after the liberalization of Indian economy in 1991. to focus on industrial pollution the Government of India has identified 19 critically polluted areas in the country and 17 industrial sub sectors which are polluting namely, cement, thermal power plant, distilleries, sugar, fertilizer, integrated iron and steel, oil refineries, pulp and paper, petrochemicals, pesticides, tanneries, basic drugs and pharmaceuticals, dye and dye intermediates, caustic soda, zinc smelter, copper smelter and aluminum smelter. The list of critically polluted areas is given in Table 1.2. out of a total of 1,551d units identified under these categories by Central Pollution Control Board (CPCB) 1,259 units have installed adequate facilities for pollution control 112 units have been closed down and the remaining are putting the polluting control schemes.

Since November 1991, World Bank with MOEF is providing finance for pollution control projects to individual industrial units for pollution prevention and control. The funds are disbursed by Industrial Development Bank of India (IDBI) and the Industrial Credit and Investment Corporation of India Limited (ICICI). From this credit line of the World Bank nearly 60 units have taken loans amounting to more than 90 million US dollars. In view of such positive response from industry, Government of India has recently negotiated second line under Industrial Pollution Prevention Project where similar funds would again be available. in the case of new units, the cost of pollution control is internationalized with the entire project cost and is financed within the overall financing package of the entire project. In general, this cost is less than 5 per cent of the total capital cost of the project except for specific industry sub-sectors where the cost may be as high as 10 per cent.

Table 1.2: Critically Polluted Areas

Sl.No.	State	Area
1.	Punjab	Mandi Gobindgarh
2.	Rajasthan	Pali
3.	Delhi	Najafgarh Drain Basin
4.	Gujarat	Vapi
5.	U.P.	Singhrauli
6.	Bihar	Dhanbad
7.	West Bengal	Durgapur
8.	West Bengal	Howrah
9.	Odisha	Talcher
10.	Tamil Nadu	North Arcot
11.	Tamil Nadu	Manali
12.	M.P.	Korba
13.	Assam	Digboi
14.	Maharashtra	Chembur
15.	A.P.	Vishakapatnam
16.	Karnataka	Bhadravati
17.	Kerala	Greater Cochin Area
18.	Himachal Pradesh	Kala-Amb
19.	Himachal Pradesh	Parwanaoo

Effective enforcement of environment legislation, increased public awareness and change in industry's perception of its social responsibility have resulted in significant increase in compliance to the various provisions of environmental legislation in the large and medium scale industries.

Status of the wastewater treatment and reuse for select industrial sectors in the country is briefly given below:

Pulp and Paper Industry

In India there are more than 379 paper mills with an annual installed capacity of around 37.78 lakh tonnes. A significant aspect of Indian paper industry is strong presence of small paper mills which contribute 50 per cent to the production of paper and paper board in the country.

The water requirement of the paper mills in India is in the range of 250-440 m^3/tonne of product and the requirement depends on the quality of paper made and extend of recycle. This water requirement is high compared to 60-120 m^3/tonne water requirement achieved in several developed countries. Combined wastewater generation from large mills range between 167 to 281 m^3/tonne of product. No significant variation in wastewater generation is observed. It is found that lignin bearing colored wastewater accounts about 30 per cent of the total effluent discharged.

A study of wastewater recycle in the paper industry has revealed that wastewater recycle varies between 8 and 48 per cent with an average recycle of 21.4 per cent, which facilitates reduction in freshwater requirement to the extent of 7 to 44 per cent. A wastewater recycle pattern in a typical Indian paper mill is given in Table 1.3.

Table 1.3: Wastewater Recycle Pattern in a Typical Paper Mill

Type of Wastewater	*Source*	*Recycle Process*	*Wastewater Recycled, per cent*
Decker wash	Unbleached pulp filter	Screening and dilution of unbleached pulp at any place and chipping when required	50-60
Alkali backwater	Alkali extraction stage filter	Diluting chlorination tower contents from 3 per cent to 1 per cent and alkali tower contents	90-95
Hypo backwater	Hypo extraction stage	Same system, alkali extraction stage and rage plant	80-90
Final washer backwater	Bleach plant	Some system and rag plant	80-85
Paper machine	Paper machine	Dilution high density pulp before feeding to paper machine and for hydropulping	50-55

Source: Comprehensive Industry Document for Large Pulp and Paper Industry, Central Pollution Control Board Delhi, COINDS/36/1991.

The Indian pulp and paper industry is adopting recycling of wastewater either without treatment or after suitable treatment. It is also observed that there is growing recycle awareness in the industry.

Recovery of Caustic Soda from Black Liquor of Small Scale Pulp and Paper Industries in Common Chemical Recovery Unit

An individual chemical recovery unit is not economically viable for small scale paper mill, hence a common chemical recovery unit for a number of paper mills is recommended option where black liquor of individual mills can be collected and processed in a central recovery plant.

The common chemical recovery unit can be installed for a group of six to eight units located in the vicinity. Initially a bench scale plant will be constructed to determine the design parameters. A bench scale study will essentially involve characterization of black liquors from different units to establish the suitability of their use as a feed stock for recovery and suggest methods to improve the liquor characteristics (to obtain higher solid content black liquor). The black liquor received from mills will be concentrated in multiple effect evaporation section. The semi concentrated black liquor then will be taken from mills will be concentrated in multiple effect evaporation section. The semi concentrated black liquor then will be taken to recovery boiler and after increasing its concentration in cascade evaporator to convert it into green liquor. This green liquor will be reacted with lime and converted into white liquor for reusing as a cooking chemical in the paper mills.

Tanneries

There are about 2000 tanneries in India. Approximately 75 per cent of these tanneries are in the cottage and small scale sector, 20 per cent in the medium sector and remaining 5 per cent in the large sector. The total wastewater discharge from Indian tanneries is estimated to be about 50 million liters per day. The tanneries in India are concentrated in clusters in the state of Tamil Nadu, West Bengal, Uttar Pradesh, Karnataka, Punjab and Maharashtra. Tanning industry is water-consuming industry, approximately 30-40 litres of water is used for processing one kg of raw hide into finished leather. Most of the tanneries in India are located near the river banks or near the natural water bodies to draw surface water. Some tanneries also use groundwater from the captive open wells. The quantity of water usage and nature of wastewater discharge in tanneries depend on type of tanning process, capacity of tannery and its production pattern.

The tanneries in India are grouped into three major categories from wastewater management.

1. Tanneries having adequate land and managerial capacity to built their own effluent treatment plant. There are about 100 large scale tanneries in the country. These tanneries have put up effluent treatment units.
2. Tanneries that are close to each other and are amenable to have a Common Effluent Treatment Plant (CETP). Common Effluent Treatment Plants are put up for these tanneries
3. Scattered tanneries not having adequate land which can neither put up individual effluent treatment plant nor be included in a CETP. The only option for these tanneries is to shift to new location so that they can join the CETP scheme.

In India more than 80 per cent tanneries adopt chrome tanning process, in this process 60 per cent of the applied chromium is taken by the leather and remaining is discharged in wastewater. Research organizations like NEERI, CLRI and IIT Kanpur have developed technologies for reduction in chromium salt utilization for chrome tanning and for recovery of chromium from the wastewater.

Recovery and Reuse of Chromium Salt from Tannery Wastewater

Lot of research work has been conducted in the past to evolve chemical methods of treatment to recovery chromium leached with sulphuric acid to produce reusable chrome liquor.

Chrome recovery system based on dewatering of chromium sludge followed by incineration to product ash which is leached with sulphuric acid to produce reusable chrome liquor.

In the second method the chrome liquor after filtration is treated with sodium carbonate to produce chromium sludge which is dewatered in rotary vacuum filter. The filter cake are redissolved in sulphuric acid to give chrome tanning solution.

Another method of chromium recovery is use of Magnesium Oxide (MgO) which has low relativity and solubility causing chromium to settle in a very compacted way. The separation from the liquor is obtained by decanting the supernatant. The sludge can be dissolved in sulphuric acid to obtain the reusable liquor.

Most of the work carried on recovery of chromium is confined to laboratory only. Not much work has been done on full scale application. An in-depth study will be conducted on all the above technologies to estimate the process parameters for design of full scale plant. A cost benefit analysis will be carried on to assess economic viability of the process. The study will also include the effect of recovered chromium on quality of leather produced and possibility of reuse the recovered chromium for number of cycles without affecting the quality.

Textile Industries

Textile industries located at Small Scale Industrial (SSI) cerntres like Pali and Balotra in Rajasthan are known to be source of high wastewater pollution loads. A total of 767units located in various areas of Pali are collectively contributing a large quantity of wastewater to the region.

NEERI has studied the pollution from these industries and prepared design packages for control of water pollution on a collective basis. Criteria as defined in Table 1.4, and the basis for treatment presented in Table 1.5, are utilized in preparing various options for such collective treatment. These options are presented in Table 1.6. the capital costs, operation and maintenance costs and specific costs were studied in detail to arrive at economically feasible option. One CETP at Pali has been commissioned successfully.

NEERI has designed nine CETPs for different types of industrial clusters. The details are presented in Table 1.7.

Recycle and Reuse of Small Scale Textile and Dying Industry Wastewater for Industrial Application

Textile processing comprises one or more of the following processes desizing/ kierning, bleaching, mercerizing, dying and printing. Additionally, carbonization is included for synthetic mixed cotton processing. Physico-chemical treatment scheme is selected for reduction of excess alkalinity, heavy metals, suspended solids colloidal material and colour followed by aerobic biological treatment to remove biodegradable organic matter.

Initially bench scale treatability studies will be conducted to identify the suitable treatment unit and to estimate the kinetic coefficients colour is an important parameter for dying industry. Dying industry wastewater will be subjected to ASP added with activated carbon. This treated effluent from modified ASP will be further treated/ polished in sand filters. This treated wastewater can be reused in the process.

Table 1.4: Criteria for the Design of CETPs as Developed by NEERI in Consultation with MEF

- ☆ Inventory of Industries
- ☆ Flow and characteristics of wastewater
- ☆ Classification of wastewater based on biodegradability (A, B, C and D)
- ☆ Design of conveyance system
  - Free from problems
  - Optimized scheme
- ☆ Treatability study
  - Bench scale
  - Pilot scale
- ☆ Stagnation of wastewater
- ☆ Pretreatment of wastewater (C and D)
- ☆ Assessment of available technologies for design of CETP to development treatment packages and optimization based on spatial distribution of CETP
- ☆ Ranking of technology options
- ☆ Reuse/recycling and resource recovery
- ☆ Disposal of treated effluents
- ☆ Cost estimation based on optimized CETP
- ☆ Cost benefit analysis
- ☆ Scheme for sharing financial burden
  - Annualized capital cost
  - O and M for different types of wastewater (A, B, C and D) based on flow and mass
- ☆ Possibilities of using cleaner technologies
  - Process
  - Water consumption
  - Raw material
  - Energy requirement
  - Consideration of waste from one industry to be used as raw material for another industry

Energy from Waste

Realizing the potential and importance of treatment of wastes and the resultant recovery of energy from these wastes, the Government of India has launched in June, 1985 a National Programme on Energy Recovery from Urban, Municipal and Industrial Wastes with a view to promoting the adoption of proper technologies as a means of improving waste management practices in the country with the goals and objectives of (i) creation of conductive conditions and environment with fiscal and financial incentive to help promote, develop, demonstrate and disseminate and utilization of wastes for recovery of energy, (ii) improving the waste management practices through the adoption of technologies for conversion of wastes into energy and (iii) promoting the setting-up of projects utilizing wastes from urban, municipal and industrial sectors.

Table 1.5: Basis for Wastewater Categorisation

WWC	*Pollutant*	*Treatment*	*Explanatory Notes*	*Pretreatment*	*Acceptability in CETP*
A	High SS	Highly settleable	Non-toxic solids highly settleable	NP	Acceptable
A	High SS	Colloidal needs coagulant	Non-toxic solids settleable after coagulation	NP	Acceptable
A	Organics	Highly biodegradable	Organic solids in solution and/or suspension, highly biodegradable	NP	Acceptable
A	Organics	Slowly biodegradable	Organic solids in solution And/or suspension, slowly biodegradable, example: benzene series	NP	Acceptable
B	Inorganics	Acid	Mineral acids produced or used in excess	N and SEP	Acceptable on neutralization
B	Inorganics	Alkali	Alkalies produced or used in excess	N and SEP	Acceptable on neutralization
B	High TDS	Precipitable	Raw or product materials are soluble in water	CP and SEP	Acceptable on precipitation
C	High TDS	Membrane Separation	Highly soluble solids need membrane separation	NIP	Require expensive collective treatment using membrane separation
D	Organics	Refractory	Naphthalene, anthracene used as raw material and/or product	NIP	Requires expensive collective treatment of ozonation and then to CETP
D	Organics	Toxic	Metal complex organo-chklorine pesticides carbonates	NIP	Requires expensive collective treatment or incineration

NP: No pretreatment; NIP: No individual pretreatment; N and SEP: Neutralisation and Solid Separation; CP and SEP: Chemical Precipitation and Solids Separation; WWC: Wastewater Category.

Table 1.6: Various Treatment Options for CETP

Option	*Treatment Alternative*
1.	Primary Settling + Aerobic Biological Process (Extended Aeration) + Sand Filter + Granular Activated Carbon Column + Disinfection
2.	Primary Settling + Anaerobic Biological Process (AFFB, D.P. – 1 day) + Cascade Aeration + Sand Filter + Granular Activated Carbon Column + Disinfection
3.	Primary Settling + Anaerobic Biological Process (AFFFB, D.P. – 0.25 day) + Aerobic Biological Process (Extended Aeration) + Sand Filter + Granular Activated Carbon Column + Disinfection
4.	Physico-chemical Process + Aerobic Biological Process (Extended Aeration) + Sand Filter + Granular Activated Carbon Column + Disinfection
5.	Physico-chemical Process + Sand Filter + Granular Activated Carbon Column + Disinfection
6.	Physico-chemical Process + Aerobic Biological Process (Extended Aeration) + Disinfection
7.	Physico-chemical Process + Anaerobic Biological Process (AFFB, D.P. – 0.5 day) + Cascade Aeration + Disinfection
8.	Primary Settling + Aerobic Biological Process (Extended Aeration) + Disinfection

AFFB: Anaerobic Fixed Film Fixed Bed Reactor.

Table 1.7: List of CETPs Designed by NEERI for Control of Water Pollution/Resource Recovery/Recycle in Cluster of Small Scale Industries

Sl.No.	State/Town	Industries in Cluster		Flow	Annualised Cost (Rs.Lakhs)	Annual (Rs. Lakhs)	Salient Features
		Type	Nos.				
1.	Gujarat, Vapi	Chemical Dyeing, etc.	750	16.0	271	114	Wastewater recycling for industrial area
2.	Himachal Pradesh, Parwanoo	Electropla-ting, engg., etc.	76	1.0	29	38	Conforming to standards
3.	Himachal Pradesh, Barotiwala	Chemical pulp and paper, etc.	34	8.4	313	413	Wastewater recycling for industrial area
4.	Himachal Pradesh, Kala-Amb	Paper, textile, etc.	22	16.0	521	715	Wastewater recycling for industrial area
5.	Himachal Pradesh, Mehatpur	Textile, Engg., etc.	111	0.37	15	27	Effluent conforming to standards
6.	Punjab, Amritsar	Textile	50	50.0	297	–	Effluent conforming to standards
7.	Rajasthan, Balotra	Textile	367	7.7	635	191	Wastewater recycling for industrial area
8.	Rajasthan, Pali	Textile	767	36.0	1900	1070	Wastewater recycling for industrial area
9.	Tamil Nadu, Salem	Sago	641	12.0	291	108	Biogas recovery and aquaculture

Biogas Generation from Industrial Wastewater

The proposed research work on anaerobic system will focus on process modifications. This work will involve estimation of factors like SRT, type of reactor, and loading rates to increase the process efficiencies towards optimal performance level. This work will also involve careful study of factors like pH, temperature to maximum the growth rate of microorganisms involved in biological degradation of organic compounds. Determination of kinetic constants for different reactor types and industrial wastewaters. Scale up formulation for different types of reactor systems will be developed as scanty information is available on these aspects.

Intensive experimentation in above areas will increase the level of technology in the treatment of industrial wastes by anaerobic process to a much higher plateau than is realized today.

Agricultural Category

India is an agricultural country. Majority of the Indian population live in rural areas and agriculture is their main occupation. According to an estimate 88 per cent of the Indians water usage is consumed by agriculture sector. As per the livestock census of 1992 our country has 204.5 million cattle 83.5 million buffaloes, 50.8 million ships, 115.3 million goats, 12.8 million pigs and 307 million poultry birds. This is one of largest livestock population in the world. The rate of chemical fertilizer consumption has increased rapidly in our country as the emphasis is on increase in food grain production. The consumption of chemical fertilizer is given in Table 1.8. the compound growth rate in agricultural production during the period 1950 to 1946 is 2.67 per cent per annum as against the compound rate of nitrogenous fertilizer consumption is 11 per cent.

Table 1.8: Consumption of Nitrogeneous Fertilizers in India

Year	*Consumption Tonnes/Year*
1951-52	5500
1961-62	249800
1971-72	1798000
1981-82	4068700
1991-92	8046300
1996-97	14930000
2000-2001	22383000

Source: i) Comprehensive Industry Document Fertilizer Industry, CPCB, New Delhi, COINDS/50/1994-95

ii) India 1998, Ministry of Information and Broadcasting, New Delhi.

Contrary to the belief, agriculture sector is a major source of water pollution. The major factors of importance are reduction in wetland area, monoculture crop production, extensive use of commercial fertilizers and pesticides, irrigation and intensified husbandry. Irrigation, which consumes maximum quantity of water, is known to cause water quality problems such as i) salinization, ii) alkalinization, iii) erosion of irrigated land and iv) increase in sediments in downstream areas like lake and dams.

There is a tendency in farmers to apply more fertilizers than the optimal rate in order to reduce a risk of being short of nutrients in good weather years. Excess chemical fertilizers and pesticides are washed off and reach the surface water or groundwater. A survey has shown alarming increasing in nitrates of groundwaters of north India.

Livestock breeding and animal husbandry is another source of water pollution from agriculture sector. The major pollution sources from animal husbandry are cattle slurry, (dirty water which contains washing from the farm yard and milking parlour) silage effluent (fermented grass used as cattle feed) and animal faeces and urine. Table 1.9 gives the BOD from animal husbandry sources. It is seen that BOD of animal sources is approximately 100 higher than raw domestic sewage. Considering the livestock population of India and BOD load, the estimated figures can be alarming.

Table 1.9: BOD of Common Animal Husbandry Waste

Sl.No.	*Effluent Type*	*BOD Range*
1.	Raw domestic sewage	300-400
2.	Cattle slurry	10,000-20,000
3.	Poultry slurry	30,000-35,000
4.	Silage effluent	30,000-80,000

There is a major and extensive built-up water pollution from agriculture. However this is going largely unattended. Popular perceptions as well as regulatory institutions have focused upon urban and industrial pollution. There are now laws governing the pollution from agriculture sector vis-à-vis no extensive studies have been conducted in our country to estimate the water pollution from agriculture sector.

Causes for Present State of Affairs

Problems of Wastewater Management in Domestic Sector

Social Awareness

Indians in general are not aware of the environmental consequences of the discharge of the untreated wastewaters. Indian urban centers are administered by municipalities or municipal corporations which are governed by body of elected members. Potable water supply and collection and treatment of wastewater is not an important agenda of municipal elections. Hence wastewater treatment is always a neglected issue and has least priority in policies of the municipalities.

Finance

Municipalities often depend upon the state/central governments for funds to construct the sewage treatment plants. These funds are either partially available or not available on time.

Bureaucracy

Administrative and bureaucratic machinery in the municipalities and in state and central governments follow a lengthy and complicated procedures for decision making which cause excessive delay in policy formation and its implementation.

Lack of Trained Manpower

It is often observed that sewage treatment plants are not equipped with qualified and trained manpower to operate them efficiently. The staff is often inexperienced.

Problems of Wastewater Management in Industrial Sector

Technology

Although technology for conventional treatment for water and wastewater is available in the country, various studies in the past have indicated need for specific technologies for specific pollution parameters control. CPCB has assessed various wastewater treatment technologies with respect to know-how, equipment's and their availability in India. The details of this study are reported in Table 1.10. Some of these technologies are already developed and demonstrated in the country while some need development of design criteria and demonstration of applicability and efficiency.

Law Enforcement

The Central Pollution Control Board (CPCB) is the national apex body for assessment, monitoring and control of water pollution. The executive responsibilities for enforcement of the Acts for Prevention and Control of Pollution of Water (1974) and of the Water (Cess) Act, 1977 are carried out through the board. Under the Environment (Protection) Act, 1986, effluent and emission standards in respect of 61 categories of industries have been notified.

State pollution control boards are formed to enforce the law constituted by central and state governments at state level. However these pollution control boards are often deficient of funds, poorly manned and have inadequate infrastructure for proper regulation of the pollution control laws.

Policy

An approach to industrial wastewater treatment should be to consider the problem in its totality and to include both preventive and curative aspects. This means that pollution control begins from the point of generation of wastewater and ends with ultimate disposal point. This concept could be broadly divided into two components, namely, inplant control and end of the pipe treatment.

Inplant control consists of measures, which will reduce the wastewater quantitatively and reduce the pollution potential qualitatively. This is achieved through appropriate selection of raw materials, cleaner technologies for production, conservation through recycle and reuse, by-product recovery and process and equipment modifications. This leads to preventive, internal systematic, integrated, conservational and low pollution system, which considerably reduce the load on end of the pipe treatment.

Table 1.10: Assessment of Wastewater Technology Status in India

Sl.No.	Process	Control Parameter	Industries Where Required	Present Status in India	Assessment of Import Needs			Notes
					Technology	Hardware		
1	2	3	4	5	6	7A	7B	8
1.	Adsorption	BOD, COD Toxic compounds Colour Odour	Dyes and Intermediates, Textile, Paper, Pharmaceuticals, Pesticides, Fruit, Canning, Petrochemicals, Rubber Chemicals, Man-made Fibres	Limited Product Range	Yes	No		Import of know how for Technology of Activated Carbon manufacture and regeneration. Import of technology for synthetic adsorbent. Initial import of special adsorbents including Activated Carbon
2.	Bio-Technology							
2.1	Conventional	BOD/COD	For all industries biodegradable and slowly-biodegradable					
2.1.1	Aerobic							
A	General			Developed	No	No	No	–
B	Rotating Biological Contactors (RBC)			Under-Developed	Yes	No	Yes	Import of Process know-how and in hardware initial import of Disc Segments only
C	Diffused Aeration			Limited Range	Yes	No	Yes	Import of know-how for manufacture of diffusers and initial import of porous diffusers of various materials
2.1.2	Anaerobic			Developed	Yes	No	No	Import of know-how for the design of reactors including anaerobic activated sludge, egg shaped digesters etc.

Contd...

Table 1.10–*Contd...*

Sl.No.	*Process*	*Control Parameter*	*IndustriesWhere Required*	*Present Status in India*	*Assessment of Import Needs*			*Notes*
					Technology	*Hardware*		
1	*2*	*3*	*4*	*5*	*6*	*7A*	*7B*	*8*
2.2	Advanced	BOD, COD						
A.	Microbial Cultures/ Enzymes			Research stage	Yes	No	Yes	Initial import of know-how and cultures/enzymes especially for recovery of single products
B.	Bioreactor			Under developed	Yes	No	Yes	Import of know-how for design of aerobic and anaerobic reactors such as deep shaft, tower, loop, upflow and downflow fixed film and fluidized bed etc. Also initial import of critical components, if necessary
3.	Disinfection							
3.1	Conventional (Chlorination etc.)	Pathogens	–	Developed	No	No	No	
3.3	Ultraviolet	Pathogens	–	Nil	Yes	No	Yes	Import of know-how and critical components
4.	Dissolved Air Floatation	Suspended Solids/Oil and Grease	Metal processing and finishing, Petroleum Refining and Petro-Chemicals, Dairy, Edible Oil, Textile Processing, Dyestuff, Tanneries, Food and Fruit Processing	Limited product Range	Yes	No	Yes	Import of know-how and critical components

Contd...

Table 1.10–*Contd...*

Sl.No.	*Process*	*Control Parameter*	*IndustriesWhere Required*	*Present Status in India*	*Assessment of Import Needs*			*Notes*
					Technology	*Hardware*		
1	*2*	*3*	*4*	*5*	*6*	*7A*	*7B*	*8*
5.	Electro-Dialysis and Electro-Deposition	Heavy metals	Metal Processing and finishing, Caustic Soda	Nil	Yes	No	Yes	Import of know-how and critical components especially membranes
6.	Evaporation and Crystallization	Toxic, Organic and Inorganic Salts	Pulp and Paper, Distillery, Metal Processing, and Finishing, Textiles, Organic Chemicals and Petrochemicals	Under Developed	Yes	No	Yes	Import of know-how for Vacuum Evaporators and critical components
7.	Incineration (Thermal Treatment)	BOD/COD Toxic Organics	Paper, Distillery, Tannery, Pharmaceuticals, Pesticides, Petrochemicals, Petroleum, Dyestuff Intermediates	Under Developed	Yes	No	No	Import of know-how especially for Effective Combustion System
8.	Ion Exchange	Heavy Metals, NH_3 Cyanides, Fluorides	Caustic Soda, Fertilizers, Manmade Fibres, Rayon, Pulp and Paper, Distillery, Pharmaceuticals, Metal Processing and Finishing, Electronics, Ferrous and Non Ferrous	Limited Product Range	Yes	No	No	Import of know-how for the manufacture of macroraticular resins and selective ion exchange resins. Initial import of macroraticular resins and selective ion exchange resins.

Contd...

Table 1.10–*Contd...*

Sl.No.	Process	Control Parameter	Industries Where Required	Present Status in India	Assessment of Import Needs			Notes
					Technology	Hardware		
1	2	3	4	5	6	7A	7B	8
9.	Ozonation	BOD, Colour, Odour, Refractory Organics, Toxic Chemicals, Pathogens	Iron and Steel Manufacturing, Dyes and Intermediates, Pharmaceuticals, Petrochemicals, Paper, Pesticides	Nil	Yes	No	Yes	Import of know-how and initial imports of Ozonators
10.	Reverse Osmosis	Dissolved inorganics	For all industries for reclamation of water	No	Yes	No	Yes	Import of know-how and import of membranes. Also import of know for suitable pump manufacture and critical components
11.	Sedimentation	Suspended Solids, Oil and grease						
11.1	Conventional	–	–	Developed	No	No	No	–
11.2	Tube/Plate Settlers	–do-	–	Under developed	Yes	No	No	Import of know-how
12.	Sludge Handling Treatment and Disposal							
	Handling	–	All industries	Yes	No	No	No	Import of know-how for the manufacture of screw pumps for high consistency sludges and air lifts pumps
12.2	Treatment							
12.2.1	Thickening	–	–	Developed	No	No	No	–

Contd...

Table 1.10–*Contd...*

Sl.No.	*Process*	*Control Parameter*	*IndustriesWhere Required*	*Present Status in India*	*Assessment of Import Needs*			*Notes*
					Technology	*Hardware*		
1	*2*	*3*	*4*	*5*	*6*	*7A*	*7B*	*8*
12.2.2	Dewatering	–	–	Under developed	Yes	No	Yes	Import of know-how for manufacture of Centrifuges, Belt Filters etc. Also import of critical components
12.2.3	Digestion	Organic Sludge	–	Under developed	Yes	–	–	Import of know-how for Reactor Design
12.2.4	Composting	Organic Sludge	–	Developed	No	No	No	–
12.2.5	Thermal Dewatering	Organic and Inorganic Sludges	–	Under developed	Yes	No	No	Import of know-how
12.3	Ultimate Disposal							
12.3.1	Smelting	Toxic, Inorganic Sludge	–	Developed	No	No	No	–
12.3.2	Incineration	Organic Sludge	–	Under developed	Yes	No	No	Import of know-how
12.3.3	Stabilization and Solidification	Toxic and Hazardous Sludges	–	Nil	Yes	No	No	Import of know-how
13.	Ultrafiltration	COD, Dissolved Polymers and Colloidals	Dairy, Oil Refinery, Wool Industry, Breweries Paper, Paints All Industries	Under developed	Yes	No	Yes	Import of know-how and critical components especially membranes
14.0	Control and Monitoring	Various Parameters	All Industries	Limited Range	Yes	No	Yes	Import of know-how for instruments for continuous monitoring and sampling systems, specific ion electrodes. Flow measurements and recording. Also import of critical components and initial import of instruments

Source: Report on Identification of Import Component of Waste Treatment Technology and Know-how, Central Board for Prevention and Control of Water Pollution, New Delhi, PROBS/32/1985-86.

End of the pipe treatment, theoretically, deals with treatment and disposal of wastewater, which are residual, after all the inplant measures are applied. In practice, however, till recent past, inplant measures were not emphasized and consequently the major pollution control efforts was concentrated on the end of the pipe treatment.

But with the scenario of large scale industrial development, the focus of the policy planners and enforcing agencies should shift towards inplant pollution control thus permitting a holistic approach to pollution.

Problems of Environmental Management in Small Scale Industry (SSI)

Management/Economics

No Investment returns on pollution control: The management of SSI's have taken the approach that pollution control is a dead-end investment. In many cases this fear is indeed realized, where the cost of installing pollution control does not add to the productivity and to the bottom-line profits of the enterprise. In addition the cost of operating the pollution control equipment cuts deeply into the normal cash flow of the unit.

Investment for pollution control higher than initial investment of plant: The cost of pollution control has traditionally been compared with the initial investment required to set up the industry. As is the case with many highly pollution small units, *e.g.* glass, lime kilns and dye-stuffs industry, the initial investment is of the order of 3 to 5 lakhs. Pollution control designed to reduce the pollutants to reasonable levels, would cost about the same amount. This causes the entrepreneurs not to install pollution control devices. However, the cost of pollution control should be compared with the annual turn-over of the industry, because in some of these cases, the turn-over could be many times the initial investment.

This fact creates the additional problem, that industry is more willing to enter protracted legal wrangling to buy time, rather than install pollution control equipment. When push comes to shove, they are willing to close down the industry and start over again.

Awareness: The awareness among the management of SSI's is fairly low with respect to technical options, finance availability in term of subsidies and tax incentives for pollution control, and the legal repercussions of non-compliance with regulations. SSI is typically not aware of notional level efforts in development and availability of non-or less-polluting technologies.

Operation of Existing Pollution Control Equipment's: Due to the shortage of electricity and water in various states, having large number of small scale industries, *e.g.* Uttar Pradesh and West Bengal, operation of the pollution control devices is severely affected. And when electricity is available, the entrepreneurs choose not to operate the control devices to save on operating costs.

Economies of Scale

Due to the size of the SSI sector, cost-effective and technically feasible pollution control becomes exceedingly difficult, if not impossible. Also due to the small size, the units oftentimes operate in an inefficient manner. For example, the small lime kilns operate at low thermal efficiencies, losing significant amount of heat as waste heat. Also, the heat content of the flues are so small that heat recovery is not possible.

In another example, hot flue gases at around 600°C from cupolas pose severe problems for removal of particulates from the flue. The option of recovering the waste heat, as steam is infeasible because there is no demand for steam in the foundry or the immediately neighbouring industries.

Finance

Even though the government provides soft loans for pollution control, with the interest rates in neighbourhood of 12-14 per cent p.a., it hardly makes the investment in pollution control cost effective.

SSI financial institutions and banks providing loans for setting up of SSI's are not aware of the environmental consequences of setting up the industry. As a result, the cost of pollution control has not been included in the feasibility studies, or required in business proposals from the entrepreneur. If environmental costs are considered, they are done apart from the cost of the project, and not internalised as a part of the project.

Technology

Know-how for cost-effective control of pollution for the SSI is not readily available. A concerted effort in developing pollution control and less polluting production technologies does not exist, and efforts of R&D institutions are sporadic and they do not solve the problems effectively. Reputed consulting engineers have not shown interest in catering to the small scale, whereas several unreliable consulting-shops have mushroomed, providing inadequate and at times grossly wrong solutions.

Regulatory/Institutional

No effect regulatory mechanism exists for regulating, controlling and monitoring the pollution status of the SSI. Both the central Pollution Control Board (CPCB) and the State PCB's are short of trained manpower and adequate funds to regulate large industries, let alone the small scale sector. Due to their very large numbers, the efforts of the PCB's, if any, have been ad hoc, and ineffective.

Vast differences occur between the levels of effort of the PCB's in various states. This encourages location of SSI's in state providing lax control. This also causes unfair treatment with respect to the environment to the SSI's located in different states, affecting their competitive position.

No database regarding the pollution from the SSI's is available to assist in development of suitable environmental management policy, and regulations. Forthcoming efforts of CPCB and financial institutions towards environmental control have only standard scratching the surface.

Problems of Wastewater Management in Agriculture Sector

The successive governments formed at center and at state have always championed the cause of farmers by advocating the policy to increase agriculture production by increasing the chemical fertilizer dosing. These governments have encouraged this policy by providing subsidized chemical fertilizers, which has led to over application of fertilizers. Any attempt to discourage this by increasing the fertilizer price will invite the wrath of powerful farmers lobby, which no government can afford due to its long-term political implications.

Similar is the case with livestock breeding. Traditional Indian values encourage to have maximum number of cattle, as it increases ones social standing in rural India. Cows and bull are considered, as holy animals and religion prohibits their slaughtering. These reasons have caused rise in livestock population of the country.

Towards Pragmatic Solutions

To affect and change the burgeoning problem of water and wastewater treatment in India, pragmatic solutions to these problems at the macro level (policy level) as well as at the grass roots levels in terms of the implementation needs to be made. One set of attempts without the other, would as in the past fail to produce the desired levels. Following are some of the possible solutions to tackle this problem.

- ✰ Adoption of low cost wastewater treatment systems for treatment of wastewater from small towns and cities. These technologies include water hyacinth pond, oxidation ponds, and application of wastewater for irrigation and constructed wetlands. These technologies requires less capital and operation and maintenance cost but require large quantity of land which is available in plenty in the vicinity of smaller towns. Valuable by products can also be recovered from these treatment schemes. For example methane, fertilizer and water for irrigation from water hyacinth pond. It is reported that domestic wastewater (3650 x 10^6 m^3) can contribute 219 x 10^3 tonne of nitrogen, 73 x 10^3 tonne of phosphate, 146 x 10^3 tonne of potash and 1460 10^3 tonne of organic matter with an economic value of Rs.2000 million per annum.
- ✰ By imparting environmental educating to the urban population by making them aware about the problems and consequences of discharging untreated wastewater into the environment and their duties towards social hygiene.
- ✰ Providing financial support to the municipalities for wastewater management either from state/central governments or from financial organizations at reasonable terms.
- ✰ Reducing the bureaucratic delays by simplifying the process of decision making.
- ✰ Providing suitable training to the treatment plant operators to efficiently run the treatment plant.
- ✰ Common Effluent Treatment Plants provide a viable option for a cluster of small industrial units which are not capable of setting up the treatment units individually on their own. Financial incentives in the form of grant should not be made available for constructing CETPs.
- ✰ The need to encourage adoption of cleaner technologies (CT) for industrial production is required. Government of India is taking steps in this direction

by setting up an Indian center for Promotion of cleaner Technologies (ICPC) with a large number of agencies operating as a Network with access to data bases within as well as outside the country. The ICPC, after becoming operation shortly, will be able to service the industry for waste management purposes in the following ways:

- Recovering wastes
- Treatment and disposal of wastes
- Waste Exchange to identify users for potential recyclable wastes.

☆ It is necessary to underline the fact that mere availability of technology is no guarantee that waste management would be undertaken in a proper manner. Skilled manpower and management package is equally important inputs for successful completion of waste management programme. Considering the quantum of waste to be handled in the public sector, it is imperative that the constitutional mechanism is suitably strengthened and also made accountable for successfully and timely completing the assigned task. This aspect, however, has not received the requisite attention and needs to be vigorously pursued along with the promotion of Research and Development effort in such areas as bio-technology intervention for waste management

☆ Another approach would be to move from CETP to Combined Effluent Treatment Plant to jointly treat domestic sewage with industrial wastes. Addition of domestic sewage from surrounding community will not only solve the problem of community waste treatment but will also improve the treatability of industrial wastewater.

☆ There are at present several fiscal incentives for installation of pollution control equipment and for shifting industries from congested areas. However, economic instruments should be amended to encourage the shift from curative to preventive measures, internalize the costs of pollution and conserve resources, particularly water. One possible suggestion is effluent change based on the nature and volume of releases to the environment. The charge will be based on the cost of treatment and the flow discharged, thus putting thrust on optimal released of wastewater. The precise choice of economic instruments adopted should be determined by the ease with which releases can be measured, as well as prospective changes in technology and market structures.

☆ At present the disincentives include penalties and fines for non compliance under the various acts as well as the Water Cess. All of these disincentives are extremely low in relation to the cost of compliance. The Water Cess, also comes nowhere close to cost of pollution prevention/control for compliance. This can be overcome by increasing the amounts on penalties and fine.

☆ In addition to the above, some market instruments could also be employed. One such instruments is the "Eco-mark" labeling system. This system, through public awareness, could provide market forces to encourage environment friendly production processes.

- ☆ Another market instrument could be rating of industries, similar to financial rating (CRISIL) or ISO-9000. A good tool would be BS-7750 the British Standard for Environmental Management Systems or ISO-14000.
- ☆ Problems associated with SSI could be solved by helping genuine SSI's especially those promoted by young and technically qualified entrepreneurs. Although adequate network of institutional finance exists, yet there is a need for co-ordinating the flow of both short and long term capital. Government should co-ordinate making capital available to SSI's, on right time and at appropriate terms. Financial institutions created for assisting SSI's should require environmental impact studies on an appropriate scale prior to project financing. Environmental expertise should be made available with the financial institutions for this purpose.
- ☆ By providing R&D encouragement; create instruments for R&D related to cleaner technologies of production and pollution control from SSI
- ☆ Multi-media Compliance represents a more holistic approach to achieving a clean environment. This approach examines the big picture-emissions, the geographic location, wind patterns, social impact – in order to come up with a plan that will be both effective and practically reliable. Release to all environmental media (*e.g.* Air emissions, wastewater discharges, and hazardous wastes) are considered simultaneously and in an integrated manner. This means that the environment permit or license for a facility address all environmental media, as do compliance inspections and enforcement responses.

  The advantages are:

  - It allows the consideration of cross media effect, *e.g.* The effect of wastewater treatment on solid wastes and air pollution generation.
  - It allows the facility operation to consider a full range of environmental improvements
  - Forward thinking decision makers in several countries have incorporated this new focus into legislation and policy with positive results.
- ☆ By promoting and implementing low cost wastewater treatment schemes for management and treatment of wastewater from agriculture and rural areas. One such example is wetland transformation. The capacity of wetlands to remove nutrients and organic loads from freshwaters has been emphasized. Natural riparian wetlands are believed to play a key role in the removal of nutrients, especially nitrates. Several studies have shown the efficiency of constructed wetlands in reduction of nitrogen in natural waters and for biomass production. The important role of wetlands in groundwater storage and recharge has also been proven in many areas, yet it is often neglected in water management planning.
- ☆ Restoration or creation of buffer strips or resparian ecotone along the water courses can arrest nutrient flow in rivers. Such areas are known to have high nitrogen removal capacity by denitrification and uptake by

macrophytes followed by harvesting and storage in the sediments. Wetlands, possibly in the form of wet buffer strips along the water course, have the same potentials.

- By implementing the schemas to popularize the use of organic nutrients in agriculture through use of compost green manure's and bio-fertilizers.
- By making a national policy of reducing the consumption of chemical fertilizers and pesticides
- Controlling the livestock population and improving the quality of breed.
- By increasing the awareness on environmental sanitation in rural areas by providing education.

Conclusion

Water and wastewater management problems in India is different from developed countries as it is linked with gross inadequacy of environmental sanitation, lack of political will, shortage of funds and, non-availability of technology and social, religious and cultural values. While taking the stock of the situation, one is left to ponder about how little has been done so far in this regard and how big is the magnitude of the problem ahead.

Chapter 2

Bio-filteration

Kiran Singh[1], R.S. Singh[1], B.N. Rai[1] and S.N. Kaul[2]

[1]IT-BHU, Varanasi – 221 005
[2]MIT-Pune

Introduction

Several control technologies are used for the destruction and recovery of VOCs (Volatile Organic Compounds) emitted from industries. In general treatment of VOCs bearing waste gases can be done in two different ways–source control and emission control. Source control involves the reduction of emissions through substitution of raw materials, recycling, and its reduction. However, these measures may reduce the quality of the product or increase costs. Secondary control involves treatment of the waste gas after it has been produced. Conventional air pollution control technologies such as absorption, adsorption, condensation, thermal oxidation and catalytic oxidation are generally not cost effective for the treatment of VOCs bearing dilute waste air streams. Besides this, the above treatment technologies produce hazardous by-products also and require the addition of chemicals or fuels that may require further treatment or disposal, thereby creating additional environmental problems. A brief account of their salient features are summarized in Table 2.1.

Bio-filteration, though the oldest of biological methods of waste control, is now emerging as a sustainable alternative for the treatment of air contaminated with VOCs. In bio-filteration the polluted air is passed through a bed of packing media covered with a layer of aerobic microorganisms. The microorganisms are immobilized on the surface of the packing media. The VOCs get transferred from air to the water adhering to the bacterial growth on the media and then get biologically metabolized (Upadhyay and Kumar, 2004). Bio-filteration is energy efficient and cost effective and also does not produce toxic end-products. The degradation of VOCs by bio-filteration

Table 2.1: Characteristics, Advantages and Drawbacks of the Main Air Pollution Control Technology (APCT) Applied to VOC Elimination (Delhomenie and Heitz, 2005).

Technology	*Principle and Characteristics*	*Performances and Limitations*	*Costs/($m^3 \bullet h^{-1}$) Air (In US $)*
Adsorption	**Transfer of VOCs to a porous solid phase, fixed or fluidized** ☆ Materials: activated carbons, zeolites and polymers ☆ Example: Activated carbon adsorbs 10–30 per cent VOCs on weight basis ☆ Doubled installations: adsorption–desorption cycles ☆ Operating temperature<50–60 ºC: ignition risks may be present	☆ Conversion: 90–99 per cent ☆ Possible recovery of VOCs (desorption) ☆ Can accept variations of flow rates and shutdown periods ☆ Treatments for adsorbentregeneration are required ☆ Moisture level of effluents: <50 per cent ☆ High pressure drop ☆ Bed poisoning problems withcertain VOCs	Investment: $15–120 Operation: $10–35
Incineration	**Thermal oxidation of VOCs** ☆ 760 < Temperature < 1200 ºC ☆ 0.3 < Residence time < 2 s ☆ VOCs concentration < 25 per cent of explosion limit ☆ Required O_2 level ~10 per cent	☆ Conversion: 98–99.5 per cent ☆ Possible energy recovery ☆ Elimination of halogenated or sulfurated VOCs with adequate,additional equipment ☆ High investment and operating costs ☆ Toxic by-products: CO, NO_x, dioxins, furans ☆ Efficiency decreases for low VOCs concentrations	Investment: $10–450 Operation: $20–150 (depends on the quantity of recovered energy)
Catalytic oxidation	**Thermal, catalytic oxidation of VOCs** ☆ 300 < Temperature 650 ºC ☆ 0.07 < Residence time < 1s ☆ Catalysts: noble metals (Pt,Pd, Rh) on supports (alumina, other ceramics), or metaloxides (Cu,Ti, Ni, Mn, etc.) ☆ Catalyst life time: 2–5 years ☆ VOCs concentration: far lower than the explosion limit ☆ Required O_2 level ~2 per cent	☆ Conversion: 90–99 per cent ☆ Less energy required than incineration and less toxicby-products ☆ Catalyst deactivation problems (clogging, poisoning, overheating) ☆ Disposal of spent catalyst ☆ Combustion by-products	☆ Fixed catalyst Investment: $20–250 Operation: $10–75 ☆ Fluidized catalyst Investment: $35–220 Operation: $15–90

Contd...

Table 2.1–*Contd...*

Technology	*Principle and Characteristics*	*Performances and Limitations*	*Costs/($m^3 \bullet h^{-1}$) Air (In US $)*
Absorption	**Transfer of the VOC to a liquid phase** ☆ Plate tower, bubble column,packed tower, atomizer ☆ Solvents: water (withadjusted pH), high boiling-point hydrocarbons, amines, etc. ☆ Counter-current operationincreases the VOC transferrate	☆ Conversion: 90–98 per cent ☆ Possible recovery and vaporization of the dissolved VOC with downstream treatments ☆ Inadequate for VOCs of low solubility ☆ Production of wastewater	Investment: $15–70 Operation: $25–120
Condensation	**Liquefaction of highboiling-point VOC (>38 ºC) via cooling and/or compression** ☆ Cooling/cryogenic systems: water (5ºC), brine (→ –35ºC), liquid nitrogen (→ –185ºC)	☆ Conversion: 50–99 per cent ☆ A recovery and vaporization waywell adapted to saturated VOCs only ☆ Disposal of condensates ☆ Problems of frost deposits	Investment: $10–80 Operation: $20–120
Membranes	**Separation of gas mixturesthrough semi-permeable membranes** ☆ Materials: polymers (hollow fibers, silicones), porous ceramics ☆ Gas flow compressed before membrane separation	☆ Conversion: 50–98 per cent ☆ VOC are concentrated 5–100 times, and vaporization (recycle)possible ☆ Selective membranes, resistantto halogenated VOC ☆ High pressure drop ☆ High operating pressures ☆ Membrane cleaning required	Not available
UV/photo-chemical oxidation	**Complete oxidation byoxygenated oxidizers (O_3,H_2O_2), and initiated by UVradiation** ☆ Possible utilization of photocatalysts (TiO_2, Fe_xO_y, etc.) ☆ Operating temperature: ambient	☆ Conversion: 90–98 per cent ☆ Moderate energetic costs ☆ Inadequate for halogenated VOCs ☆ Deposit of oxidation by-products on the catalysts surface (cleaning) ☆ Complex systems	Not available

Contd...

Table 2.1–*Contd...*

Technology	*Principle and Characteristics*	*Performances and Limitations*	*Costs/($m^3 \bullet h^{-1}$) Air (In US $)*
Biotechnologies (detailed later)	**Biocatalytic oxidation of VOCs** ☆ Three configurations: biofilters (most frequent), biotrickling filters and bioscrubbers ☆ Biocatalysts: microorganisms (bacteria, fungi) ☆ 30 s < Residence time < several min ☆ Operating temperature: 20–40°C ☆ Filter-bed life time: 3–5 years	☆ Conversion: 80–95 per cent ☆ Moderate installation andoperating costs ☆ Low maintenance ☆ Strict control of biological parameters (pH, temperature, moisture level, nutrients, etc.) ☆ Large space required for biofilters ☆ Pressure drop problems	Investment: $10–70 Operation: $3–10

is very effective at low concentrations, *i.e.*, less than 1000 ppmv. This technology has been successfully used for removing a wide range of VOCs including mal-odorous pollutants such as ammonia, mercaptans, and sulphurous compounds. There have been many commercial bio-filters using compost-type material installed in Europe and the United States over the past 20 years. Their applications to date also include several industries like, chemical and petrochemical industry, oil and gas industry, synthetic resins, paint and ink, pharmaceutical industry, waste and wastewater treatment, soil and groundwater remediation, and odor abatement applications. Bio-filters and biotrickling filters are nearly similar and the basic difference between them is that in biotrickling filter nutrients are continuously trickled through the porous bed but in bio-filters nutrient circulation is occasional. Therefore bio-filteration technology is cheaper than biotrickling filters.

Bioscrubber is essentially a two stage unit in which absorption occurs in first stage and biodegradation occurs in the other in which microbes are suspended. Bioscrubbers are usually applied for the treatment of readily soluble VOC (alcohols, ketones etc.) having concentration less than 5 g m^{-3}. The major problem associated with bioscrubbers is the generation of excess sludge and liquid waste. The guide lines for selecting treatment technology based on concentration and flow rates are given in Figure 2.1.

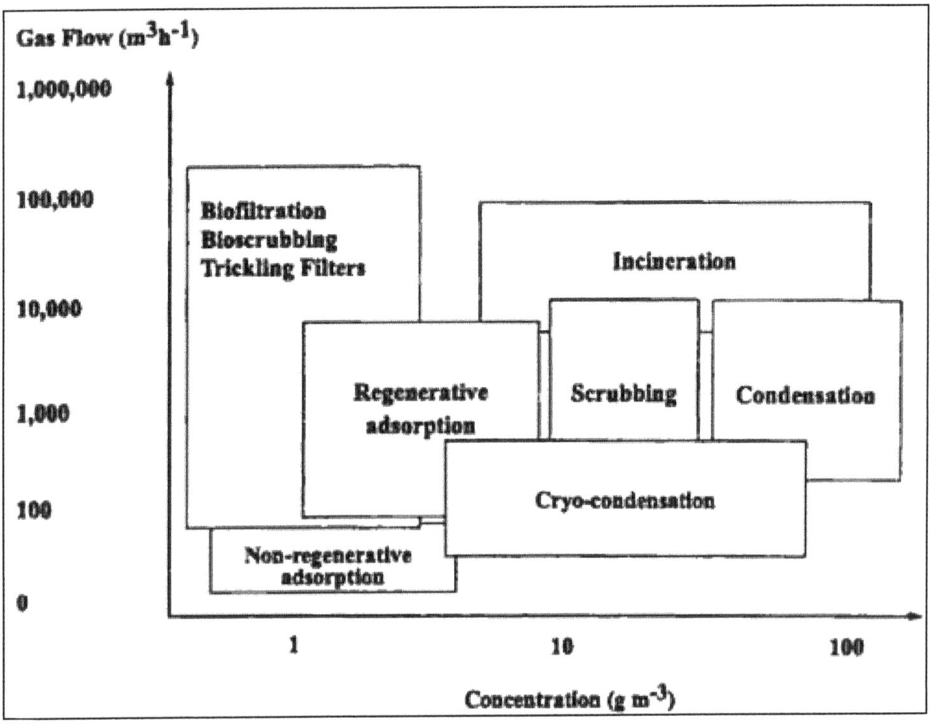

Figure 2.1: Guidelines to Choose VOCs Control Technologies, Based on Flow Rate and Concentration.

Bio-filteration Technology

For the treatment of malodours compounds bio-filteration process was proposed for the first time in the year 1923. Since 1950's the bio-filteration process is being used to treat odorous compounds emitted from wastewater treatment plants. This process is also used to remove odour generated from solid waste processing, animal rendering plants, composting operations, and breweries. For the treatment of hazardous air pollutant bio-filteration process has been widely and successfully used in Germany and the Netherlands since 1970's. In US, bio-filteration process was first used to treat odours resulting from H2S gases in sewage works. There are more than 600 bio-filters operating both in Germany and Netherlands and the technology is also widely spreading in United States. A schematic diagram of a typical bio-filter is shown in Figure 2.2.

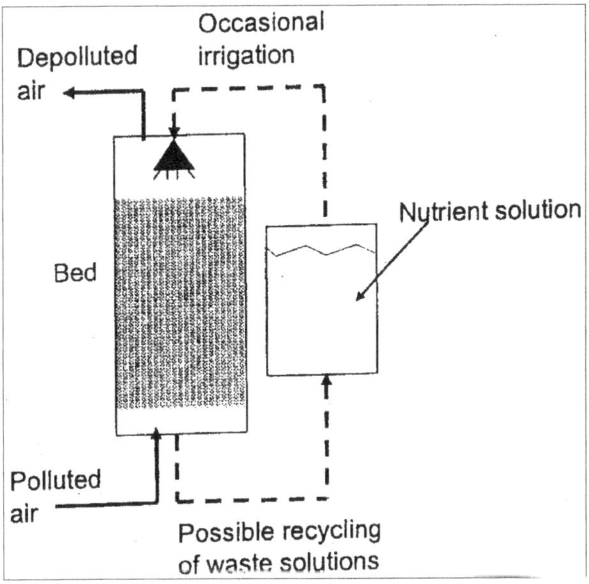

Figure 2.2: Schematic Diagram of a Bio-filter.

In India a few research centers like BITS Pilani, IIT Roorkee, National Environmental Engineering Research Institute (NEERI), Nagpur, Regional Research Laboratory (RRL), Trivendurm, Banaras Hindu University, Varanasi, have been doing work in this area. Giri *et al.* (2010) studied the degradation of dimethyl sulfide in the bio-filter packed with compost and wood chips. Singh *et al.* (2006) and Kiran *et al.* (2010), examined various filter media like wood charcoal, yellow gram stalk, maize cobs and polyurethane foam. Raghuvanshi and Babu, (2009) used a mixture of coal and matured compost for the degradation of methyl ethyl ketone.

Many researchers have reviewed the research on bio-filteration (Leson and Winer, 1991; Bohn, 1992; Singhal *et al.*, 1996; Swanson and Loeher, 1997; McNevin and Barford, 2000; Cohen, 2001; Pandey *et al.*, 2002; Groenestijin and Kraakman, 2005; Delhomenie and Heitz, 2005). Leson and Winer (1991) reviewed the history and current status of bio-filteration discussed the principles and applicability of bio-filters for a wide range of specific emission and sources. Bohn (1992) summarized the work done by other researchers and also provided the brief description of commercial bio-filters running successfully at various locations in Europe and United States.

Bio-filteration Process

Bio-filteration is nothing but bio-oxidation of contaminated air by beds of organic or inorganic media containing microorganisms that have the ability to minimize pollutants aerobically. Bio-filteration utilizes a supported media such as lavarock, peat, compost, agricultural by-products, soil, etc. for microbial growth to remove odours and organic contaminants for air streams. The filter bed is kept wet by humidifying the influent air to maintain sufficient moisture for maintaining a biologically active layer (biofilm) surrounding the media. The contaminated air stream is passed through the packing media and the oxygen present in the contaminated air helps in maintaining aerobic conditions within the bio-filter unit. As the contaminated gas passes through the filter bed, the gaseous contaminants are transferred from gas phase to water phase and get absorbed in the biofilm layer. The contaminants are now available to the microorganisms as food and energy source to support microbial life and growth. This degradation process is an aerobic heterogeneous biochemical process controlled either by mass transfer or biochemical reaction or by both. The removal efficiency of bio-filters varies with media characteristics (*e.g.* particle size, surface area, and porosity), carbon and energy source, nutrients, pH and temperature. The removal of pollutants during bio-filteration process largely depends upon the presence of microorganisms in the filter bed capable of biodegrading the contaminant of interest. The capacity of microorganisms to degrade pollutants decreases with the increase in inlet load.

Biotransformation process can be expressed as:

$$\text{VOCs} + O_2 \xrightarrow{\text{Microorganism}} \text{Biomass} + CO_2 + H_2O + \text{Heat} \quad (1)$$

The biodegradation of VOCs present in air stream begins with the transfer of pollutants from gas phase to aqueous phase and then to biofilm. These pollutants then get consumed by the microbial cells as the source of carbon and energy and get converted to CO_2, water and new biomass. Figure 2.3 shows the mechanism of bio-filteration.

The transfer of pollutants from gas phase to aqueous phase is not a rate limiting step. The partition between gas and aqueous phase is generally described by Henry's law, which is

$$P = HC_w, \quad (2)$$

where, P is the partial pressure of contaminant in the gas phase (Pa), C_w is the equilibrium concentration of contaminant in the water phase (mol/l_{water} or g/l_{water}), and H is Henry's law constant (Pa l/mol or Pa l_{water}/g).

Pollutants with large Henry's law constants are difficult to degrade and require longer residence times than those with small Henry's law constant. Generally the

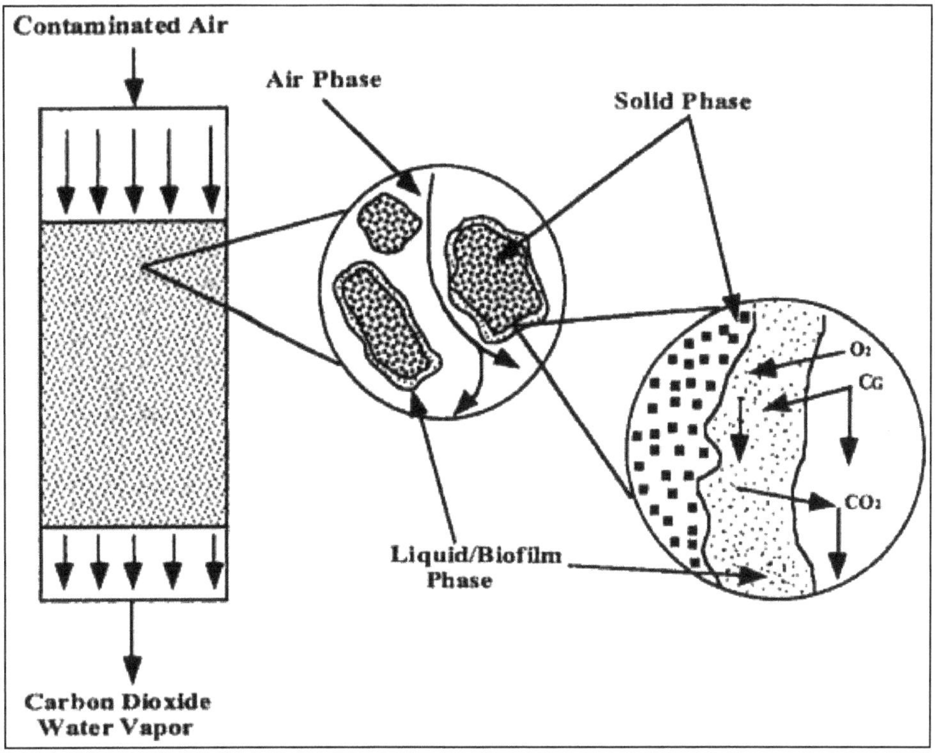

Figure 2.3: Internal Mechanisms of a Bio-filter.

Contaminated air (C_G) passes through the filter bed medium (compost, peat, soil, etc.) with (O_2) and sorbs into a microbial biofilm/liquid phase attached to the filter medium. Microbes convert the contaminant to carbon dioxide (CO_2) and water. (Devinney *et al.*, 1999).

elimination capacity of a bio-filter declines with increasing Henry's law constant. The reversible adsorption of pollutant on the surface of filter media can be described by an equilibrium isotherm (Linear Freundlich, Langmuir, etc.) The transport of pollutants in the biofilm occurs by molecular diffusion following Ficks law.

The destruction of contaminants occurs in the biofilm, containing immobilized microorganisms, surrounding the filter media. The microorganisms degrade the gaseous contaminants and produce water, carbondioxide, some inorganic acids, intermediate products from the incomplete biooxidation of VOC, such as acetic acid, and microbial biomass. The development of biofilm is influenced by several factors such as gas flow rate and type of filter media. Filter media supports the biomass and provides moisture, nutrients, etc. for the microbial reaction. The biofilm thickness usually varies from tens of micrometers to more than 1 cm, with an average value of 1 mm or less. The microbial activity increases with the thickness of the biofilm, up to a level termed as the 'active thickness'. Above this value, the diffusion of nutrients becomes a limiting factor, thus differentiating the 'active' biofilm from the inactive one. Once the rate of adsorption of pollutants becomes faster than the rate of oxidation, the bio-filter becomes overloaded, which decreases the removal efficiency.

Microbiology

Microorganisms are the most critical component of the bio-filter, as because of the microbial metabolism, the adsorbed pollutants are degraded into carbon dioxide and water and cell biomass. In the bio-filters around 1 billion microorganisms are present per gram of organic material. A complex microbiological community exists in the bio-filter. A broad variety of bacteria, yeast, fungi, protozoa, nematodes, algae and other higher organisms are generally observed in bio-filters. For the proper growth of microorganisms and for the optimal contaminant degradation microbes require a suitable environment. The most important environmental factor for microbial function is the moisture in the contaminated air stream entering the bio-filter and the filter media. Initially for evaluation of the microbial population dynamics cell counting method were used but now a days, molecular biological techniques like denaturating gradient gel electrophoresis (DGGE) of polymerase chain reaction (PCR) amplified 16S rDNA fragments, fluorescence *in situ* hybridization (FISH) and single-strand conformation polymorphism (SSCP) are applied. With the help of these techniques entire microbial community existing in the filter bed can be assessed and the most dominating microbe can be found out. In terms of biomass density, a bio-filter contains between 10^6 and 10^{10} cfu[e] of bacteria and actinomycetes (Delhomeni and Heitz, 2005).

The inoculation of the filter bed was done on the basis of the type of filter media used and the biodegradability of volatile organic compound. Generally natural filter media do not require any inoculation, since they have indigenous microbes. Most researchers preferably use packing media like compost and soil as they do not need any inoculation. During the course of bio-filteration the microbes get acclimatized and then the most resistant microbial population dominates the bed. Highly soluble VOCs degrade very easily and a general source such as activated sludge having a wide spectrum of microorganisms is often sufficient for inoculation. For treatment of the VOCs which are less soluble such as halogenated hydrocarbons like TCE etc, soil or water samples from site of contamination with the pollutant of interest or consortia that are enriched in the laboratory on the pollutants of interest, or pure cultures degrading the pollutant of interest are obtained either from culture collection or isolated from the mixed consortia, or leachate of the bio-filter treating the pollutant of interest. By using such sources for inoculation of the filter bed the acclimation time can be minimized. Fungi generally grow at a slower rate in comparison to the bacteria but they are capable of degrading a broad variety of pollutant and can withstand with more adverse conditions. For toluene degradation bacterial consortia was used by many researchers (Singh *et al.*, 2010; Kiran *et al.*, 2010). But now a day's fungi have also used as an inoculum for toluene degradation by many researchers (Gracia-Pena *et al.*, 2001; Maestre *et al.*, 2007; Aizpuru *et al.*, 2005). Microorganisms can survive for fairly long periods when the bio-filter is not loaded up to 2 months, if sufficient nutrients are available from the filter material. Steele *et al.* (2005) observed difference between initial communities and community after acclimation due to restructuring of microbial communities after acclimation. The organisms which are capable of removing different type of odorants and VOCs are listed in Table 2.2.

Table 2.2: Microorganisms Identified during the Degradation of VOC by Bio-filteration

Pollutant	*Microorganisms*
Benzene	*Pseudomonas* sp.
Benzene	*Alcaligenes xylosoxidans*
Benzene	*Cladosporium sphaeraspermum, Exophila lecaniicorni, Phanerochaete chrysosporium*
BTX	*Phanerochaete chrysosporium*
Butylacetate	*Cladosporium resinae, Exophiala lecanii-corni, Phanerochaete chrysosporium*
Diethyl ether	*Graphium* sp. *ATCC 58400*
Ethylbenzene	*Phanerochaete chrysosporium*
Ethylacetate	*Rhodococcus fascians*
Methylisobutylketone	*Phanerochaete chrysosporium*
Phenols and Chlorophenols	*Pseudomonas putida, Pseudomonas fluorescence, Azotobacter, Flavobacterium, Bacillus* sp., *Achromobacter* sp., *Rhodococcus* sp., *Xanthomonas, Aspergillus,Penicillium, Neurospora, Deberomyces, Pichia, Trichosporon, Oospora*
Styrene	*Exophiala jeanselmei, Rhodococcus pyridinovorans, Phanerochaete chrysosporium* ATCC 34541
Toluene	*Pseudomonas pseudoalcaligenes, Paecilomyces variotii* CBS115145, *Cladosporium sphaerospermum, P. putida*
Trichloroethane	*Pseudomonas putida, Methylosinus trichosporium* OB3b, *Burkholderia cepacia* G4, *Pseudomonas mendonica* KR-1, *Nitrosomonas europaea, Pseudomonas putida* F1
Xylene	*Pseudomonas pseudoalcaligenes, Phanerochaete chrysosporium*
Hexane	*Aspergillus niger, Fusarium* sp.
α-pinene	*Pseudomonas* sp., *Nocardia* P18.3; *Bacillus pallidus* BR425
Sulphurous odourants	*Hyphomicrobium* sp., *Thiobacillus thioparus,Thiobacillus ferroxidans, Pseudomonas* sp., *Xanthomonas* sp., *Aeromonas* sp., *Basidomycetes* sp.
Formaldehyde	*Gliocladium deliquesecens, Paecilomyces variotii*

Source: Delhomenie *et al.*, 2002; Pandey *et al.*, 2002; Kennes and Veiga, 2004.

Pollutants are not always degraded independently by microorganism. Several sparingly soluble VOCs such as high chlorinated aliphatics (*e.g.* trichloroethylene (TCE)) are degraded by bio-filteration using co-metabolism *i.e.* certain other easily degradable hydrocarbon (*e.g.* toluene) helps in degrading TCE. On the other hand certain pollutant in mixed stream works as inhabitant for another one. Continuous exposure of microbes to high pollutant concentrations during bio-filteration processes makes them less effective as the bacteria lose their ability to degrade pollutant. This also enhances the growth of non-pollutant degrading bacteria. In the bio-filteration process change in the filter bed temperature during the degradation of VOCs indicates the intensity of microbial activity.

Factors Affecting Bio-filter Performance:

Nature of Pollutant

Pollutant's solubility and biodegradability have a great influence on the overall efficiency of bio-filteration. The adsorption of pollutants into the biofilm largely depends on their solubility. High water solubility, low vapor pressure and hence low Henry's law constant are the desirable properties. Alcohols, aldehydes, ketones and some simple hydrocarbons are highly to moderately water soluble. On the other hand chlorinated hydrocarbons are less water soluble. The degradation of slightly soluble compounds at low concentrations in the bio-filter coloumn is limited by the mass transfer rate into the biofilm. However, soluble compounds introduced at higher concentrations, are easily absorbed into the biofilm, but in this case biodegradation rate is the limiting factor. To increase the solubility of hydrophobic pollutant now a days some additives like silicone oil is being added.

Bio-filter Media

The ability of bio-filter to remove pollutant largely depends on the selection of the packing material as bio-filter medium. The packing medium can briefly be described as solid phase on which adhesion of the microorganisms takes place, resulting in the development of a biofilm around it, which grows as a result of pollutant degradation. The choice of bio-filter medium is crucial in order to maintain long-term efficient bio-filter operation. The desirable properties which are considered during the selection of an appropriate packing material are its specific surface area, density, porosity, pH, water holding capacity, buffering capacity, and elemental composition. Presence of these properties ensures intrinsically active bio-filter medium with large effective surface area, minimum pressure drop, and uniform distribution of air throughout the bed without channeling. Bio-filter medium also serves as living space and reserve of substrate for microorganisms, as humidity reservoirs and as mechanical support for the maintenance of the internal structure of the filter bed. Filter materials can be broadly classified into two categories *i.e.* natural and synthetic filter materials. Soil, compost, peat, wood chips, biomass, coal are the most commonly used natural filter materials and Pall rings, granular activated carbon (GAC), polyurethane foam, polystyrene, etc. are commonly used synthetic media. Natural materials show compaction problem and may deteriorate over long term operations so some inert materials like poly vinyl alcohol, ceramic, perlite, vermiculite, glass beads, and polypropylene are blended with them to increase their strength. Nutrient supply is also a critical parameter for the microbial activity, so nutrients must be added during the operation. A synthetic filter material PVA/peat/KNO_3 composite bead are also developed which contains nitrogen and phosphate bearing compounds as nutrients.

Natural Filter Media

These filter media have most of the desirable properties required for achieving good removal efficiency such as the presence of intrinsic nutrients, indigenous microflora, large specific surface area (from 300 to 1000 $m^2 m^{-3}$), large water retaining capacity, low bulk density, ease of availability and low cost. However, there are some disadvantages also associated with the use of natural medias. Most of these packing materials have a common disadvantage of being strongly subjected to aging phenomena, resulting in bed shrinkage. This problem can be prevented by adding bulking inert agents such as polystyrene sphere, lava particles, glass beads, porous clay and ceramic.

Compost

Compost is the most widely used bio-filter medium because it has most of the desirable properties required for a good medium. A wide variety of raw materials and processes are used to produce compost based media. These include municipal solid waste, industrial sludge, garden waste, animal manures and food and agricultural residues. The only disadvantage of compost is compaction of the bed, aging effect and problem of stability during long operation. Compost inherently contains broad spectrum of microbial population so inoculation is not necessary. But for rapid system start up sometimes addition of microbial inocula may be useful. There is also little requirement of external supply of nutrient in compost as compared to other materials used as bio-filter media. Active compost has more available nitrogen than finished compost, but temperature in active compost may exceed 60°C. However, the airflow through the bio-filter will normally reduce the temperature into mesophilic range. In order to overcome the limitations of compost bed bio-filter now a days compost are used in combination of some other natural and synthetic media.

Agro Wastes

Use of agricultural waste as a filter media in bio-filter column is very attractive from cost and environmental point of view. In recent years, there has been increasing trend towards more efficient utilization of agro-industrial residues. There are six characterized agricultural by-products as potential bio-filter media- peanut shells, rice husk, coconut shells, cane bagasse, maize stubble and yellow gram stalk.

Peat, Coal and Soil

Peat is also an excellent medium for bio-filters. Its low pH – 3-4 is an attractive feature. It provides more surface area, better flow characteristics and less pressure drop during operations. Major problem associated with peat bed is moisture control. It also requires supplemental nutrient and buffering. Soil is a cost effective packing medium having indigenous microbial population. It is structurally stable and permits use of deeper beds. Its hydrophilic nature permits ease of moisture control. The problems associated with soil media are high pressure drop and its solubility in water. It is suitable for situations having large land area.

Synthetic Filter Media

These man made packing materials such as activated carbon, polyurethane foam, pall rings, and composite beads are some examples of the materials which have been successfully used as bio-filter media. The synthetic media have several advantages over natural media such as high strength, low pressure drop, durability, better flow characteristics, large specific surface area, and easy removal of biomass. The disadvantages of these media are long acclimation period and need for inoculation.

External supply of nutrient is also required but research is going on to develop new media with nutrients added on to their surface using appropriate physico-chemical techniques.

Granular Activated Carbon

During the last decade, GAC has been used successfully as a good packing material because of the advantage of rapid pollutant adsorption, large surface area and porosity and the resultant improved the degradation capacity of bio-filters.

Polyurethane Foam

Polyurethane foam has been found to be very useful media in bio-filteration. Due to open structure and high porosity, low density, easy inoculation procedure and ability to absorb water, the pressure drop over the packing remains low at relatively high gas velocity. The large specific area, less aging effect, and physical strengths make this media suitable for long time operation.

Pall Rings

Pall rings are used as bio-filter packing medium which prevents media-bed plugging. However, startup is relatively long presumably because of poor biofilm establishment on the surface. The relatively low specific surface area is also a disadvantage for achieving a high elimination capacity. Pall rings are made of plastic and so its use as packing material may be beneficial because of its stability, low cost and high porosity.

Temperature

Temperature regulation is very important in bio-filteration as the biological activity is temperature-dependent. General operating temperature range is 14°C to 40°C, which is suitable for the growth of mesophilic microorganisms. Bio-filter performance is in general limited by the biological reaction rate or by the mass transfer rate. The temperature has a different effect on either of these limitations; consequently, effect of temperature on the performance varies depending on the particular application. These effects may counterbalance each other. The influence of temperature on microbial activity has been well studied and in the suboptimum range (15-35°C) it can in general be described by the Arrhenius equation. The temperature beyond the optimum range will cause a decline in the performance and even cause death of the microorganisms. However this is favorable for thermophilic organisms. The metabolic rate of the thermophilic bacteria increases with temperature, doubling with each 10°C increase up to 65-70°C. Due to increased metabolic activity of microbial population higher removal efficiency and elimination capacities can be achieved. During the thermophillic operation of bio-filteration less biomass is produced and thus clogging of bed does not occur and thus the pressure drop through the bed is minimized. While a warmer bio-filter generally supports more active organisms, the physico-chemical effects of higher temperatures are usually unfavourable. For most gases with the rise in temperature the Henry's law coefficient increases; so that less of the compound will be dissolved in the water. This will reduce the sorption of the compound. Therefore, the transfer of the contaminant from the air to the microorganisms will be less effective.

Moisture Content

Among the other parameters affecting bio-filter performance (filter bed material, additional nutrients, temperature, pH, etc.), water availability within a bio-filter must be carefully controlled to ensure high microbial activity. The microorganisms need minimum water content for optimal growth and activity. Too little moisture content causes drying of the bed and leads to channeling, resulting in non-uniform gas distribution. Also the microorganisms get deprived of water which causes significant reduction in biodegradation rate. Excess water has been shown to cause a reduction of bio-filter porosity and formation of anaerobic zones. Along with this it will also increase the backpressure due to the reduced void volume, and channeling of the gas within the bed. Loss of water from the bed occurs due to (i) evaporation to an unsaturated influent gas stream, (ii) bio-oxidation caused by bed heating, (iii) metabolic water requirements, and (iv) heat exchange with the environment. Excessive moisture content reduces the transfer of oxygen and thus decreases the biological activity. To maintain optimum moisture content of the filter bed, the polluted air stream should be humidified before introducing in the bio-filter column. For maintaining the moisture content it is often advantageous to have downflow bio-filter. Since most of the drying occurs at the entrance to the filter bed, drying at the top is easily handled through direct water addition and flow from the top to the bottom.

pH Value

Most microorganisms prefer a specific pH-range. Thus changes in the pH of the filter material will strongly affect their activity which in turn will affect the performance of the bio-filter. The microorganisms involved in the degradation of VOCs generally grow in the pH range of 6.5-7.5. The pH in compost filters is typically between 7 and 8.5, a range preferred by bacteria and actinomycetes. Highly acidic condition generally inhibits microbial activity and most bacteria favor neutral conditions. In low pH bio-filters fungi are to be the dominating species. Sulfur oxidizing bacteria, however, are well adapted to acidic conditions. During microbial degradation, acidic by-products may be produced. Therefore, to maintain the pH of the filter bed, buffering agents may have to be added. Slaked lime ($Ca(OH)_2$), lime (CaO), limestone ($CaCO_3$), crushed oyster shell ($CaCO_3$), dolomite [$CaMg(CO_3)_2$], marl (soil + $CaCO_3$), and slag ($CaSiO_3$) are used to neutralize the acidic condition. In addition to controlling pH by buffering materials, periodic washing or back washing of bio-filters are also done.

Pressure Drop

For economical operation of bio-filteration process the pressure drop across the bio-filter column should be minimum since an increase in pressure drop requires more enrgy and can result in air channeling through the media. Pressure drop across the bed depends on the volumetric flow rate of gaseous stream, media porosity and bed depth. Pressure drop is directly related to the moisture content in the media and the media pore size. Increased moisture and decreased pore size result in increased pressure drop. Consequently, filter media selection and watering is critical to bio-filter performance and energy efficiency. Due to accumulation of biomass, pressure increases which can be minimized by periodic washing of the bed with water. This will help removing the excess biomass from the filter column and thus the proper aeration of the filter bed can be maintained.

Acclimation

Bio-filters require an acclimation period following startup to reach optimum pollutant removal capacity. This period generally ranges from days to weeks depending upon system and pollutants being treated. Low loading rates may be necessary during this period. During acclimation period appropriate pollutant-degradation enzymes get induced within the microbial population and microorganisms grow to established the biofilm on the media surface. The acclimation period can be shortened by inoculating the filter with organisms already acclimated to and actively degrading the pollutants present in the gas stream. Cultures isolated from already running bio-filters or activated sludge plants can be used for inoculation purposes.

Operation under Transient Conditions

Most operating bio-filters are subjected to varying feed conditions due to process shut downs, startup and surges in influent pollutant concentrations. During process shut down or low inlet pollutant concentrations, microbial population starves for the food and undergoes endogenous decay, resulting in decreased pollutant degrading activity. The longer the period of non-operation, longer is the duration of acclimation after restart-up to restore full activity. If the process shut down and transients are too frequent, microorganisms adapt to the situation and restart-up acclimation periods is shorter. Continuation of air flow through the filter during process shut downs helps in maintaining aerobic conditions to reduce destruction of aerobic micro-organism film. Unstable or transient feed conditions during startup are the most detrimental because they can resist the formation of microbial growth and result in excessively long acclimation periods. Sudden increase in influent pollutant concentration is also detrimental to bio-filter performance, leading to pollutant breakthrough, toxic inhibition or even death of the microorganisms in the bio-filter. In bio-filters treating high loads the biomass will grow rapidly if nutrients are available. This may sometimes lead to clogging and pressure drops. To overcome such a situation backwashing is done.

Gas Flow Rate and Pollutant Loading Rate

Gas flow rate is one of the most important hydrodynamic parameters in the bio-filteration process because it quantifies the amount of pollutant entering the bio-filter column per unit time. The increase in gas flow rate will increase the inlet load and consequently the diffusion flux through the biofilm. It will also decrease the time of contact between the pollutant and the microbial growth on the filter material. The decrease in the pollutant residence time will cause a reduction in the removal efficiency. The high gas flow rate is also responsible for stripping water from bed which tends to desiccate the bio-filter. Direction of flow may also have a role in bio-filteration. Better distribution of nutrients may be under downflow condition but energy requirement to operate bio-filter will be higher. The typical range of residence time is 30-60 s. The gas flow rate through the bed is an important parameter in deciding the loading rate of the bio-filter. The typical range for the gas flow rate through the bed is 0.3-1.6 m^3/min-m^3.

Oxygen Supply

Oxygen supply is necessary for the operation of bio-filters because the predominant microorganisms used in bio-filteration are aerobic, and require oxygen for metabolism. Aerobic heterotrophic bacteria present in filter beds require at least 5-15 per cent oxygen at the inlet gas stream to survive. The oxygen concentration is in most polluted gases is in general several order of a magnitude higher than the pollutant concentration. However, because of the low oxygen water solubility, bio-filter performance may be limited by mass-transfer of oxygen into the biofilm and/or by diffusion in the biofilm. The elimination capacity is higher when partial pressure of O_2 is higher in gas thus demonstrating that at normal oxygen content, VOC removal is limited by oxygen availability.

Nutrient Supply

Microorganisms utilizes VOCs as their food source by bio-oxidizing it to CO_2 and water. In order to maximize the pollutant elimination capacity, nutrients like nitrogen, potassium, phosphorus, calcium, zinc, selenium, iron, manganese, magnesium, sulfur and sodium should be present at a concentration high enough to maintain an active, growing culture. Nutrient solution is also added to the bio-filter column for attaining the optimum moisture content. In general, the nutrient requirement is qualitatively the same as the elemental composition of biomass. Although organic packing materials contain the necessary nutrients for the growth of microorganisms, but the performance of toluene-degrading bio-filters improves when an extra source of ammonium and phosphate is provided to the reactor. Therefore a good packing material for VOC treatment should have low inherent nitrification activity level and high slow-releasing nitrogen content rather than high soluble nitrogen content.

Bed Depth

The depth of bio-filter medium ranges from 0.5 to 2.5 m. A depth of approximately 1m appears to be most common, to allow sufficient residence time while minimizing filter land area requirements. Some manufacturers recommend the use of multiple layers of bio-filter media, since these will need less land area for high loading rates. Very high filter depths are restricted due to the fact of high pressure head and potential for compaction of bed at the bottom which further increases pressure drop and causes channeling.

Waste Gas Pre-Treatment

The microbial communities in bio-filters can be poisoned by the presence of toxic contaminants, excessive concentration of the contaminant, or excusion in environmental conditions like pH, temperature, and moisture content. In order to meet the basic requirements for optimal operation of the bio-filter, waste gas conditioning is often required. A sufficient supply of oxygen and humidity, and an acceptable range of temperature and pH levels in the filter bed, are indispensable for the survival of the microbial community present in the bed. High particulate loads in the waste gas can adversely affect the operation of a bio-filter by clogging the air distribution system and the filter material itself. Pretreatment options can include humidification for temperature and humidity control and/or the use of devices for particulate control, such as wet scrubber or a wet electrostatic precipitator.

Clogging and Channeling

In the bio-filter column clogging and channeling occurs due to the overgrowth of biomass generally at the higher loading rates, as a result the performance of bio-filter is altered due to less availability of specific surface area required for better transport of oxygen, water and VOCs. Therefore, the excess biomass should be periodically removed from bio-filter columns for long term operation. Several techniques such as backwashing is used but it is found that this technique is only suitable in case of small-scale bio-filters having cross-sectional areas of 0.004-0.015 m^2. Modified backwashing combined with air sparging or chemical treatment can also be done to avoid clogging and channeling. Agitation of filter media and change in mode of operation of bio-filter from the upflow to downflow can help in reduction in excess growth of microorganism.

Mathematical Modeling

Mathematical modeling is an important tool to improve the process design and performance of the biological treatment systems. The pioneer and most commonly used model was proposed by Ottengraf and van den Oever (1983). They proposed a steady-state model and used the concept of a bed of solid filter particles coated with a biologically active aqueous layer called biofilm. They assumed a plug flow model for the flow of contaminated gas through the filter bed. Oxygen was assumed to be in excess, thus oxygen limitation on kinetics was ignored. The biofilm thickness was assumed to be small compared to the diameter of the support particles and a constant value for the film thickness was used throughout the reactor despite the changes in the VOC concentration along the bio-filter column. Pollutants are removed from the gas phase by absorption followed by simultaneous diffusion and a zero order reaction in the biofilm. Analytical solution of this model was presented in terms of dimensionless concentration. Biokinetics of bio-filter can be explained with the help of most commonly used modified Monod model. Monod model is assumed for growth as a function of existing concentrations of biomass and the concentrations of contaminant:

$$\frac{dX_{act}}{dt} = \mu X_{act}, \quad \mu = \frac{\mu_{max} C_{bf}}{K_S + C_{bf}}; \qquad \frac{dX_{act}}{dt} = Y\frac{dC_{bf}}{dt}$$

where X_{act} is the biomass density, μ is the growth rate, μ_{max} is the maximum value of the growth rate, K_S is the Monod or half-saturation constant, and Y is the biomass yield. For high values of C_{bf}, the growth rate is constant, and growth rate is approximated by zero-order kinetics and similarly for low values of C_{bf}, growth is linear with contaminant concentration, and hence first-order kinetics may be presumed. However, when the model includes sufficient detail to show biodegradation rates as a function of depth within the biofilm, concentrations will range from the Henry's equilibrium value at the surface of the biofilm to zero at the maximum depth of penetration, so it is likely that both regimes will be encountered and the full form of the Monod equation will be needed. Often the appropriate values for K_S and μ_{max} are uncertain. Both values are strongly dependent on the conditions under which they are determined and most data in the literature are from experiments performed on microorganisms in stirred, well-aerated suspensions, rather than in biofilms and are highly variable. Thus, these parameters are often fitted to the bio-filter data obtained with the experiment being modeled.

Design and Operating Factors

Operation and performance of bio-filter is generally reported in terms of removal efficiency, or pollutant elimination capacity as a function of pollutant loadings of gases or the empty bed retention time (EBRT). These terms are defined as:

$$\text{Removal Efficiency (RE)} = \frac{C_{in} - C_{out}}{C_{in}} \times 100 \text{ (per cent)}$$

$$\text{Elimination capacity (EC)} = \frac{C_{in} - C_{out}}{V} \times Q \ (\text{g m}^{-3}\,\text{h}^{-1})$$

$$\text{Empty Bed Retention Time (EBRT)} = \frac{V}{Q} \ (s \ or \ \text{min})$$

$$\text{Pollutant loading (L)} = \frac{C_{in}}{V} \times Q \ (\text{g m}^{-3}\,\text{h}^{-1})$$

where, C_{in} and C_{out} are the inlet and outlet pollutant concentrations, respectively, V is the packed bed volume (m^3) and Q is the air flow rate ($m^3\,h^{-1}$).

Most of the researchers used EBRT as the parameter to evaluate the performance of bio-filter at different gas flow rates. For a constant volume of bio-filter bed, EBRT and gas flow rate have linear relationship. In most of the operational bio-filters, the EBRT ranges from 15 seconds to several minutes. Generally EBRT considered in the range of 0.5 for readily degradable compounds to 10 minutes for recalcitrant compounds.

Selection of EBRT value for bio-filter also depends on other operating conditions such as the VOCs concentration and biodegradability level. VOCs concentrations introduced beyond the toxicity threshold value tend to inhibit the microbial growth and activity. The actual appropriate bed height depends on media type and expected pressure drop.

A typical elimination capacity versus pollutant loading curve is shown in Figure 2.4. It is usual to report the performance as a function of the load, *i.e.* product of inlet concentration and air flow rate, rather than the concentration. This enables comparison of systems of different sizes operating under different conditions.

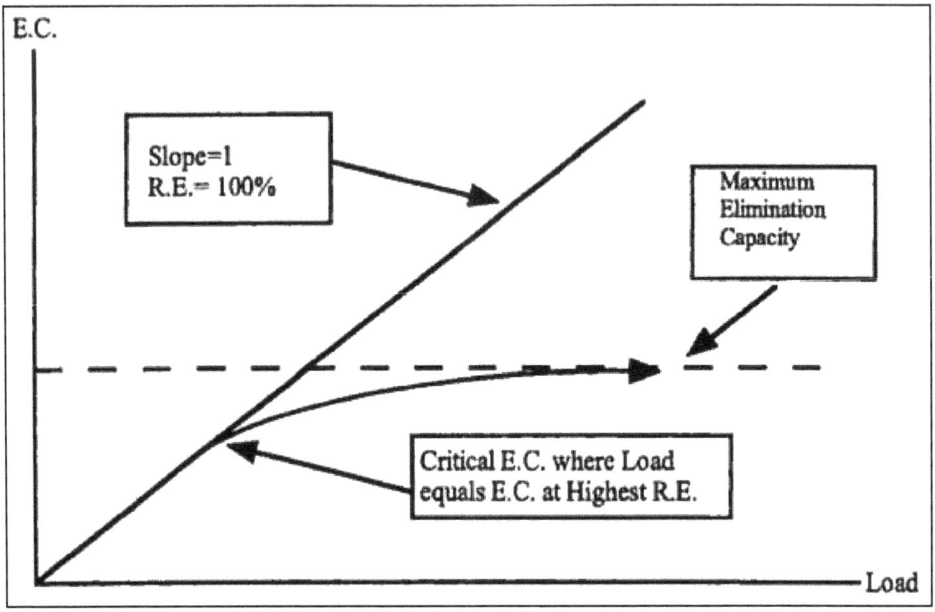

Figure 2..4: Typical Elimination Capacity vs. Load Curve. Elimination capacity is always Equal to or less than the load. The ratio between elimination capacity and load is the removal efficiency of the system.

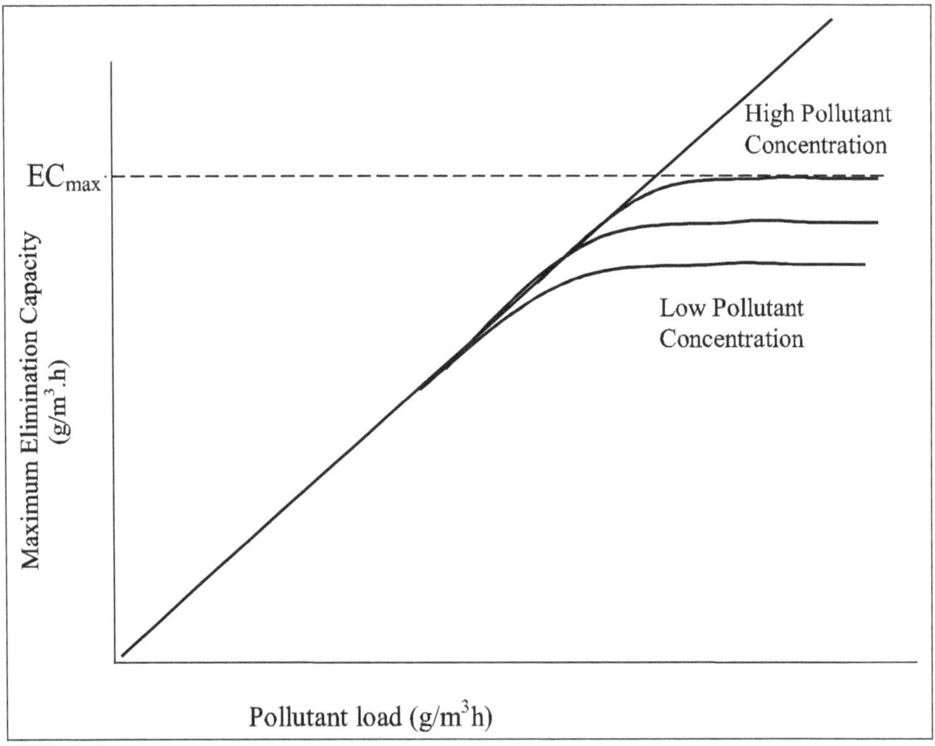

Figure 2.5: Elimination Capacity and Load Curve for a Typical Bio-filter.

Figure 2.5 reveals that there are essentially three operating regimes:

1. Low loading, (also called first order regime): The elimination capacity and loading are identical and the pollutant is completely removed. The bio-filter is operated well below its maximum elimination capacity. The performance increases proportionally with the loading.
2. Intermediate range: Breakthrough of the pollutant occurs with higher inlet concentration or higher air flow rates, the elimination capacity increases, but to a lesser extent than the loading.
3. High loading, (also called zero order regime): The bio-filter is operated at its maximum elimination capacity. Increase in the loading doesn't result in further increase in elimination capacity, the removal efficiency decreases. As indicated on the figure, if the concentration is very low, the maximum elimination capacity is reduced accordingly. Above behaviour can also be explained with help of the controlling mechanism in the bio-filter. At low loadings of pollutant, overall process is controlled by mass transfer and at high loading reaction in the biofilm becomes the controlling mechanism. For evaluating bio-filter performance, one should consider both the maximum elimination capacity and the removal efficiency.

The relationship between off-gas flow rate, required residence time, and the corresponding reactor volume is the most crucial aspect in the bio-filter design, since it strongly affects space requirements and capital cost of a bio-filter. The elimination of a single pollutant in a well-functioning bio-filter generally follows the concentration profile shown in Figure 2.6. Removal rates depend on the compound and the media type in the bio-filter. It is reported in units of g/kg of dry media/day. The rate of removal is nearly linear with distance into the media or with Empty Bed Residence Time (EBRT) at higher concentrations. Generally, the lower-molecular –weight, less-complex compounds are more easily degraded and more quickly removed in a bio-filter.

At lower concentrations, the rate of removal decreases and follows a power function. Lowering the polluted air flow rate (Case-B) by increasing the filter bed area increases the effective residence time and improves performance per unit of bed height, thus causing a steeper concentration profile. However, it also requires more filter volume per unit of air flow.

Advantages and Disadvantages of Bio-filteration

Bio-filteration has several advantages over conventional treatment technologies:

1. The main advantage of using bio-filteration over other conventional control technologies are lower capital and operation and maintenance costs, low chemical usage, and no need of combustion source. Further, it does not transfer pollutants from one phase to another phase but it completely destroys the pollutants and produces non-harmful products.
2. Bio-filteration units can be designed to physically fit into any industrial setting. A bio-filteration unit can be designed in any shape and size. These

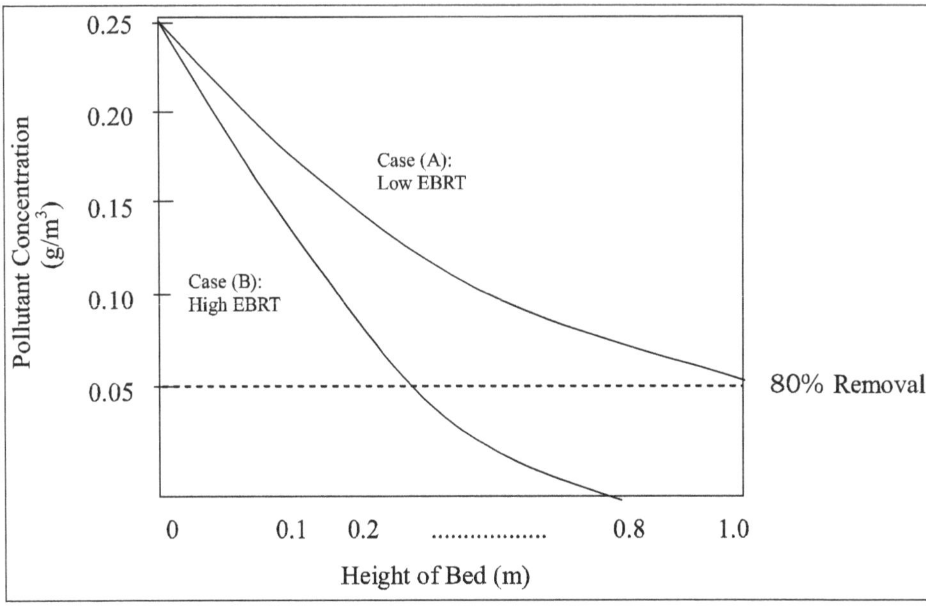

Figure 2.6: Removal of Pollutant in a Bio-filter as a Function of Empty Bed Residence Time (EBRT).

can be in open field with the piping and delivery system underground or in-door. In addition, bio-filters can be designed with stacked beds to minimize space requirements and multiple units can be run in parallel.

3. Bio-filteration is versatile enough to treat odorants like hydrogen sulfide, toxic compounds, and VOCs simultaneously, from the emission of various industries. The treatment efficiencies of these constituents are above 90 per cent for low concentrations of contaminants (<1000 ppm) and large volumetric flow rates.
4. Different media, microbes and operating conditions can be used to enhance the performance of bio-filteration system for many emission points.

Inspite of all their advantages, bio-filters also have several disadvantages:

1. Fluctuating inlet stream characteristics can be detrimental to the bio-filter's microbial population and overall performance and also risks violating emission discharge restrictions. The inability to maintain consistent removal efficiency can be a major limitation unless full support of the regulatory authority and community is achieved.
2. Bio-filteration cannot successfully treat some organic compounds, which have low adsorption or degradation rates. This is especially true for chlorinated VOCs. With the help of this technology only the biologically assimilable gaseous pollutants can be controlled.

3. Contaminant sources with high chemical emissions would require large bio-filter units or open areas to install a bio-filteration system.
4. The presence of microorganisms in the bio-filter media has raised concern over their potential release into the treated air and resultant exposure of workers on-site and individuals off-site to pathogens.
5. The most frequent problems were caused by changes in the bio-filter media due to dry-out, rapid degradation and clogging. These problems result in excessive pressure drops and gradual accumulation of acidic by-products. The bio-filter bed must not dry out (because of insufficient moisture addition) or flood (because of excessive moisture addition). If the bed dries out, the microorganism will become inactive and biodegradation will rapidly decrease. If the bed floods, the gas flow will be impeded and consequently the biological activity will decrease.
6. Compounds that cause either a pH decrease or poisoning of bacteria will cause temporary or irreversible loss of biodegradation activity.
7. With the continuous supply of nutrient, bio-filter accumulates large amount of biofilm, which leads to clogging. At present, clogging of the reactor over long run is one of the major obstacles for proper functioning. Clogging increases pressure drop across the reactor thereby decreases pollutant removal rate (Okkerse *et al.*, 1999; Cox and Deshusses *et al.*, 1999a). Clogging of air distribution systems, rapid corrosion of ductwork and concrete parts, emissions of odorous byproducts and flooding of media have also been observed.
8. Acclimation periods for the microbial population may take weeks or even months.

Observing the overall performance of different technologies available, it can be said that bio-filteration has the capacity to serve as a viable VOCs control technology.

Conclusion

Bio-filteration is an attractive technique for removal VOCs as well as odour from various emissions. The biological mechanism involved in the treatment of wastewater is time proved, the gaseous pollutant treatment are almost similar and forces a new engineering conception and has led new application in the field of environmental biotechnology. This new concept offers scope and challenges in the development of novel bio-filter, support media, nutrient requirement, specific organisms and scale up to large systems. There are also potential possibilities in application of genetic engineering for development and exploration of more effective microorganisms to degrade mixed streams of different types of pollutants.

References

1. Aizpuru, A., Dunat, B., Christen, P., Auria, R., Garcia-Pena, I., Revah, S., 2005. Fungal bio-filteration of toluene on ceramic rings. *J. Environ. Eng.-ASCE.* **131**, 396–402.

2. Cohen, Y., 2001. Bio-filteration–the treatment of fluids by microorganisms immobilized into the filter bedding material: a review. *Bioresource Tech.* 77, 257-274.
3. Devinney, J.S., Deshusses, M.A., and Webster, T.S., 1999. *Bio-filteration of Air Pollution Control.* CRC Press. Boca Raton. FL.
4. Delhomenie, M.C., Bibeau, L., Heitz, M., 2002. A study of the impact of particle size and adsorption phenomena in a compost-based biological filter. *Chemical Engineering Science* 57, 4999 – 5010.
5. Delhomenie,M.-C., Heitz, M., 2005, Bio-filteration of air: a review, *Critical Reviews in Biotechnology*, 25, 53-72.
6. García-Peña, E.I., Hernández, S., Favela-Torres, E., Auria, R., Revah, S., 2001.Toluene bio-filteration by the fungus *Scedosporium apiospermum* TB1. *Biotechnol Bioeng*. 76, 61–69.
7. Groenestijn, J.M., van, Karaakman, N.J.R., 2005. Recent developments in biological waste gas purification in Europe. *Chemical Engg. Journal* 113 (2-3), 85-91.
8. Giri, B.S., Mudliar, S.N., Deshmukh, S.C., Banerjee, S., Pandey, R.A., 2010. Treatment of waste gas containing low concentration of dimethyl sulphide (DMS) in a bench-scale bio-filter. *Bioresource Technology* 101 (7), 2185-2190.
9. Kennes, C., Veiga, M.C., 2004. Fungal biocatalysts in the bio-filteration of VOC-polluted air. *J Biotechnol* 113, 305–319.
10. Kiran Singh, R S Singh, B N Rai and S N Upadhyay, 2010, Bio-filteration of Toluene Using Wood Charcoal as the Bio-filter Media, *Bioresource Technology*, 101, 3947-395.
11. Leson, G., Winer, A. M., 1991. Bio-filteration: An innovative air pollution control technology for VOC emissions. *Journal of the Air and Waste Management Association* 41 (8), 1045-1054.
12. McNevin, D., Barford, J., 2000. Bio-filteration as an order abatement strategy. *Biochemical Engg. Jl.* 5, 234-242.
13. Maestre, J.P., Gamisans, X., Gabriel, D., Lafuente, J., 2007. Fungal bio-filters for toluene bio-filteration: Evaluation of the performance with four packing materials under different operating conditions. *Chemosphere* 67 (4), 684-692.
14. Ottengraf, S.P.P., van den Oever, A.H.C., 1983. Kinetics of organic compound removal from waste gas with a biological filter. *Biotechnol. Bioeng*. 25, 3089-3102.
15. Pandey, R.A., Padolay, K.V., Subbarao, T.V., Mudlier, S.N., 2002. Biological methods for deodorization of industrial emissions- an overview. *Journal of I.A.E.M.* 29, 158-175.
16. Raghuvanshi, S., Babu, B.V., 2009. Experimental studies and kinetic modelling for removal of methyl ethyl ketone using bio-filteration. *Bioresource Technol.* 100, 3855-3861.

17. Singhal, V., Singla, R., Walia, A.S., Jain, S.S., 1996. Bio-filteration – An innovative air pollution control technology for H_2S emission. *Chemical Engg. World* 31 (9), 117-124.

18. Swanson, W.J., Loeher, R.C., 1997. Bio-filteration: fundamentals, Design and Operations principles, and applications. *J. Environmental Engineering* 123 (6), 538-546.

19. Steele, J.A., Ozis, F., Fuhrman, J.A., Devinny, J.S., 2005. Structure of microbial communities in ethanol bio-filters, *Chemical Engg. Jl.* 113 (2-3), 135-143

20. Singh, R.S., Agnihotri, S.S., Upadhyay, S.N., 2006, Removal of toluene vapour using agro-waste as bio-filter media, *Bioresource Technol.* 97, 2296-2301.

21. Singh Kiran, Singh R.S., Rai B.N., Upadhyay S.N., 2010, Bio-filteration of toluene using wood charcoal as the bio-filter media, *Bioresource Technology* 101, 3947–3951.

22. Upadhyay, S.N. Kumar, V., 2004. Biological control of air pollutant, first edition. *Consice Encyclopedia of Bioresource Technology*. The Haworth Reference Press, New York.

Chapter 3

Biochemical Removal of Nitrogen from Industrial Effluents: A Full Scale Plant Performance and Stability

L. Szpyrkowicz[1] *and S.N. Kaul*[2]

[1]*University of Venice, Venice, Italy*
[2]*CSIR-NEERI, Nagpur*

Introduction

Biological nitrification/denitrification is a well-known cost-effective method of domestic or mixed wastewater treatment. If used to solve problems of industrial effluents, the performance of the plant is often unsuccessful, with periodic block of nitrogen removal. The present paper concerns the performance of nitrification/denitrification during a 1.5-year period in a plant treating tannery wastewater. Operational parameters are evaluated on a day-to-day basis and their role, along with pH, temperature, TKN, NH_3 concentration and other analytical parameters, which characterize wastewater quality at different points in the treatment scheme, is discussed in reference to process rates. As shown, biological nitrification/denitrification of tannery wastewater is extremely difficult to conduct in a full-scale plant and there may be many reasons for the periodic failure of the system. Only a slight deviation of the optimal pH in nitrification or denitrification basins, a drop in temperature or a high influent nitrogen content are factors, whose magnitude would not affect the performance of a plant treating domestic wastewater, but during treatment of industrial effluents they can readily cause system failure. Most often, however, it is the concomitant occurrence of several causes that leads to an unsuccessful performance or inhibition of one or more biochemical reactions. In such a complicated situation, sludge age can be the key parameter in controlling the plant. Keeping it high, far above the value related to the specific growth velocity for a given temperature, can be a useful tool for preventing problems in plant performance, particularly when high fluctuations in the concentration of inflow solid make a proper sludge management difficult. Furthermore, on-line monitoring of wastewater characteristics at key points of the plant can give useful information on the inhibition risk.

Basic Considerations

The term 'nitrification' denotes oxidation of ammoniacal nitrogen into nitrites and nitrates by a biochemical route. It is performed by specific autotrophic micro-organisms, whose behaviour is significantly different from those of the heterotrophic type. The process is conducted in two steps:

$$2NH_4^+ + 3O_2 \xrightarrow{\text{Nitrosomonas}} 2NO_2^- + 2H_2O + 4H^+ \quad (1)$$

$$2NO_2^- + O_2 \xrightarrow{\text{Nitrobacter}} NO_3^- \quad (2)$$

The nitrifying bacteria consume ammonium for their cellular synthesis and as a source energy. The quantity of this ion used for biomass production of nitrifiers constitutes an insignificant fraction (about 2 per cent [(1)]) of the total quality processed, which results in this quantity generally being neglected in the mass balance for nitrogen. Since generally the conversion of ammonium into nitrite is slower than the transformation of nitrite into nitrate, it is assumed that the nitrification kinetics is governed by the first step of the process. In the absence of any inhibition, a Monod expression for biomass growth can be applied.

$$\mu = \mu_{max} \frac{\left[NH_4^+\right][DO]}{K_4^+ \left[NH_4^+\right] K_O + [DO]} \quad (3)$$

where,

μ: Specific growth rate (T^{-1});

μ_{max}: Maximum specific growth rate (T^{-1});

$[NH_4^+]$: Bulk ammonium concentration (ML^{-3});

K_N: Semisaturation constant for ammonium (ML^{-3});

K_O: Semisaturation constant for dissolved oxygen (ML^{-3});

[DO]: Bulk concentration of dissolved oxygen (ML^{-3}).

From equation[3], the importance of dissolved oxygen in the process nitrification is evident. The value of the semisaturation constant for oxygen is generally assumed to be equal to 0.5 mg dm^{-3} [(2)]. For concentrations higher than 2 mg ^{dm}-3, dissolved oxygen is no longer considered to be a limiting factor.

Both the specific growth rates and semisaturation constants are very sensitive to temperature and pH variations. The influence of temperature on the magnitude of the specific growth rate is given by [2].

$$\mu_{max}[T] = \mu^{20}_{max} 1.123^{(T-20)} \quad (4)$$

A similar exponential equation describes the dependence of the semisaturation constants on temperature.

Temperature can also affect the overall process of nitrogen elimination by diminishing the efficiency of the removal of ammonium by ammonification during the biomass production [3] or by increasing the inhibition or toxicity effects of the constituents of the wastewater [4,5].

Regarding the influence of pH on the variation of μ_{max}, it is generally assumed that it has no impact if the pH ranges between 7.2 and 8.5, but for lower values the specific growth rate diminishes; according to:

$$\mu[pH]_{max} = 7.2Q^{(pH - 7.2)} \quad (5)$$

where, Q = sensitivity coefficient, equal to 2.35 [2].

The variation of the semisaturation constant K_N as a function of pH is not documented in the literature.

In the case of inhibition of nitrification, the Monod-type equation may remain valid, but the value of the maximum specific growth rate will be much lower. It may also happen that the hypothesis made for the specific growth rate, as being a function of the substrate concentration only, may no longer be true. In this case, other types of equations, based on enzymatic kinetics, are applied. The Haldane equation, which describes the process kinetics inhibited by the substrate itself, is most commonly used in this case [6]:

$$\mu = \frac{\mu_{max}\left[NH_4^+\right]}{K_N + \left[NH_4^+\right] + \left[\frac{NH_4^+}{K_i}\right]^2} \quad (6)$$

where K_i = inhibition constant (ML^{-3}).

Results and Discussions

The research consisted of the long-term observation (about 600 days) of the performance of a full-scale common effluent treatment plant, processing tannery wastewater with a mean flow of 30000 m^3 day^{-1}, generated by a cluster of tanneries. The plant, based on a single-sludge process. The wastewater was sampled at different points of the plant to assess the removal efficiency of various treatment stages and analyzed for pH, redox potential, Chemical Oxygen Demand (COD), Biochemical Oxygen Demand (BOD), total and settleable solids, chlorides, sulphates, sulphides, total chromium, nitrates, nitrites, Total Kjeldahl Nitrogen (TKN), Dissolved Oxygen and total volatile solids. The analyses were performed following Standard Methods [7].

The mean values of the parameters analyzed at different sampling points are reported in Table 3.1. As can be seen, the wastewater was characterized by an elevated content of organic substances of high biodegradability (BOD/COD = 0.74) a high concentration of nitrogen, present mainly as ammoniacal nitrogen and a high concentration of chlorides. Suspended solids were also present in significant quantity.

Table 3.1: Mean Values of the Characteristics of the Effluent at different Sampling Points [(mg dm^{-3}), Except for pH, Sedimentatable Solids (cm^3 dm^{-3}) and Redox (mV)].

Parameter	*Sampling Point*			
	Inlet into Denitrification	*Secondary Sedimentation Tank*	*Recycle into Denitrification*	*Inside Nitrification Basin*
pH	8.0	7.85	7.5	7.2
COD	5422	185		
COD filtered	2736		228	183
BOD	2030	13		
Total solids	2582	81	6.7	6.9
Sedimentation solids	38.06	2.4		
Chloride	3538	2964		
Sulfate	1615	1817		
Cr tot	103			
Fe tot	20.0			
$N\text{-}NO_{2-}$		1.8	0.6	0.3
$N\text{-}NO_{3-}$		22.8	12.7	18.5
$N\text{-}NH_{3-}$		34.1		40.0
TKN	493.0	38.5		
Sulfide	121.2	0		
redox	–438	76		
P tot		1.0		
DO				3.1
Phenol			756	616
Alkalinity				24.5

The day-by-day variation of the characteristics of the wastewater entering the plants during the whole period of the study was very high. The TKN and COD ratio is of particular importance for the single sludge process, considering the need to assure the organic load during denitrification at a level sufficient to accomplish biochemical conversion of nitrates to gaseous nitrogen by means of heterotrophic bacteria, which use organic matter as a source of energy and carbon. An example of the denitrification reaction is the following one, which employs methanol as an external carbon source and the concentration of different forms of nitrogen in the treated wastewater. As can be seen, the performance of the plant was very unstable, with long periods of insufficient nitrification and the consequent discharge of ammonium and/or oxidized nitrogen compounds present in the final effluent. During the periods when nitrification performed well, the rate of this process was equal to 0.02 kg N kg^{-1} MLVSS day^{-1}, which is about five-fold lower than the rates observed at a pilot plant treating the same kind of wastewater [8].

$$6NO_3^- + 5CH_3OH \longrightarrow 3N_2 + 5CO_2 + 7H_2O + 6OH^- \quad (7)$$

During the periods when nitrification caused problems, the final effluent was also often characterized by increased COD, indicating a decrease in removal of the organic load (Table 3.2). It may be that this was caused by a lack of chemical oxygen in the denitrification basin, resulting in the lower efficiency of this section.

Table 3.2: Global Efficiency of Nitrogen and COD Elimination at the Plant

Period	*Removal Efficiency (Per cent)*			
	Mean Value		*St. Dev.*	
	N	*COD*	*N*	*COD*
Problems in nitrification				
01.01.93 – 02.03.93	43.46	89.74	10.01	1.17
15.06.93 – 31.07.93	50.00	90.21	17.34	2.02
01.10.93 – 14.03.94	44.18	88.44	14.99	2.49
24.03.94 – 31.03.94	92.98	87.40	2.00	3.36
20.04.94 – 09.05.94	83.87	88.93	6.38	3.92
Nitrification in function				
03.03.93 – 14.03.93	99.80	91.07	0.74	2.11
01.09.93 – 30.09.93	100.00	92.46	0.00	1.14
15.03.94 – 23.03.94	98.34	87.52	1.32	2.15
01.04.94 – 19.04.94	96.75	89.53	2.18	7.04
10.05.94 – 23.05.94	99.32	87.63	1.52	2.97

The plant exhibited significant fluctuation of pH in nitrification and denitrification sections. It is of a particular interest to note, comparing these data that the high pH (particularly above 7.7) can be associated with low rates of nitrification due to inhibition by the increased level of free ammonia. In fact, on the increase of the pH, the equilibrium between ammonium ions and ammonia shifts towards this last compound. At the pH of 8.0 and the temperature equal to 20°C, about 5 per cent [9] of ammoniacal nitrogen is present in the form of free ammonia. This means that in wastewater bearing a TKN content as high as 300 mg dm^{-3}, the concentration of free ammonia can be approximately equal to 15 mg dm^{-3}.

While examining possible reasons for deficiency in nitrification, dilution rate should be taken into particular consideration. To avoid a wash-out of the biomass, the plant should operate at a dilution rate lower than the critical value:

$$D = \frac{F_o}{V} < D_{crit} \tag{8}$$

where,

D: Dilution rate (T^{-1});

F_0: Nominal flow rate (L^3T^{-1});

V: Volume of the reaction basin (L^3).

This critical dilution rate is equal to the specific growth rate μ for the plants without the sludge recycle. For the plants, which operate with sludge recycle, the critical dilution rate, calculated on the basis of the sludge mass balance is given by [10]:

$$D = \frac{\mu}{1 - \frac{F_r}{F_o}\left(\frac{X_o}{X_1} - 1\right)} \tag{9}$$

The specific growth rate being a function of several operational parameters and wastewater characteristics, it is clear that the critical dilution rate also changes with variation of these parameters and an optimal value cannot be indicated. Furthermore, in the case of nitrification of industrial wastewater, the combined effect of temperature and inhibiting factors can occur. This can lead to a total or partial block of nitrogen removal, even under conditions of relatively high temperature.

In the study, critical dilution rates were calculated assuming the μ_{max} value for nitrifiers at a temperature of 20°C equal to 0.3 day^{-1} [2]. This value was further corrected, considering that the temperature of the wastewater measured in the basin differed from 20°C and the actual pH often was not optimal. Equations (4) and (5) were used for calculation purposes. It has to be stressed that the plant should be operated at dilution rates below and not close to the values indicated as critical, considering that an eventual inhibition was not taken into consideration while determining the critical values. Inhibition, if present, would affect the specific growth rate and, as a consequence, the critical dilution rate would be lower.

The critical dilution rates calculated for those days of the study for which data were available which shows the actual daily dilution rates at which the plant operated. As can be seen, in some periods, the critical and actual dilution rates were very close or the plant even exceeded the critical value. These periods corresponded to winter temperatures, which manifested in January-February of the first year study and November-January of the second year (respectively the periods between 1 and 60, and 300 and 420 days on the time scale). Under these conditions the plant nitrified poorly, which confirms the validity of the dilution rate approach.

There exists a vast list of different substances, that can cause an inhibition of nitrification [9]. It includes organic compounds used in the tanning industry, *e.g.* surfactants or substances preventing the development of fungi *(e.g.* chloromethacreosol). Particularly, noxious are cationic surfactants, which can block nitrification at a concentration as low as a few μg dm^{-3} [11,12]. Anionic surfactants are less dangerous, but their inhibitory effect also manifests itself at low concentrations, in the range of few mg dm^{-3}.

Another inhibiting factor is salinity, 50 per cent inhibition of nitrification was observed when the NaCl concentration was equal to 0.4 eq dm^{-3} [13].

The greatest danger of inhibition is, however, associated with the same compounds which are used by nitrifying bacteria as substrates. In fact, both steps of nitrification can be inhibited by free ammonium (threshold levels reported in the literature are 0.1 mg dm^{-3}, respectively for Nitrobacter 0.1 mg/L and Nitrosomonas 10.0 mg/L [9] and undissociated nitric acid (0.2 mg dm^{-3} for Nitrobacter [14].

If nitrification is conducted at a higher temperature, the risk arises of an accumulation of nitrites. This effect is not caused by a partial or total block of the second step of nitrification, but derives from a variation of the kinetics of the two steps, the first one being more sensitive to the temperature rise than the second. Thus there exists a critical temperature level, above which the transformation of nitrites into nitrates is slower than their production [4, 5]. In the case of the "single sludge" process such a situation leads to the discharge of nitrite-rich final effluent.

Accumulation of nitrites in the wastewater can also influence alkalinity and pH. The decrease of pH can be observed if the "reserve" of alkalinity is insufficient and lower than its stoichiometric need, resulting from the mass balance calculated using the equations (1), (2) and (7). The decrease of pH causes a shift of the equilibrium between dissociated and undissociated forms of nitrites, leading to an increase in the quantity of free nitric acid. Since this compound is a strong inhibitor of the first step of nitrification, the whole process can become completely blocked. This is what happened between days 270 and 300 of the study.

Other possible causes of inhibition of oxidation of nitrites into nitrates include the oxygen concentration being lower than the minimum value equal to 0.5 mg dm^{-3} [15] (a situation which occurred on 380 – 400 day) or a presence of organic substances in the nitrification basin at a high level. This can result in the predominance of heterotrophic over autotrophic bacteria, with a consequent competition for oxygen. Under such conditions, the maximum specific growth rate of nitrifying bacteria diminishes and the value of the semisaturation constant for oxygen increases [15].

The parameter, which is of a paramount importance in biological wastewater treatment plants, is sludge age. This is calculated from the sludge mass balance, as a ratio between the quantity of the biomass in the activated sludge basin and the quantity wasted daily. Considering that the inflowing wastewater contains solids, their quantity should be subtracted from the total mass of solids present in the system. Thus the following equation is used to calculate the sludge age SRT:

$$SRT = \frac{C_{MLSS}\,V - C_S\,F_o}{Q} \tag{10}$$

where, C_{MLSS} and C_s – concentration of solids in the activated sludge basin and in the influent, respectively (ML^{-3}); Q – quantity of MLSS wasted daily (MT^{-1}) other notations as above.

From the above equation, it results that if the quantity of the solids arriving at the biological unit is not carefully determined or considered at all, it can cause a significant discrepancy between the real and calculated value of the sludge age. It was observed the quantity of solids arriving at the nitrification/denitrification section of the plant was very variable and was often in the range of the wasted quantity. This made the calculation of the sludge age an impossible task, as even negative values were obtained. This difficulty indicates the importance of correct determination of the concentration of the solids, while determining the daily quantity of the sludge to be wasted and, generally, in the management of the nitrification plant.

Another way to calculate the SRT is to consider volatile instead of total solids in the mass balance. In such an approach, the inflowing solids can be neglected, assuming that they represent the organic matter, which is hydrolysed and further decomposed in the biological basin; thus this quantity constitutes a "feed" for the biomass. It has to be outlined that this assumption may not be in practice totally fulfilled. The apparent SRT values calculated for the plant using this approach varied widely between 8 to 40 days. The loss in nitrification was observed for the values below 10 days (days 175 to 200 on the time scale). Considering that in this period the temperature in the basin was relatively high (above 19°C), the SRT of 10 days could have been regarded safe for nitrification of municipal wastewater, but resulted too short for industrial effluents. Thus, while treating industrial sewage, it is advisable to assure the stability of the plant performance to operate at a much higher SRT than the value estimated from the specific growth rate of nitrifiers.

Conclusion

Analysis of causes of the failure in the performance of the full-scale plant leads to the conclusion that the lack of nitrification observed at the plant cannot be explained by one single cause. The whole range of factors that influence the development of the nitrifying biomass has to be considered. Among the factors, which can be cited as the most frequent are: exceeding of the critical value of the dilution rate, pH too low (below 7.0) or too high (above 8.0), resulting in generation of inhibition by nitric acid and free ammonia, respectively, insufficient oxygenation (dissolved oxygen concentration below 1 mg dm^{-3}). In the majority of the cases more that one cause occurred.

To ensure that the stable and satisfactory performance of the plant, it is advisable to install of sensors to monitor continuously ammonium, nitrites, nitrates, pH, temperature and dissolved oxygen. Caution should also be paid to sludge management practice, to ensure the possibility of defining the sludge age at which the plant operates.

In such a complicated situation, sludge age can be the key parameter in controlling the plant. Keeping it high, far above the value related to the specific growth velocity for a given temperature, can be a useful tool for preventing problems in plant performance, particularly when high fluctuations in the concentration of inflow solids make a proper sludge management difficult.

References

1. Van Handel, A.C. and Marais, G.v.R. (1981), Research Report W 42, Dept. of Civil Eng., Univ. of Cape Town.
2. Water Research Commission by the University of Cape Town, City Council of Johannesburg and the National Institute for Water Research of the CSIR (1984), Theory, Design and Operation of Nutrient Removal Activated Sludge Processes, Pretoria, Rep of South Africa.
3. Oleszkiewicz, J.A. and Berquist, S.A. (1988), *Wat. Res.*, **22**, 1163-1171.
4. Randall, C.W. and Buth, D. (1984), *J.WPCF*, **56**, 1039-1044.
5. Randall, C.W. and Buth, D. (1984), *J.WPCF*, **56**, 1045-1049.
6. Rozich, A.F. and Castens, D.J. (1986), *J.WPCF*, **58**, 220-226.
7. APHA. (1989), Standard Methods for examination of water and wastewater, 17th ed., American Public Health Assoc., Washington D.C.
8. Szpyrkowicz, L., Rigoni-Stern, S. and Zilio Grandi, F. (1991), *Wat.Res.*, **25**, 1351-1356.
9. Sedlak, R. (1991), *Phosphorus and Nitrogen Removal from Municipal Wastewater*, 2nd ed., Lewis Publishers, New York.
10. Bailey, J. and Ollis D. (1986), *Biochemical Engineering Fundamentals*, Mc Graw-Hill, Singapore.
11. Munaò, F., Di Pietro, A., Scoglio, M.E., Grillo, O.C., Anzalone, C. and Minolfi, P. (1990), *ACQUA-ARIA*, 769-774.
12. Munaò, F., Di Pietro, A., Scoglio, M.E., Grillo, O.C., Anzalone, C. and Minolfi, P. (1991), *ACQUA-ARIA*, 465-470.
13. Dollan, J.F., O'Neill, M.J. and Horan, N.J. (1990), *J.IWEM*, **4**, 457-468.
14. Anthonisen, A.C. (1978), *J.WPCF*, **48**, 835-842.
15. Hanaki, K., Wantawin, C. and Ohgaki, S. (1990), *Wat. Res.*, **24**, 297-302.

Chapter 4

Titanium Dioxide Photocatalysis: A Candidate for Advanced Wastewater Treatment Technology

***N.N. Rao*[1] *and S.N. Kaul*[2]**

[1]*CSIR-NEERI, Nagpur*
[2]*MIT-Pune*

ABSTRACT

Heterogeneous photocatalysis is about absorption of photons from UV or sunlight by semiconductors and harvesting the charges that are formed. The phenomenon can be tapped to produce electricity or hydrogen fuel. Alternatively, some specific changes may be brought in selected organic compounds. The search for newer applications of photocatalysis is continuing. The current interest in photocatalysis is mainly due to its ability to destroy a variety of organic pollutants present either in water or in air. Stable semiconductors such as TiO_2 and ZnO absorb UV light from sunlight or from artificial light sources and destroy a variety of organic compounds with remarkable efficiency. The present state of research in this area is to develop photocatalysts and photoreactors that promise much higher efficiencies in order to handle large volumes of wastewater. In this chapter, we discuss the environmental applications of TiO_2 photocatalysis. We initially take recourse to well known principles and mechanisms of photocatalysis and gradually diverge into purification of water and wastewater, air, disinfection/sterilization, photocatalysts and catalytic media and design of photoreactors, highlighting our own contributions.

Keywords: *Photocatalysis, Environmental pollution, Photoreactors.*

Introduction

Acceleration of a photoreaction by the presence of a catalyst is photocatalysis. The catalyst may accelerate the photoreaction by interaction with the substrate in its ground state or excited state, or with a primary photoproduct, depending upon the mechanism of photocatalysis. In photocatalysis, no energy is stored. There is merely an acceleration of a slow event by a photon-assisted process. There are many types of well characterized solid state materials that belong to metal oxide, chalcogenide, and perovskites families that act as photocatalysts.

The subject of photocatalysis began in 1970s. Titanium dioxide, a simple oxide material, has emerged as an excellent photocatalyst material for environmental purification. Many scientific studies have proven titanium dioxide as a chemically stable UV responding material of great potential to decontaminate water, air and solids. The photocatalytic process is employed to break down and destroy many types of organic pollutants. It has been used to purify drinking water as well as air streams, destroy bacteria and viruses, remove metals from waste streams, and break down organics into simple components of water and carbon dioxide.

Photocatlysis

Basic Band Picture, Suitable Semiconductors, Generation of Reactive Species, and Mechanism of Photocatalytic Oxidation Process

Crystalline solids are categorized as conductors, semiconductors, or insulators according to their ability to conduct electrical current. The valence band (VB) is the energy level occupied by the outermost electrons in a solid. The conduction band (CB) is the highest energy level in an atom. Because it is unoccupied by electrons at the ground state, electrons from the valence band can move into it and are free to flow within the solid. The valence and conduction bands are often separated by an energy barrier called the band gap. In a conductor, the valence and conduction bands overlap, and electrons can easily move to the conduction band without any energy input. Within that band, they can flow freely from atom to atom in the solid. In insulators and semiconductors, the valence band (which is completely occupied by electrons) is separated from the conduction band (which is empty) by the band gap energy zone. In order for the electrons to move to the conduction band and subsequently move from atom to atom within the solid, an external energy input equal to or greater than the band gap energy must be applied. The difference between an insulator and a semiconductor is a matter of degree. As a rule of thumb, a material is considered to be an insulator if the band gap energy is greater than about 3.5 V and to be a semiconductor when it is less than 3.5 V.

Semiconductors are categorized as n-type or p-type depending on whether they contain an excess of electrons or holes within their lattice. An *n*-type semiconductor has an excess of electrons and a p-type semiconductor has a deficiency of electrons in the crystal lattice. For example TiO_2 is considered to be an *n*-type semiconductor because its crystal structure is oxygen deficient, causing an excess of electrons within the crystal lattice. Copper (I) oxide (Cu_2O) is considered p-type because it is copper deficient causing an excess of holes within the crystal lattice. Semiconductors can be altered from n-type to p-type or vice versa by the insertion of impurities (doping) called electron donors or electron acceptors into the crystal lattice.

Photoactive catalysts are semiconductors primarily composed of metal oxides. Table 4.1 lists several different photoactive semiconductors; their band gap energies vary widely ranging from 2.2 V to 3.5 V. The equivalent wavelength is the wavelength of a photon that has energy equal to the band gap energy. The table shows that photo activation of the anatase crystalline form of TiO_2, for example, requires light of wavelength less than or equal to about 413 nm. Semiconductors typically used in solar energy applications and oxidation processes are categorized as n-type semiconductors because the main charge carriers are electrons. Titanium dioxide is most commonly used because it is very photoactive and stable in comparison to other semiconductors and is very fascinating material.

Table 4.1: Band-Gap and Threshold Wavelengths for Oxide Semiconductors Suitable for Photocatalytic Environmental Applications

Semiconductor	*Band-gap, Eg (eV)*	*Threshold Wavelength, nm*
TiO_2	3.0	413
ZnO	3.2	385
SnO_2	3.5	346
$SrTiO_2$	3.2	385
WO_3	2.8	437
Fe_2O_3	2.2	515
In_2O_3	2.7	450
CdO	2.1	528

Generation of Reactive Species and Mechanism of Photocatalytic Oxidation

A semiconductor is characterized by band structure; a filled VB (valence band) separated by E_g (energy gap) from a vacant CB (conduction band) as shown in Figure-1. The energy level at the bottom of the conduction band (LUMOs) reflects the reduction potential of the photoelectrons, whereas the uppermost level of the valence band (HOMOs) is a measure of the oxidizing ability of the photo-holes. The flatband potenial, V_{fb}, fixed by the nature of the material and by the proton exchange equilibrium, determines the energy of the two charge carriers at the interface. Hence, reductive and oxidative processes for adsorbed couples with redox potentials more positive and more negative than the v_{fb} of the conduction and valence bands, respectively, can be driven by surface-trapped e^- and h^+ carriers. Figure 4.1 illustrates a fraction of the complex sequence of events that may take place in a semiconductor photocatalyst.

The band gap defines the wavelength sensitivity of the semiconductor to irradiation. Photo-excitation with light of energy greater than or equal to E_g ($h\nu \geq E_g$) promotes an electron from VB into CB leaving a hypothetical hole (h^+) in VB (Figure-

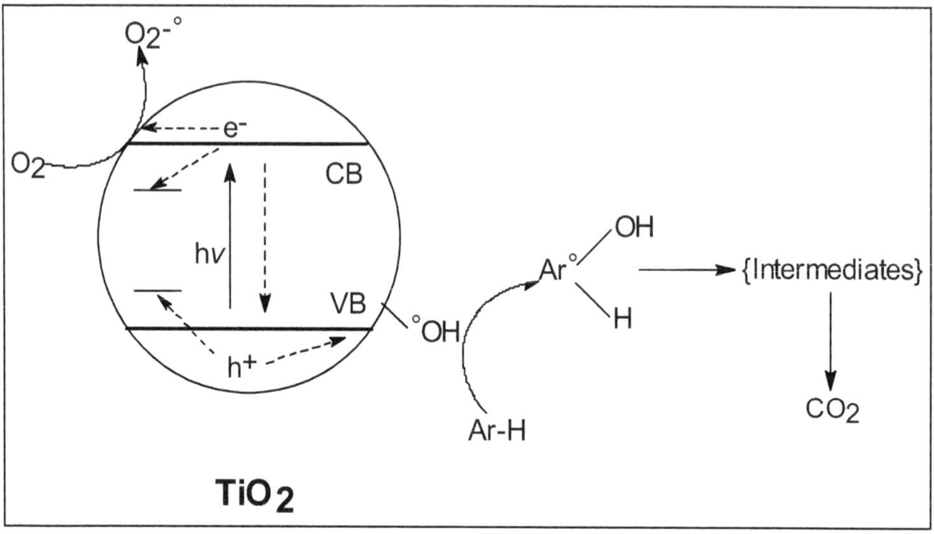

Figure 4.1: Sequence of Photophysical and Photochemical Events taking Place upon Irradiation of a TiO_2 Particle with hν Greater than E_g Together with Secondary Reactions to Total Mineralization to CO_2.

4.1). A hole can also be identified as a chemical entity, *e.g.* a bound O^- lattice radical or a surface associated OH radical in a typical metal oxide semiconductor. This hole can cause oxidation of an adsorbed substrate *(e.g.*OH or H_2O) to OH radicals through electron transfer at the interface. On the other hand, the photogenarated electron in CB reduces a suitable acceptor (A, like H^+ or O_2) and generates reactive H_{ads} (or H_2) or $O_2^{-\bullet}$ (superoxide radical anion). It is imperative that in order to find appreciable generation of reactive chemical species (such as •OH or $O_2^{-\bullet}$), one must efficiently separate the photogenerated electrons and holes by kinetically favoring the formation of their chemical equivalents. The Schottky barrier at semiconductor/liquid junction performs this task partially. In a typical TiO_2/UV process, thc substrate to be oxidized by $O_2^{-\bullet}$or •OH is usually an organic pollutant chemical generalized as RX. By virtue of 3.0 -3.2 eV band gap (413-380 nm) illumination of TiO_2 results in the generation of h^+ with oxidizing power +2.8 V and e^- with reduction power of –0.2 V (based on the positions of VB and CB respectively at pH 7). It is interesting to note that the oxidizing power of h^+ in TiO_2 is equivalent to that of the hydroxyl radical. The usual mechanistic steps of TiO_2/UV process are:

$$TiO_2 + h\nu \rightarrow TiO_2\,(e^-, h^+) \quad (1)$$

$$TiO_2\,(h^+) + RX \rightarrow TiO_2\,(RX^+)_{ads} \quad (2)$$

$$TiO_2\,(h^+) + H_2O/OH^- \rightarrow TiO_2\,(OH)_{ads} + H^+ \quad (3)$$

$$TiO_2\,(e^-) + O_2 \rightarrow TiO_2\,(O_2^{-\bullet})_{ads} \quad (4)$$

$$TiO_2\,(RX^+)_{ads} + TiO_2\,(OH)_{ads} \rightarrow TiO_2\,(R\text{-}OH)_{ads} + X^- \quad (5)$$

$$TiO_2\,(R\text{-}OH)_{ads} + OH \rightarrow CO_2 + H_2O \quad (6)$$

Both adsorbed H_2O/OH^- and molecular oxygen mediates the PCO process as shown in steps (2)- (6). The steps (5) and (6) are too simplified; but they constitute a series of attacks by powerful OH radicals to finally disintegrate R-X to CO_2 etc. Heteroatoms in the pollutant molecule are transformed into their corresponding anions such as SO_4^{2-}, NO_3^-, PO_4^{3-} as per the generalized reaction in (7)

$$C_xH_yN_zO_pCl_qP_rS_t \rightarrow CO_2 + H_2O + NO_3^- + PO_4^{3-} + SO_4^{3-} + Cl^- \quad (7)$$

Charge Carrier Formation vis-a-vis Recombination of Photo-Generated e^- and h^+

In the simplest case of an intrinsic semiconductor, absorption of a photon (hv) of band gap energy E_g or greater excites an electron (e_{cb}^-) to the conduction band, while this leaves an electronic vacancy (called a hole, h_{vb}^+) in the valance band. A potential exists between the interior bulk solid and the external surface of the solid under this condition. This is due to depletion of conduction-band electrons in the space-charge region. The resulting gradient causes the photo-excited hole and electron to migrate to the exterior surface. In the absence of an electrical circuit, at steady state, the electron and the hole arrive at the surface simultaneously. These electron-hole pairs at the surface can recombine, producing only luminescence, or can react with adsorbed species on the surface to produce redox reactions.

These separated charge carriers may also recombine, migrate to the surface while scanning several shallow traps. The traps may be identified as anion vacancies and/ or Ti^{4+} for the electrons; and oxygen vacancies or other defect sites for the hole. Both charge carriers scan the surface visiting several sites to reduce adsorbed electron acceptors (A_{ads}) and to oxidize adsorbed electron donors (D_{ads}). This occurs in competition with surface recombination of the surface trapped electrons and holes (e^-_{st} and h^+_s) to produce light emission and/or phonon emission. Oxygen on the particle surface acts as an electron acceptor whereas OH^- groups and H_2O molecules are available as electron donors to yield the strongly oxidizing •OH radicals. Trapping of electrons and holes in pristine naked TiO_2 colloids of nanometer size takes place in less than 30ps. At concentrations of organic pollutant substrates normally found in the environment (a few tens of mg L^{-1}) the •OH radicals are the primary oxidants. In the case of an aromatic substance, the corresponding. •OH adducts (a cyclohexadienyl radical) that ultimately breaks down into a variety of intermediate products on the way to total mineralization to carbon dioxide.

Reaction Pathways-Hydroxyl (•OH) and Oxygen (O_2)

The primary oxidant responsible for the photoassisted catalytic oxidation of organic compounds in aqueous solutions is the highly reactive hydroxyl radical. The mechanisms of hydroxyl radical formation have been discussed in the literature (Matthews 1984, Izumi *et al.*, 1980; Okamoto *et al.*, 1985; Sakata *et al.*, 1984; Pelizzetti 1990; Serpone *et al.*, 1986; Kraeutler and Bard 1978; Norton *et al.*, 1988; Turchi and Ollis 1990; Al-Ekabi and Serpone 1988).

Two pathways have been proposed for the formation of hydroxyl radicals. Matthews (1984) proposed that the major pathway for hydroxyl radical formation is through the reaction of the valance-band hole with either water (H_2O) or hydroxide ions (OH^-) adsorbed on the catalyst surface in the presence of photon energy ($h\nu$). However, Okamoto *et al.* (1985) proposed the following alternative pathway, suggesting that hydroxyl radical formation can also be promoted by the reaction of the conduction-band electron with adsorbed oxygen, $O_{2\,(ads)}$, to produce hydrogen peroxide (H_2O_2) from super-oxide ions ($O_2^{-\bullet}$): where $HO_2^{\cdot}$ is the perhydroxyl radical and HO_2^- is the perhydroxyl ion. Thus, Okamoto *et al.* (1985) and Hsiao *et al.* (1983) believe that oxygen plays an important role in photocatalytic reactions. Surface adsorbed oxygen not only reacts to produce hydroxyl radicals but also traps the conduction-band electrons, preventing electron-hole recombination. Hsiao *et al.* (1983) conducted experiments using aqueous suspensions of TiO_2 to determine the effect of PCO on dichloromethane and dibromomethane in both the presence and absence of O_2. They found that the rate of disappearance of these compounds was significantly slower when O_2 was not present in solution and that the participation of O_2 is a kinetically important step. Similar results were reported by Okamoto *et al.* (1985) for the oxidation of phenol using an aqueous suspension of TiO_2. Several researchers have since then reported that the molecular oxygen is necessary for complete oxidation of organic compounds (Matthews 1984; Kraeutler and Bard 1978; Turchi and Ollis 1990; Al-Ekabi and Serpone 1988; Rao and Natarajan 1994).

It is possible, thermodynamically, that many organic compounds in solution can be oxidized by direct contact with the valance band holes. The question whether h+ or •OH is the continuing source of curiosity in the photocatalysis field. However, this pathway is unlikely because the concentration of H_2O or OH^- is much higher than the concentration of organic compounds. Nevertheless, holes may become important when the coverage of organic compounds is very high. Izumi *et al.* (1980) conducted PCO experiments using aerated pure organic solvents (no water present) and found that only partial oxidation occurred with a very slow production of carbon dioxide (CO_2). In another study, Cunningham and Srijaranai (1988) conducted photoassisted catalytic oxidation experiments with aqueous TiO_2 slurries using iso-propanol and deuterated iso-propanol. They found that the substitution of deuterium for hydrogen on iso-propanol had no effect on the initial reaction rate and that direct reaction with the valance-band hole was not significant. Cunningham and Srijaranai (1988) conducted experiments where deuterium oxide (D_2O) was used instead of water as the solvent and found that the initial reaction rate decreased by a factor of 3 as compared to where H_2O was used as the solvent. The observed decrease in the rate when D_2O was used as a solvent was due to the higher electron volts (eV) required for generating the ${}^{\bullet}$OD radical as compared to the ${}^{\bullet}$OH radical. Their results showed that the rate-limiting reaction step is the formation of the hydroxyl radical (${}^{\bullet}$OH) or the deuterium radical (${}^{\bullet}$OD).

Turchi and Ollis (1990) summarized studies that have used electron spin resonance (ESR) to identify and verify the existence of hydroxyl radicals in near-UV illuminated aqueous solutions of TiO_2. Dimethyl Picryl Hydrazide (DMPH) is commonly used to trap ${}^{\bullet}$OH. The DMPH${}^{\bullet}$-OH spin trap complex displays a four-finger symmetric resonance with 1: 2: 2: 1 intensity ratio. Turchi and Ollis (1990) pointed out that the radical-spin trap complex is detected in the aqueous phase but stated that it is not clear whether the actual trapping occurs on the catalyst surface or in solution. The perhydroxyl radical ($HO_2^{\bullet}$) and $O_2^{-\bullet}$ were also detected in these studies.

In summary, experimental evidence supports the fact the hydroxyl radical is the primary oxidant responsible for the mineralization of organic compounds, and its formation is principally due to the reaction of the valence-band hole with water or surface OH^- groups. Although it can be stated that hydroxyl radicals are produced on the catalyst surface, the reaction pathway of the radicals in mineralizing organic compounds is still unclear. Several detailed reaction pathways have been proposed for the photoassisted catalytic oxidation of organic compounds in the aqueous phase (Izumi *et al.*, 1980; Okamoto *et al.*, 1985; Sakata *et al.*, 1984; Pelizzetti 1990; Serpone *et al.*, 1986; Kraeutler and Bard 1978; Ollis *et al.*, 1984; Childs and Ollis 1981; Matthews 1987a,b; Hashimoto *et al.*, 1984; Herrmann and Pichat 1980; Pruden and Ollis 1983; Fox *et al.*, 1988). As pointed out by Turchi and Ollis (1990), there are many possible reaction pathways between the radical species and the organic molecules. The adsorbed hydroxyl radical may interact with the catalyst solid or react with an adjacent adsorbed or aqueous-phase organic molecule or other radical species. Alternatively, the hydroxyl radical may diffuse into the aqueous phase and react with an adsorbed or aqueous-phase organic molecule or other radical species. The pathway may be some combination of all of these possibilities. The actual reaction pathway for the degradation of organic compounds is still unknown.

Water Treatment-Advanced Oxidation Processes

The water supplies, formerly of good quality, are now potentially hazardous to mankind and aquatic life in particular. Extensive industrial development and insincere use of methods for disposal and storage of polluted effluents have resulted in progressive deterioration of water quality in many places. A number of contaminant chemicals have been found to enter ground and surface waters through routine industrial, domestic and agricultural operations. Conventionally, varieties of chemicals in effluents are removed by physicochemical and biological methods. However, these methods are associated with some severe problems like sludge disposal and regeneration of adsorbents and thus can not offer permanent solution to the pollution hazard. Particularly, the biological methods are indifferent to some toxic and hazardous chemicals and do not ensure desired removal efficiency. Photocatalysis may be chosen as a pre-treatment method to induce molecular transformations for enhanced biological degradation (Step A, Figure 4.2) or as post-treatment method for ensuring the standards compliance (Step B, in Figure 4.2).

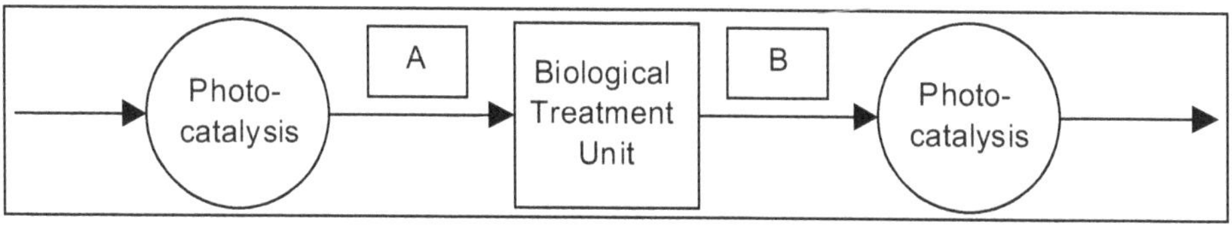

Figure 4.2: Role of Photocatalysis in Industrial Wastewater Treatment.

Based on the water purification, self-cleaning, air cleaning and self-sterilizing properties of TiO_2/UV system a number of process applications have been investigated. The selected applications are listed in Table 4.2.

Table 4.2: Selected Applications of Photocatalysis

Property	*Category*	*Application*
Self-Cleaning	☆ Materials for residential and office buildings ☆ Indoor/outdoor lamps ☆ Materials for roads/ subways	Exterior tiles, kitchen and bathroom interior furnishings, plasic surfaces, building stones and curtains, Tunnel wall and lamps, Cloth for hospital and spray coatings for cars
Water Purification	Drinking water	River water, groundwater, lakes, fish feeding tanks, drainage water etc.,
Air cleaning	Indoor air cleaners Outdoor air purifiers	Room/factrory store rooms air cleaner, Photocatalysts for highway tunnel air cleaning etc.,
Self-Sterilizing	Hospital and others	Tiles to cover the floor and side walls, public rest rooms, bath rooms etc.,

Process Description – Photocatalytic Oxidation

Photocatalytic oxidation involves the use of photoactive n-type semiconductor powders and near-UV light. When these semiconductors are illuminated in water, a redox environment is established that can cause the oxidation of organic compounds. In most cases, the crystalline anatase form of titanium dioxide has been used as the semiconductor because of its high activity and stability. Laboratory studies (Okamoto 1985; Al-Ekabi and Serpone 1988; Ollis *et al.*, 1984; Childs and Ollis 1981; Matthews 1987a,b; Herrmann and Pichat 1980; Pruden and Ollis 1983) have demonstrated that a wide variety of organic compounds, such as chlorinated alkanes and alkenes, polychlorinated phenols, aromatics, aldehydes, and organic acids, can be oxidized using near-UV or solar- illuminated TiO_2 aqueous suspensions. Parachlorobenzene as well as 2,4-dichlorophenoxyacetic acid undergo rapid degradation with TiO_2/UV technique (Rao *et al.*, 1993). Details in respect to each class of compounds are published (Rao and Natarajan, 1994). Surfactants can also be destroyed under solar or simulated sunlight in the presence of TiO_2. The authors applied this technique to treat water containing commercial soap/detergent formulations, which comprise mixtures of surfactants (Rao and Dube, 1996). The technique may also be applied to remove various metal impurities also. Thus, the authors (Ramachandriah *et al.*, 1996) were able remove copper from industrial plating sludge and achieved about 80 per cent removal as Cu/TiO_2 in 2h with 400W Hg Lamp and in 6h under natural sunlight. Rao and Dube (1996) have also developed compartmentalized photoelectrosynthetic cell constituting an illuminated TiO_2 photoanode and a Pt cathode and found H_2 evolution at cathode at the expense of organic pollutants as sacrificial donors. Organic pollutants can act as sacrificial electron donors in PEC cells which may be used to generate hydrogen and to destroy water soluble organic pollutants simultaneously, both of which are useful reactions. The system appears to possess some practical interest as it combines energy and environment related reactions. The authors reported photodegradation of 2,4-D, a herbicide with a COD removal efficiency of 90 per cent and above (Rao and Dube, 1995). In addition to some TiO_2 catalysts from Aldrich, Degussa etc., the Indian commercial TiO_2 also shows significant photoactivity upon reduction in H_2 atmosphere at 450°C. Rao and dube (1997) have also reported TiO_2-catalyzed photodegradation of Reactive Orange 84 and alizarin red S biological stain. While complete color removal took place within 30 min., about 4h are required to remove COD to the extent >90 per cent. The photocatalytic degradation of Reactive Orange 84 in dye house effluent using single pass photo reactor is reported (Rao and Dube, 1998). In this system, completely colorless treated solutions were obtained at <5-ml/min flow rate. Simultaneously some COD is also reduced. The reactor was so designed to retain catalyst particles in the reactor itself. Both color and COD are reduced from a dye containing wastewater using photocatalytic method (Dhodapkar *et al.*, 2000). Through a detailed study of photodegradation of surfactants using MO_3/TiO_2 catalysts (M=Mo or W), the independence of rate constant on the organic reactants has been examined (Rao and Dube 1996). The TiO_2 photocatalysis is indiscrete and does not significantly depend on the nature of organic reactant. The use of photocatalysis for inducing biodegradability in 2-CP solutions has been reported (Rao *et al.*, 2003). The 2-CP solutions were readily detoxified as C-Cl bond is broken, and the treated solutions show increase in BOD. Some new intermediates formed were found more biodegradable. The aspect of photocatalytic destruction of organic pollutants and various methods to enhance the efficiency of photocatalytic process has been reviewed recently (Rao and Kaul 1999; Dhodapkar *et al.*, 1999).

Table 4.3: List of Compounds that have been Destroyed Using PCO

Name of Compound	*Name of Compound*
Chloroform	Cyanide
2-chlorophenol	4-c: hlorophenol
1,2-dibromomethane	1,2-dichloroethane benzoic acid
Biphthalates	Carbon tetrachloride
Chloromethane	Nitrobenzene
Methanol	Tetrachloroethene
Dichloroacetic acid	n-propanol
2-propanol	1,1,1-trichloroethane
2,4-dichlorophenol acetic acid	Ethyl acetate
p-dichlorobenzene	o-dichlorobenzene
2-naphthol	Sucrose
Vinyl bromide	Difluorodichloro- methane
Hydrogen sulfide	

Table 4.3 lists many of the compounds that have been successfully destroyed using this process. Most were completely mineralized to such compounds as carbon dioxide, water, and hydrochloric acid (HCI). Several of the compounds listed in the

table are important to the water industry. However, studies have shown that some nitrogen-containing compounds, such as atrazine, are difficult to mineralize with this process. For example, Pelizzetti *et al.* (1990) reported that atrazine was photocatalytically oxidized to cyanuric acid.

Now, development of PCO for destroying organic contaminants is in the pilot-scale stage and several pollution equipment companies and research laboratories are moving toward its commercialization. For example, one company recently received funding from the USEPA for a pilot-plant demonstration of its PCO system. In the process developed by this company, TiO_2 is coated onto a fiberglass mesh, which is wrapped around near-UV lamps and placed in a stainless-steel reactor. The contaminated water is then pumped through jacket so that it is exposed to the illuminated TiO_2. The contaminants are oxidized (Al-Ekabi Hussain 1990). In another process, developed and pilot tested by researchers from Sandia National Laboratories, a linear parabolic sun- tracking solar collector concentrates sunlight onto a UV - transmitting reactor containing TiO_2 powder suspended in contaminated water (Al-Ekabi Hussain 1990). The reactor is capable of mineralizing 5 ppm of trichloroethane (TCE) to 5 ppb in about 4 minutes at a rate of 42 L/min.

The PCO process is an emerging water and wastewater treatment technology. However, major engineering strides must be taken if this process is to become viable and cost effective.

Photoacvity

For water treatment, the primary goal of the photocatalytic oxidation process is to produce an oxidizing environment that will allow organic compounds to be destroyed or mineralized in a cost-effective manner. Meeting this goal depends on these process variables: (1) the catalyst type; (2) the incident light intensity and wavelength; (3) the reactor configuration; and (4) the physical and chemical properties of the water matrix. Providing a cost-effective process requires optimizing at least the first three variables, although understanding the impact of the physical and chemical properties of a given water matrix is necessary in the overall process operation. Optimization of those properties (such as pH and temperature) is usually not cost effective for drinking water treatment.

Several factors can influence the efficiency of electron-hole separation at the surface of a photoactive semiconductor and the subsequent oxidation-reduction reactions at the solid-liquid interface. Some of the important factors, which include crystal morphology, are surface character, substitution doping, and surface modification.

TiO_2 as a Photocatalyst, Photocatalysts Configuration and other Materials-Related Aspects

In terms of criteria for choosing an appropriate semiconductor photocatalysts, the most important appears to be factors related to the location of the band edges at the surface. That is, the conduction and valence band edges have to be located at energies (potentials) such that reactions of hydroxyl radical formation using adsorbed H_2O and O_2 may be photodriven. Hence, the valence band has to be placed at potentials that are at least +2.85 V (with respect to the standard hydrogen electrode). Only then, will the photogenerated holes that have sufficient energy to oxidize water. Similarly, if O_2 were used as the electron acceptor the conduction band would have to lie at a value negative of the standard potential for the reduction of O_2. In addition to surface energetics, the stability of the semiconductor particle to photo corrosion is a crucial factor. Finally, in terms of practical considerations, cost is also an issue. Titanium dioxide has been by far the most popular photocatalyst. Both anatase and rutile modifications have been used, although commercial samples (*e.g.*, Degussa P-25) often contain a mixture of the two. The energy band gaps of anatase (3.23eV, 384nm) and rutile (3.02 eV, 411nm) combine with the valence- band positions to create a favorable situation for the photogeneration of highly energetic holes at the interface. However, anatase is superior to rutile for photocatalytic applications. First, the conduction band location for anatase is more favorable for driving conjugate reactions involving electron. Other variant reasons have been given. For example, the poorer photocatalytic activity of rutile was attributed to its high e^--h^+ recombination rate and its lower oxygen photoadsorption capacity. A more recent study concludes that very stable surface peroxo groups can be formed at the anatase during photo oxidation reactions but not at the rutile surface. The implication is that the oxidation of organic compounds such as 4-chlorophenol proceeds through an indirect pathway involving these surface species. The decrease noted in the photocatalytic activity when titania (prepared by precipitation method) is annealed at temperatures higher than 600 °C appears to have a similar mechanistic origin in the extent of hydroxylation of the oxide surface. Other semiconductors have also been employed, and a representative listing appeared in Table 4.1. However, none of these appear to match (at least to date) the attributes of TiO_2.

The band gap values as listed are for the 'bulk' semiconductors or for colloidal particles of several hundred-nanometer sizes. Size qantization, however can cause sizable shifts in these band gap energies. For example, TiO_2 micro crystallites as small as 2 nm have been prepared in Nafion and clay inter layers with a corresponding band gap energy as high as 3.95 eV. Similar trends have been observed for Fe_2O_3, ZnS and CdS. Very high activities have been observed for Fe_2O_3 nano-sized particles (relative to bulk Fe_2O_3 powder) for model photo reactions such as the photo decomposition of saturated carboxylic acids.

The vast majority of the semiconductor photocatalysts listed have rather high band gaps. Large band gap semiconductors in general tend to be more stable against photo corrosion. Oxidation of many pollutants, especially organic species, requires high potential, with the results that the valence band location at the semiconductor/ electrolyte interface has to be rather positive as exemplified by TiO_2 and CdS.

The ability of a photocatalyst, such as TiO_2 to perform oxidation-reduction reactions can be significantly influenced by its morphology. The two most common crystalline forms in which TiO_2 exists are anatase and rutile. Kraeutler and Bard (1978) found that the anatase form of TiO_2 was significantly more reactive than the rutile form in the photocatalytic decarboxylation of acetic acid-sodium acetate mixtures. Sclafani *et al* (1990) evaluated the photo assisted catalytic oxidation rate of phenol in an aqueous solution using seven commercially available TiO_2 powders, six primarily anatase and one rutile. They found a wide range of phenol degradation activity for the anatase phase samples, whereas the rutile form was not active.

Sclafani *et al.* (1990) conducted further experiments to evaluate the impact of calcination temperature and time on phenol destruction. Two differently prepared TiO_2 powders were used. One method precipitated titanium trichloride ($TiCl_3$) in an aqueous solution of 25-wt per cent ammonia, and the other method precipitated titanium tetrachloride ($TiCl_4$) in pure water. Both methods produced the anatase form of TiO_2. In the experiments, calcination temperatures ranged from 120 to 800°C and calcination times from 3 to 336 hours. For TiO_2 prepared by the first method (using $TiCl_3$), the phenol degradation activity was low and increased slightly with calcination temperature up to about 650°C. At this temperature, TiO_2 crystallinity was 70 per cent anatase and 30 per cent rutile. The calcination time did not appear to influence the catalyst activity. For TiO_2 prepared by the second method (using $TiCl_4$), the phenol degradation activity was significantly higher. A significant rutile phase (42 per cent) was formed at 400°C and increased steadily to 100 per cent at about 650°C. The highest oxidation rate for phenol was reported at 550°C, where the TiO_2 contained about 25 per cent anatase and 75 per cent rutile. The results also showed that the surface area of the TiO_2 decreased with increasing calcination temperature and that increasing calcination times decreased the anatase: rutile ratio.

Similar results were reported by Abrahams *et al.* (1985). In addition, Abrahams *et al.* (1985) found that the particle size decreased with increasing calcination temperature. These results demonstrate exceptional variability in the photoactivity of TiO_2 due to morphological changes. The changes depend on the starting material and method of preparation.

Changing the crystal morphology of the semiconductor by substituting other transition metal ions into its crystal lattice can increase its photocatalytic activity. Substitution doping has been tested in solar energy applications for photo dissociating water (Kreautler and Bard 1978; Borgarello *et al.*, 1981; Kiwi 1983; Pelizzetti and Visca 1983). For example, Borgarello *et al.* (1981) evaluated the impact of niobium (NbS^+) doping on both the anatase and rutile forms of TiO_2. For the anatase form, they found that increasing the NbS^+ doping level from 0.07 wt per cent to 0.4 wt per cent increased the activity by about 20 per cent, whereas for the rutile form, the activity increased dramatically. Substituting the titanium (IV) ions (Ti^{4+} ions) in the TiO_2 with the higher valence NbS^+ ions introduces extra valence electrons into the lattice and causes the NbS^+ ions to behave as electron donors. Their behavior extends the spectral response toward visible light wavelength changing the band gap of TiO_2 (Pelizzetti and Visca 1983) and allows the electrons to migrate more easily to solid-liquid interface (Kiwi 1983). However, Malati and Wong (1984) showed that substituting chromium (III) ions (Cr^{3+} ions) for Ti^{4+} ions had the opposite effect. The Cr^{3+} ions act as electron acceptors and retard electron migration toward the solid-liquid surface. These investigators also showed that TiO_2 doped with Cr^{3+} extends the spectral response toward visible light further than NbS^+ does; however, lower quantum efficiencies were observed.

For a photo active semiconductor particle immersed in an aqueous medium, the production of hydroxyl radicals at the exterior particle surface is influenced by the electron-hole separation at the surface. For a given incident light intensity and wavelength, minimizing the charge recombination at the catalyst surface will help maximize the production of hydroxyl radicals available for oxidation reactions, increase the quantum yield, and increase the efficiency of the overall process. In the presence or absence of oxygen, efficient charge separation at the surface of the particle can be carried out by depositing transition metals, such as platinum (Pt), and other metal oxides, such as ruthenium oxide (RuO_X), on the particle surface (Kreautler and Bard 1978; Borgarello *et al.*, 1981; Kiwi 1983; Pelizzetti and Visca 1983; Nakabayashi *et al.*, 1983; Duonghong *et al.*, 1981; Sakata and Kawai 1983). The conduction-band electrons are attracted to transition metals, whereas the valence-band holes are attracted to the metal oxides. The result is a high degree of electron-hole separation. The use of band gap semiconductors such as TiO_2, ZnO, ZrO_2, CdS etc, and their various modified forms as the gaseous and aqueous phase photocatalysts has been known. Further it has long been known that certain materials such as noble metals (Pt, Pd, Au, Ag) and some metal oxides (RuO_2, WO_3 and SiO_2) facilitate electron transfer and prolong the length of time that electrons and holes remain segregated.

Sakata and Kawai (1983) evaluated the rate of hydrogen evolution for an ethanol-water solution containing TiO_2 loaded with several different transition metals at 3wt per cent. The TiO_2 loaded with Pt, palladium (Pd), and rubidium (Rb) showed significantly higher rates of hydrogen evolution than did TiO_2 alone. They also compared the hydrogen evolution rate from the photo dissociation of water for TiO_2 alone, Pt-TiO_2 (weight ratio 10: 100), and RuO_2-TiO_2-Pt (5: 100: 10). The rate of hydrogen evolution for RuO_2-TiO_2-Pt was about 4 times greater than for Pt- TiO_2, which was about 190 times greater than forTiO_2 alone. The quantum yields for bifunctional catalysts such as RuO_2-TiO_2-Pt have also been reported to be higher than for TiO_2 alone. Sakata and Kawai (1983) reported quantum yields of hydrogen production of 50 per cent for water-alcohol mixtures containing RuO_2-TiO_2-Pt and illuminated with monochromatic light of 380 nm; for mixtures containing TiO_2 alone, quantum yields were only 3 per cent. For a similar water-alcohol mixture under monochromatic light of 310 nm, Duonghong *et al.* (1981) reported quantum yields of hydrogen production of about 30 per cent for a RuO_2-TiO_2-Pt catalyst.

Surface modification (Muggli and Ding 2001) of TiO_2 with sulfate has been reported to be very useful in the photocatalytic oxidation of volatile organic compounds such as heptane, trichloroethylene, ethanol, acetaldehyde and toluene. The coverage of all organics on SO_4^{2-}/TiO_2 were significantly higher than on P-25 TiO_2, room temperature PCO is faster with P-25. The SO_4^{2-}/TiO_2 may be preferable to Degussa P-25 for PCO at elevated temperatures. On the otherhand, the authors (Dhodapkar *et al.*, 2003) have recently shown that carbonate present in wastewater could inhibit photodegradation strongly. The adsorption of organic compounds is strongly limited by preferential adsorption of CO_3^{2-} that also forms very stable Ti-CO_3^{1-} surface complexes.

Advantages of the TiO_2/UV Process

Some of the advantages of the TiO_2/UV method over other AOP processes are listed below:

- It does not require the use of expensive oxidants (like H_2O_2 or O_3). Instead, it uses cheap and abundant oxygen.
- The oxide photocatalysts, especially TiO_2, are available at a low price (RS 175-200 per kg) and can be recycled on a technical scale due to its stability in aqueous media.
- The light required to activate the catalysts may be a natural UV component of sunlight in addition to low and medium pressure Hg vapor lamps which generate UV-B and UV-A light. Many of these lamps are commercially available. The use of TiO_2/UV process with UV-A domain light opens the possibility of using concentrated solar light.
- The TiO_2/UV based oxidation is powerful and indiscriminate leading to mineralization of a majority of organic pollutants.

Design of Photocatalysts and Photoreactors

Three types of Hg lamps *viz.*, low, medium and high-pressure lamps are used as UV light sources in a typical photocatalytic experiment. However, medium pressure Hg lamps (MPML) are more frequently used. The MPML used commonly for photocatalytic studies generate large amounts of thermal radiation at relatively higher temperatures, with electric to UV energy conversion efficiency of < 15 per cent. The heat must also be used for driving thermocatalytic reactions.

It is generally recognized that only a very thin layer on the photocatalyst surface can actually be excited to enter photocatalytic reactions. For most active photocatalysts, the physical thickness of this layer or skin does not exceed few microns. This is because UV radiation is completely absorbed within a skin only few microns thick on the exposed photocatalyst surface, in contrast to thermal radiation that can penetrate deep into the supported catalyst. The fact that most target species can also be adsorbed into the deep layers of the porous photocatalytic media encourages the use of multifunctional catalysts capable of utilizing both heat and light emitted by medium and high pressure UV lamps. Thus, a multipurpose catalyst can comprise a base material that acts as both a thermocatalyst as well as support structure for the photocatalyst. Alternatively, a dual catalyst may be used that can function as both thermocatalyst and photocatalyst, simultaneously.

Assuming band gap energy of 3.1 eV for TiO_2, a threshold wavelength of about 400 nm is obtained. Thus, TiO_2 will absorb light having a wavelength equal to or lower than this value. While these wavelengths are much longer than the excitation wavelengths needed for the UV/H_2O_2 and UV/O_3 processes, the match of the TiO_2 absorption profile with the solar spectrum is unfortunately rather poor. However, solar photocatalysis would be better candidate in terms of economics. Attempts to extend the TiO_2 spectral response into red through dye sensitization has relied on initial excitation of the dye followed by subsequent electron transfer from the excited state dye molecules into the TiO_2 conduction band. Such an approach should specially be suited while aiming photocatalytic treatment of dye containing wastewater.

Recombination of Charge Carriers-Photoreactor Design

Once holes and electrons are photo generated they move about the crystal lattice freely. Thus a major portion of the electrons and holes get a chance to recombine and only a small portion reach the surface whereby they react with adsorbed species to form reactive radicals. An important factor in controlling the rate of e-h recombination on the photocatalyst surface is the size of the photocatalyst particles. The smaller these particles are, the shorter the distance that charge carriers must travel to reach the surface and larger the exposed catalyst surface area is. Photocatalysts of a few nanometer diameter and BET surface areas of >100 m^2/g are commercially available (ST-01 and ST-31 grades titania produced by Ishihara Sangyo Kaisha, Ltd. of Japan).

The rate of recombination of holes and electrons is a function of the catalyst surface irradiance. The greater is the surface irradiance, the greater is the rate of recombination. It is generally recognized that hydroxyl radical generation is rate-limiting step. The rate of surface reactions will then be equal to $r = k_{(C+d)}[h^+_{(VB)}]$. The rate of hole formation is $k_{aqi,}$ where qi is catalyst surface irradiance (quanta/s/cm^2). The rate of e-h recombination is then $k_e[h^+_{VB}][e^-_{cb}] = k_e[h^+_{vb}]^2$. When qi is high, a large number of electrons and holes will be generated, and $r = k\,(q_i)½$. At low values of qi when surface concentration holes is relatively small, the recombination term will be negligible and $r = k_a q_i$. The surface irradiance value at which the reaction rate transition from q_i to $(q_i)½$ is called Egerton-King threshold ($q_{EK} = 2.5 \times 10^{15}$ quanta/s/cm^2 at λ = 335, 365 and 404 nm). The q_{EK} for two commonly used UV light sources (low and medium pressure Hg lamps) is approximately equal to 1.95 mW/cm^2 (for λ = 254 nm) and 1.36 mW/cm^2 (for λ = 365 nm), respectively. In order to limit the rate of recombination of electrons and holes and maximize the photorector performance, it is necessary to limit the catalyst surface irradiance to levels at or below the Egerton-King threshold. The rate of surface reactions, r, is proportional to $(q_i)^m$, where m varies between ½ and 1. To increase the rate of surface reactions for target pollutants, it may be necessary to allow q_i to exceed q_{EK} under certain conditions. Therefore, under practical situations, the requirement for an efficient utilization of the photogenerated charge carriers must be balanced against the need for optimum rate of the surface reactions involving primary and secondary reactants. This requires a careful photoreactor design that allows uniform irradiance over all photocatalyst surfaces at a level that is close to q_{EK} as possible and optimum rate of conversion of surface-borne target species to desirable final products.

Just like radiation and heat transfer, transport of the primary reactants to and final products from the catalyst surface affect the photo process' performance. The reactor engineering is closely coupled to the choice and configuration of the media and the type of light source used. A proper photoreactor design should provide for uniform irradiance on all catalytic surfaces as well as effective mass transport to and from active sites on the photocatalyst surface. In general, photoreactor designs fall into three categories (Table 4.4).

Table 4.4: Types of Photoreactors

Type	*Properties*	*Reactor Types*
Category I	Most photoreactors possess good mass transfer but generally poor radiation field characteristics	Photocatalyst coated monolith, photocatalyst coated panel, and baffled annular photoreactor.
Category II	Poor mass transfer but uniform catalyst surface irradiance	Annular photoreactor designs.
Category III	Poor mass transfer and non-uniform catalyst surface irradiance	Externally lit annular- photoreactor, catalyst coated on the inner wall.

Mass transfer of species is generally improved by generating and enhancing turbulence in the flow. Many articles such as ribs, fins, pleats, beads, chips, flaps, strips, coils, baffles, baskets, wires etc., have been conceived, used and patented for generating mixing and turbulence. However, due care should be given when using such articles for enhancing turbulence so that the radiation field characteristics are not altered. In view of this, it is advantageous to design the photoreactors in such a way that the bulk of catalyst resides on the reactor wall. This limits the number and proximity of internals, in general, baffles receive more radiation compared to walls. Thus, most internals would be undesirable in the context of uniform radiation field that is always desired.

Catalytic Media

At least six distinct types of catalytic (photocatalyst/support) media arrangements have been in use for related applications. The type-0 to type-5 media have been described in Table 4.5.

Choice Based on the Physical Form of the Photocatalyst

Semiconductor materials may be available in three forms: powders, single crystals, and thin films. For practical applications, single crystals can be safely excluded as serious candidates for remediation processes both in terms of economic and technical perspectives. Semiconductor thin films are attractive for use in photovoltaic devices but are inadequate for photocatalytic waste treatment applications unless special efforts are made to tailor their surface morphology. This leaves us with the option of reactors containing slurries and suspensions of semiconductor powders. These do offer the advantage of high surface dispersion that entails higher particle- substrate encounter frequency. However, photocatalysts recovery after use becomes an issue of major practical concern. Immobilized photocatalysts in the form of highly microporous particles that are attached to a solid support represent an effective compromise incorporating to a degree, the positive features of powder based slurries and suspensions. Thus TiO2 has been immobilized on beads, hollow tubes, Vycor glass, woven fabric, silica gel, optical fibers and sand. Porous thin films of TiO_2 have been synthesized thermally fusing ZnO or TiO_2 particles onto conducting glass or by dispersing the powder on polymeric binder. Several immobilization approaches have been evolved, most of them are support specific. A novel method for immobilization of the titania catalysts on a metallic support (SS

Table 4.5: Types of (Photo)catalytic Media

Type	*Properties*	*Remarks*
Type-0	A suitable catalyst such as titania is used in colloidal form without any support or base materials. There exists a sub-category to Type-0 media. This employ, in addition to titania, an adsorbent material that absorbs the target compounds. The contaminant-loaded adsorbent is then separated from the fluid and brought into contact with aqueous slurry of a suitable photocatalyst. This is implicit that whenever adsorption target species is rate limiting, pre -adsorption of contaminant followed photocatalysis is a more suited method. However, this is generally not the case. It is desirable to simplify the treatment process by eliminating additional adsorbent in favour of multifunctional catalytic media.	Type-0 catalytic media need to be separated and recovered after the photocatalytic reaction is over. If the activity persists, it may be reused.
Type-1	In this type of photocatalyst/support arrangement, the catalyst such as TiO_2 is immobilized or bonded onto a ceramic, glassy *(e.g.*, fiber glass mesh, woven glass tape etc.,) or metal oxide (*e.g.* silica gel), metallic *(e.g.* Stainless steel) or synthetic polymeric support substrate *(e.g.*, plastic, PE fabric etc,)	The substrate has no function other than providing physical and structural support for the photocatalyst.
Type-2	In type 2 media configuration, impregnated glassy mesh/matrix or porous ceramic monolith or beads, metallic and metal oxide substrates (in the form of plates, beads etc) are employed as the photocatalyst support to which titania is bonded utilizing a method known as the "sol-gel" technique. In this technique, titania colloid is generated by controlled hydrolysis of a titanium precursor salt. The titania from colloid is coated on supports by dip-thermal cycles are spin coating technique.	The substrate has no function other than providing physical and structural support for the photocatalyst.
Type-3	Type-3 catalyst/support configuration is a variation of the Type-2 media that involves synthesis and use of metal oxide aerogels, most prominently, SiO_2 aerogels doped or co-gelled with other transition metal oxides such as titania to produce photochemically active catalyst/support material. The preparation comprises two steps; a condensed metal oxide intermediate is formed. From this intermediate, aerogels are prepared having desired density, clarity and UV transparency, moisture and stability.	The competetive adsorption of organics at TiO_2 as well as unreactive SiO_2 sites may lead to incomplete mineralization of pollutants.
Type-4	In type-4 photocatalyst/support media, a photocatalyst *(e.g.*, doped and undoped modifications of TiO_2, CdS etc., is deposited by bonding or cementing onto fabric of a modified or unmodified natural or synthetic polymer material. Examples for polymers include: (natural) wood, paper, kozo, gampi, kraft lignin, woven cotton, kenaf, linen, wool etc.	There is upswing research in this area, as naturally available supports are cheap and devoid of catalyst- support interactions, unlike Type-3 catalytic media

Contd...

Table 4.5–*Contd...*

Type	*Properties*	*Remarks*
Type-5	Type-5 media includes the broad field of moderate temperature (150-350°C) thermal oxidation catalysts. A sub-class of relevance to the present subject is supported transition metal oxide catalysts and cation modified zeolites as dual function sorbent/catalyst media. Such catalysts are useful for removing dilute VOCs from air at room temperature, and then it acts as a catalyst at higher temperature to both desorb and oxidize trapped VOCs. Due to their microporous crystalline structure, various forms of zeolites have been widely used as commercial adsorbents. Further, noble or base metal supported titania or zirconia (1-wt per cent Pt, 5-wt per cent V, 5-wt per cent La, 89 per cent TiO_2) with added promoters such as Mo, W, or V, are useful for catalytic oxidation of organic nitrogen containing compounds.	An excellent choice if dual function adsorbent/catalyst media is needed.

304) was developed by using the chemically inert and photo stable fluroresin –PTFE (Teflon) as a binder (Kuo, 2000). The conditions with respect to baking temperature, baking time and the ratio of resin/TiO_2 were optimized by applying response surface methodology (RSM). The resin bonded titania photocatalyst showed twice more photo activity. Rao and Dube (1995) have been able to immobilize TiO_2 on to polyester fabric using polyvinyl alcohol as binder. Similarly, the authors (Rao and Dube, 1995) have experimented with Ti-TiO_2, prepared by thermal as well as flame oxidation of Ti sheet. Sol-gel technology can also be used for fabricating porous films onto suitable supports. The physicochemical and photocatalytic properties in aqueous solutions of the supported TiO_2 films prepared by different sol-gel methods have been studied (Guillard 2002). The properties are influenced by the nature of the stabilizing agents (*e.g.*, acetic acid, acetylacetone etc.,); the nature of organo titanium precursors (methoxo-, ethoxo-, isopropoxo- or ethoxo-isoporopxo-) and solvent. The migration of cations native to supports into thin TiO_2 films drastically reduces the photoactivity of the supported catalysts. The migration is facilitated at higher temperatures and longer calcination times.

Unfortunately, immobilized photocatalyst reactor configurations resulted in lower photocatalytic activity relative to their slurry counterparts. This is traced to substrate mass transport problems crucial with immobilized catalysts. This area of research is now awaiting a break through in the development of supported form of TiO_2 with a hundred fold higher photoactivity for oxidation of organics unselectively.

Photocatalysis with Sunlight

Most of the solar photochemical processes are based on the utilization of high-energy short-wavelength photons (300-400 nm) to promote photochemical reactions. In solar heterogeneous photocatalytic detoxification process, the near-UV part of the solar spectrum (wavelength shorter than 380 nm) is used to photoexcite a semiconductor catalyst. Although scientific research on solar detoxification has been conducted for at least last three decades (Blake 1999), industrial/commercial applications, engineering systems and engineering design methodologies have only been developed recently (Blanco *et al.*, 2000; Malato *et al.*, 2002).

The hardware needed for solar photochemical applications is similar to that used for solar thermal applications. The original solar photoreactors were based on line-focus parabolic trough concentrators and Non-concentrating collectors (Figure 4.3). The first engineering scale solar photochemical facility for water detoxification in Europe was developed by CIEMAT (25) using 12-axis PTC. The sunlight was reflected by the parabola onto the reactor tube at the focus through which contaminated water is circulated. The reactor receives large amount of radiation energy per unit volume and the optical efficiency > 50 per cent is obtained (Blanco *et al.*, 1991; Blanco and Malato 1992). Apart from a variety of organic contaminants, industrial wastewaters are treated using the PTC reactors. The main disadvantages are (*i*) use only direct radiation, (*ii*) collectors are expensive, and have low optical efficiencies at least for TiO_2 applications.

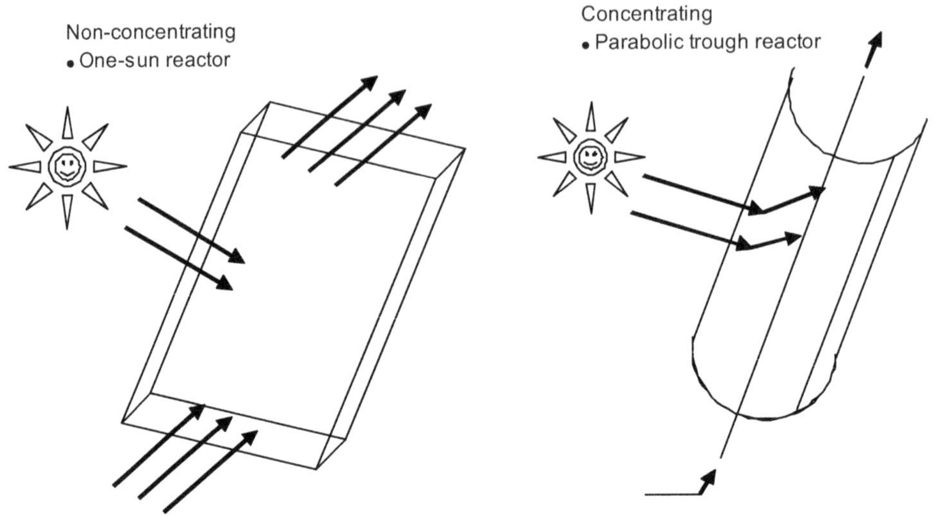

Figure 4.3: Design Concepts for Solar Water Detoxification Photoreactors.

On the other hand, non-concentrating solar reactors are in principle cheaper than PTCs as their components are simple and easy to maintain and also do not have expensive tracking devices etc. There are three types of non-concentrating reactors:

1. Free-falling film in which the process fluid falls slowly over a tilted plate with a catalyst attached to the surface, which faces the sun and is open to atmosphere (Goslich *et al.*, 1997; Blanco *et al.*, 1994);
2. Pressurized flat plate which consists of two plates between which the fluid circulates using a separating wall (Well *et al.*, 1997; Dillert *et al.*, 1999; Bahnemann *et al.*, 1999);
3. Solar Ponds; small, shallow on-site pond reactors (Gimenez *et al.*, 1997, 1999).

The fluid flow in non-concentrating solar reactor is usually laminar. This presents with mass transfer problems and vaporization of reactants. In this sense, tubular reactors do have decisive advantage.

Compound Parabolic concentrator is a hybrid reactor possessing characteristics of both trough and one-sun reactors. They are static collectors with involute reflective surface around a cylindrical reactor tube. They are simple and cost-effective. The reflector design allows almost all the UV-irradiation reach the reactor aperture where the tube reactor is placed.

Some chlorophenols (Malato *et al.*, 1997; Dillert *et al.*, 1999; Minero *et al.*, 1993) have been subjected to photocatalytic degradation in a two axis solar pilot-plant of PTC design (260 L, 20mg/L chlorophenol, 0.2gTiO_2/L and 32 m^2 collector surface area. The mineralization treatment efficiency is found to be 0.015, 0.012 and 0.03g

TOC $h^{-1}m^{-2}$ for PCP, DCP and CP respectively. The mean intensity of solar UV light was considered $30WUVm^{-2}$. When applied to real wastewater containing pesticides, about 20-30 mg TOC L^{-1} was reduced using $56kJL^{-1}$ solar UV-energy (Blanco *et al.*, 1994; Malato *et al.*, 2000; Malato *et al.*, 2002). Considering mean solar UV intensity at Almeria as 18.6 $W_{UV}m^{-2}$, approximately 1000 m^2 collector is estimated for treating 5000 m^3 of wastewater per year.

Titanium Dioxide Photocatalysis: Self-Cleaning Building Materials

Cleaning of building materials like tiles, facades and glass panes causes practical problems of inaccessibility as well as consumption of energy and chemicals. Cleaning based on lowering of contact angle using soap, detergents is of late outsmarted by thin film coatings of photocatalyst materials on the surfaces to be cleaned. Thus, very thin films of photoactive TiO_2 shone with UV light helped to lower the contact angle to as low as <1°. These coatings are superhydrophilic and water lies flat on the surface in sheets instead of forming droplets. The superhydrophilicity is reatianed for approximately two days even after the UV-light is put off. In addition, the UV irradiated TiO_2 has the ability to oxidize and decompose many types of bacteria, organic and inorganic materials. This effect is used for cleaning the tiled walls and free them from bacteria (Fujishima *et al.*, 2000). Thus in a reasonably well lit room these coatings can decompose a hydrocarbon layer of approximately 1 m thickness every hour (Fujishima *et al.*, 2000). Some of the applications reported in the literature are TiO_2-containing paper, microfilms, self –cleaning TiO_2 subway tunnel lamps and flow-through water purification. Further, self-cleaning window glass has been developed, as well as TiO_2-containing, self-cleaning paint. The TiO_2 photocatalysts can also be applied for killing bacteria. Based on this self-sterilizing surfaces can be prepared (Kikuchi *et al.*, 1997).

Conclusion

The photocatalytic process using TiO_2 and UV light may be considered as feasible for the treatment of wastewater containing hazardous contaminants for which biological waste treatment plants are improbable at medium or low pollutant concentrations. The use of sunlight instead of artificial light for this process would dramatically lower the process costs and thus provide a major step towards industrial applications. The feasibility of solar photocatalytic process for decolorization of textile dye wastewater has been demonstrated at NEERI as well as at many other places in the world. Since the rate of the process is linearly dependent on UV-photon flux, the associated investment is also dependent on the photon collector surface. Thus, it is needed to conduct further research and development studies to design appropriate photocatalysts and photoreactors so that optimum utilization of sunlight is achieved towards decontamination of large volumes of industrial wastewater.

References

1. Abrahams J, Davidson R S, Morrison C L. (1985). Optimization of the properties of titanium dioxide. *J. Photochem*. 29, 353.

2. Al-Ekabi H (1990). Catalytic oxidation of waste organics. *Chemical Engineering*, 97 (9), p.27.

3. Al-Ekabi H, Serpone N. (1988). Kinetic studies in heterogeneous photocatalysis. 1. Photocatalytic degradation of chlorinated phenols in aerated aqueous solutions over TiO_2 supported on glass matrix. *J. Phys. Chem.*, 92, 5726.

4. Bahnemann D, Dillert R., Dzengal J., Goslich R., Sagawe G., Schmacher H.-W, Benz V.W., *J. Adv. Oxid. Tech*. 4 (1999)11

5. Borgarello E, Kiwi J, Pelizzetti E, Visca M, Gratzel M. (1981). Sustained water cleavage by visible light. *J. Am. Chem. Soc.*, 103, 6324.

6. Blake D M., 1999. *Bibliography of work on the photocatalytic removal hazardous compounds from water and air*, National Technical Information Service, US Department of Commerce, Springfield, VA, USA, Update No.3

7. Blanco J., Malato S., 2000. *Solar detoxification*, Natural Sciences, World Solar Programme, UNESCO, 1996-2005 (http: //www.unesco.org/science/wsp).

8. Blanco, J., Malato, S., Sanchez M., Vidal A., Sanchez B., 1991. In: M. E. Arden, S. M. A. Burley, Coleman M. (Eds.) *Proceedings of the ISES Solar World Congress*, Denver, USA, Pergamon Press, Oxoford, p.2097

9. Blanco J., Malato S. (1992). In: *Proceedings of the 6th Int. Symp. On Solar Thermal Concentrating Technology*, Mojacar, Spain, CIEMAT, Madrid, p. 1183

10. Blanco J., Malato S., in: D.e. Klett, R. E. Hogan, T. Tanaka (Eds.) *Proceedings of the Solar Engineering* 1994, ASME, USA, 1994, p.103.

11. Blanco J., Malato S., Bahnemann D, Bockelmann D., Weich grebe, D., Carmona F., Martinez, F. in: O. Popel, S. Fris, E. Shchedrova (Eds.) (1994). In: *Proceedings of the 7h Int. Symp. On Solar Thermal Concentrating Technology*, Moscow, Russia, Institute for High Temperature of Russian Academy of Science (IVTAN), 1994, p. 468.

12. Childs L P., and Ollis D F. (1981). Photoassisted heterogeneous catalysis. Rate equations for oxidation of 2-methyl-2-butyl alcohol and Isobutane, *J. Catal.* 67 (1), 35.

13. Cunningham J., and Srijaranai S. (1988). Isotope effect for hydroxyl radical involvement in alcohol photo-oxidation sensitized by TiO_2 in aqueous suspension., *J. Photochem. Photobiol. Chem.*, 43, 329.

14. Dhodapkar R. S., Rao, N. N., Pande S. P., and Kaul S. N. (1999). Photocatalytic destruction of organic pollutants in water, In *Adv. in Ind. Wastewater Treatment*, Tech Sci. Publ. Chapter 26, 453-466.

15. Dhodapkar R. S., Rao N. N., Pande S. P., and Kaul S. N. (2000). Photocatalytic route for reduction of color and chemical oxygen demand from dye containing wastewater, *Environment Conservation Journal*, 1 (1): 13-20

16. Dube, S., and Rao N. N. (1996). Rate parameter independence on the organic reactant: a study of adsorption and photocatalytic oxidation of surfactants using MO3/TiO2 (M=Mo or W) catalysts, *J. Photochem. Photobiol. A: Chem*, 93, 71.

17. Duonghong D., Borgarello E., and Gratzel M. (1981). Dynamics of light induced water cleavage in colloidal systems. *J. Am. Chem. Soc.*, 103 (16), 4685.
18. Fox M. A., Ogawa H., Pichat P. (1988). Regioselectivity in the semiconductor mediated photooxidation of 1,4-pentane diol. *J. Org. Chem.*, 54 (16), 3847.
19. Fujishima A, Tata N. Rao and D. A. Tryk. 2000. Titanium dioxide photocatalysis, *J. Photochem. Photobiol. C: Photochemistry Reviews* 1 (2000)1.
20. Gimenez J., Curco D., Marco P. (1997). *Water Sci. Technol.* 35 (4), 207.
21. Gimenez J., Curco D, Queral M. A. *Catal. Today*, 54 (1999)267.
22. Goslich R., Dillert R., Bahnemann D. (1997). *Water Sci. Technol.* 35 (4), 137.
23. Guillard C., Beaguraud B., Dutriez C., Herrmann J. M., Jaffrezic H., Jaffrezic-Renault N., Lacroix M. (2002). Physicochemical properties and photocatalytic activities of TiO_2-films prepared by sol-gel methods. *Appl. Catal. B: Environmental*, 39, 331.
24. Hasimoto K., Kawai T., Sakata T. (1984). Photocatalytic reactions of hydrocarbons and fossil fuels with water., *J. Phys. Chem.* 88 (18), 4083.
25. Herrmann J M., and Pichat P. (1980). Heterogeneous photocatalysis: oxidation of halide ions by oxygen in ultraviolet irradiated aqueous suspension of titanium dioxide, *J. Chem Soc. Faraday Trans. I*, 76 (5), 1138.
26. Hsiao C. –Y, Lee C. L. and Ollis D. F. (1983). Heterogeneous photocatalysis: Degradation of Dilute Solutions of Dichloromethane, Chloroform, and Carbon Tetrachloride with illuminated TiO_2 Photocatalyst., *J. Catal.*, 82, 418.
27. Izumi I. W., Dun W., Wilbourn K. Fan F., and Bard A. J. (1980). Heterogeneous photocatalytic oxidation of hydrocarbons on platinized TiO_2 powders, *J. Phys. Chem.* 84 (24), 3207.
28. Kiwi J., In *Energy Resources through Photochemistry and Catalysis* (Ed. M. Gratzel) New York, Academic Press, 1983.
29. Kikuchi Y, Sunada K., Iyoda T., Hashimoto K., Fujishima A. 1997. Photo catalytic bactericidal effect of TiO_2 thin films: dynamic view of the active oxygen species responsible for the effect. *J. Photochem. Photobiol. A.Chem*: 106 (1997) 51.
30. Kraeutler B., and Bard A. J. (1978). Heterogeneous photocatalytic decomposition of saturated carboxylic acids on TiO_2 powder. Decarboxylation route alkanes. *J. Am. Chem. Soc.*, 100, 2239 and 5985.
31. Kuo W. S. (2000). Preparation and photocatalytic activity of metal-supported resin-bonded titania. *J. Environ. Sci., Health* A35 (3), 419.
32. Malati M. A. and W. K. Wong. (1984). Doping TiO_2 for solar energy applications. *Surface Tech.* 22, 305-322.
33. Malato S., Gimenez J., Richter C., Curco D., Blanco J, (1997). Low-concentrating CPC Collectors for photocatalytic water detoxification: comaparison with a medium concentrating solar collector, *Water Science and Technology*, 35 (4), 157

34. Malato.S., Blanco J., Richter C., Vincent M. *Solar Energy* 56 (5) 1996, 401.

35. Malato S., Blanco J., Vidal A., and Richter C. (2002). Photocatalysis with solar energy at a pilot-plant scale: an overview, *Appl. Catal. B: Environmental* 37, 1-15.

36. Matthews R. W. (1987). Solar-electric water purification using photocatalytic oxidation with TiO_2 as a stationary phase. *Solar Energy* 38 (6), 405.

37. Matthews R. W. (1987). Photooxidation of organic impurities in water using thin films of titanium dioxide., *J. Phys. Chem.* 91 (2), 3328.

38. Matthews R. W. (1984), Hydroxylation eactions induced by near ultraviolet photolysis of aqueous TiO_2 suspensions. *J. Chem. Soc. Faraday* I, 80 (2), 457.

39. Muggli D. S., and Ding L. (2001). Photocatalytic oxidation using SO_4^{2-}/TiO_2., *Appl. Catal. B: Environmental*, 32, 181.

40. Nakabayashi S., Fujishima A., and Honda K. (1983). Experimental evidence for hydrogen evolution site in photocatalytic process on Pt/TiO_2., *Chem. Phys. Lett.*, 102 (2), 464.

41. Norton A. P., Bernasek S. C., Bocarsly A. B. (1988). Mechanistic aspects of the photo oxidation of water at the n-TiO_2/aqueous interface. Optically induced transients as a kinetic probe, *J. Phys. Chem.* 92 (21), 6009.

42. Okamoto K., Yamamoto Y., Tanaka H., Tanaka M., Itaya A. (1985). Heterogeneous photocatalytic decomposition of phenol over TiO_2 powder. *Bull. Chem. Soc. Jpn.* 58, 2015.

43. Ollis D. F., Hsiao C.-Y., Budiman L. and Lee C. L. (1984). Heterogeneous photoassisted catalysis. Conversions of perchloroethylene, dichloroethane, chloroacetic acids and chlorobenzenes. *J. Catal.* 88 (1), 89.

44. Pelizzetti E., and Visca M. 1983. Examples of photogeneration of hydrogen and oxygen from water. In Energy *Resources through Photochemistry and Catalysis* (Ed. M. Gratzel) New York, Academic Press.

45. Pelizzetti E., Maurino V., Minero C., Carlin V., Pramauro E., Zerbinati O., Tasato M. L. (1990). Photocatalytic degradation of Atrazine and other s-Triazine herbicides, *Envi. Sci. Technol.* 24 (100), 1559.

46. Pruden A. L., Ollis D. F. (1983). Photoassisted heterogeneous catalysis: The degradation of trichloroethene in water., *J. Catal.* 82 (2), 404.

47. Ramachandriah G., Thampy S K., Narayan P K., Chuahan D K., Rao N N., Indushekhar V. K. (1996). Separation and concentration of metals present in industrial effluent and sludge samples by using electrodialysis, coulometry and photocatalysis. *Separation Science and Technology*, 31 (4): 523-532.

48. Rao N. N. and Dube S. (1995). Application of Indian commercial TiO_2 powder for the destruction of the organic pollutants: photocatalytic degradation of 2,4-dichlorophenoxy acetic acid (2,4-D) using suspended and supported catalysts. *Ind. J. Chemical Technol.*, 2, 241.

49. Rao N. N., and Dube S. (1995). Application of Indian commercial TiO_2 Powder for destruction of 2,4-dichlorophenoxy acetic acid using suspended and supported TiO_2 catalysts. *Ind J. Chem. Technol.* 2: 241-248.

50. Rao N. N. and Dube S. (1996). Photocatalytic degradation of some commercial soap/detergent products using suspended TiO_2 catalysts. *J. Mol. Catal. A: Chem.* 104: L197-L199.

51. Rao N. N., Dube S., and Natarajan P. (1993). Photocatalytic degradation of some chlorohydrocarbons in aqueous suspensions of MO_3/TiO_2 (M = Mo or W). *Trace Metal Environment*, 3: 695-700.

52. Rao N N., Dube S. (1996). Photoelectrochemical generation of hydrogen using organic pollutants as sacrificial electron donors. *Int. J. Hydrogen Energy*, 21 (2), 95-98.

53. Rao N. N., Dube S. (1995). Application of Indian commercial TiO_2 powder for destruction of organic pollutants: photocatalytic degradation of 2,4-dichlorophenoxyacetic acid (2,4-D) using suspended and supported TiO_2 catalysts, *Ind. J. Chem. Technol.* 2, 241-248.

54. Rao N. N., Dube S. (1997). TiO_2-catalyzed photodegradation of reactive orange 84 and alizarin red S biological stain. *Ind. J. Chem. Technol.* 4, 1-6.

55. Rao N. N., Dube, S. (1998). Photocatalytic degradation of Reactive Orange 84 (RO84) in dye house effluent using single-pass reactor, *Studies in Surface Science and Catalysis*, 113, 1045-1050.

56. Rao N. N., and Kaul, S. N. (1999). Methods to enhance the efficiency of photocatalytic process, *Bull. Catal. Soc. India*, 9 (1 and 1), 81-91, 1999,

57. Rao N. N., Kaul S. N. (2000). Photocatalytic activity of H_2 reduced Indian commercial TiO_2: photodegradation of 2,4-dichlorophenoxyacetic acid, *Bull. Catal. Soc. India*, 10 (1 and 2), 95-112.

58. Rao N. N. (1996). Environmental applications of photocatalysis: a new member of advanced oxidation processes, *TERI Information Digest on Energy (TIDE)*, 6 (3).

59. Rao N. N., Dubey A., Jain Rajesh, Khare P., Mohanty S., Kaul S. N. (2003). Photocatalytic degradation of 2-chlorophenol: a study of kinetics, intermediates and biodegradability, *J. Hazard. Mater.*, 101 (2).

60. Rao N. N. and Natarajan P. (1994). Particulate models in heterogeneous photocatalysis. *Current Science* 66 (10), 742.

61. Sakata T., Kawai T., Photosynthesis and photocatalysis with semiconductor powders. In *Energy Resources through Photochemistry and Catalysis* (Ed. M. Gratzel) New York, Academic Press, 1983.

62. Sakata T., Kawai T., Hashimoto K. (1984). Heterogeneous photocatalytic reactions of organic acids and water: New reaction paths besides the Photo-Kolbe reaction, *J. Phys. Chem.*, 88, 2344.

63. Sclafani A., Palmisano L., and Schiavello M. (1990). Influence of the preparation methods of TiO_2 on the photocatalytic degradation of phenol in aqueous dispersion. *J. Phys. Chem.*, 94 (2), 829.

64. Serpone N., Borgarello E., Harris R., Cahill P., Borgarello M., Pelizzetti E. (1986). Photocatalysis over TiO_2 supported on a glass substrate. *Solar Energy Materials* 14,121.

65. Turchi C. S., and Ollis D. F. (1990). Photocatalytic degradation of organic water contaminants: Mechanisms involving hydroxyl radical attack, *J. Catal.* 122 (1),178.

66. Well M., Dillert R H G., Bahnemann D., Benz V. W., Mueller M. A., J. *Solar Energy Engg.* 119 (1997)114.,

Chapter 5

Pilot Plants for Treatment of Phenol and Cyanide Bearing Wastewaters: An Overview of Experiences

R.A. Pandey, T.V. Subbarao and S.N. Kaul

CSIR-NEERI, Nagpur

ABSTRACT

Completely Mixed Activated Sludge (CMAS) process for treatment of phenolic wastewater and cyanide containing wastewater was investigated separately on pilot scale for ascertaining the feasibility of the developed processes for large scale commercial applications. A number of problems related to siting, climatic conditions, design, construction, development of biomass and operation of these systems were encountered during these investigations. Some of the problems alongwith mitigation measures adopted during investigation are presented and discussed in this paper.

Keywords: *Pilot plants, Microorganisms, Climatic conditions, Phenol and Cyanide bearing wastewaters, Treatment.*

Introduction

Biological processes are gaining more importance in the field of environmental management since these are viable, cost-effective and environment-friendly through which one can achieve the goal of sustainable development. The scope of the application of biotechnological processes, with reference to environmental issues, has acquired greater attention and, therefore, it has been utilized for various purposes; *viz.* environmental monitoring, restoration of environmental quality, resource/

residue/waste recovery/utilization/treatment, and substitution of non-renewable source base with renewable resources. These processes pass through several stages of development prior to their commercialization. Some of these stages for development of such systems like characterization of samples with which biotechnological systems are likely to be attached, isolation and characterization of microorganisms to be used in the system, selection of environmental, hydro-dynamic and nutritional conditions for the potential isolates, batch experimentation for evaluation of kinetic coefficients for selection of suitable bioreactor systems, design, fabrication, installation and performance evaluation of the reactor systems on bench scale and on continuous feed basis for generation of process design and scale up data for pilot plant, and also design, fabrication and installation of pilot plant for generation of design data for a full scale commercial plant. Today, there appears to be a trend towards the omission, under certain circumstances, of the pilot plant stage in the development of new processes. However, there is little doubt that it would be techno-economically feasible to transfer any new biotechnological process directly from the laboratory scale to large scale application, provided that unlimited money is available, so that designer could be allowed to use high factors of safety and users could, if necessary, meet the cost of overcoming various teething troubles, since live entities would be involved in the system. It is obvious that especially the processes based on biotechnology for environmental management require pilot plant investigation before commercialization, since the strength and quality of wastewater varies considerably due to the changes in the input of raw materials of various origin, and local environmental conditions for the microbial activity.

A second approach to the development of biotechnological process is based on experimental studies followed by the application of model theory for scaling up of the process for full scale application. It is a well-known fact that small scale experiments are not always reliable for large scale application. The aim of model theory is to predict the scale up effect and determine the conditions under which the performance of a model gives a reliable forecast of the prototype performance. The most crucial and important stage in the development of such biotechnological processes is the pilot plant. The pilot plant trials give an idea of the cost effectiveness of the process as well as technical feasibility of the developed system for large scale application. This paper outlines some of the experiences encountered while operating two pilot plants for the aerobic treatment of phenolic wastewater and inorganic cyanide containing wastewater separately, using Completely Mixed Activated Sludge (CMAS) process.

Materials and Methods

Pilot Plant for the Treatment of Phenol Bearing Wastewater

Extensive laboratory experiments on bench scale indicated that phenol in a simulated wastewater could be degraded and stabilized to the end products by using activated sludge process. Based on the laboratory results, a pilot plant was installed for treatment of live wastewater generated from a Low Temperature Coal Carbonization (LTC) industry. The dimensional details of the pilot plant are presented in Table 5.1, while front view of the units is shown in Figure 5.1. The pilot plant

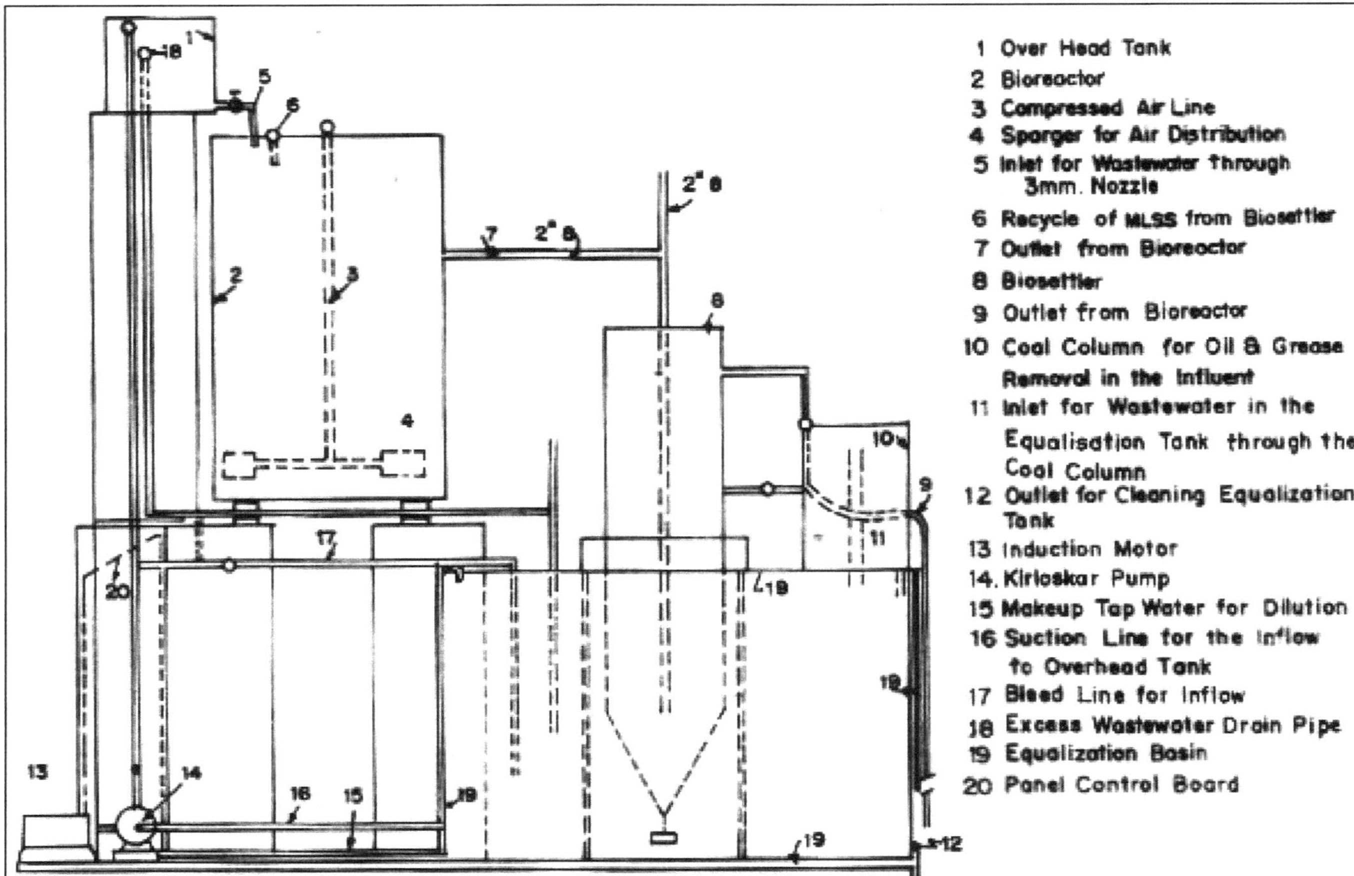

Figure 5.1: Front View of the Pilot Plant Assembly for Detoxification of LTC Wastewater.

consisted of a feed preparation tank, an aeration tank with diffused aeration system, conical settler, and arrangement for recirculation and wastage of the sludge using suitable pumps [1]. The starter seed for the process was a mixed culture of *Pseudomonas* sps. and *Candida tropicalis*. The LTC wastewater containing high concentration of phenols, ammonia, cyanide and thiocyanate was used in these studies.

Table 5.1: Dimensional Details of the Pilot Plant for Detoxification of Phenols in Low Temperature Carbonization Wastewater

Item	*Specification*
Gross volume of the bioreactor	1.5 m^3 (1 m x 1 m x 1.5 m)
Working volume	1 m^3 (1 m x 1 m x 1 m)
Equalization tank	2.440 m^3 (2.1 m x 1.25 m x 1 m)
Constant overhead dosing tank	0.080 m^3
Influent dosing pump	1 HP Kirloskar Centrifugal pump, 3 Phase (2 Nos.)
Sludge recirculation (recycle/wastage) pump	1/20 HP, Austin Pump (1 No.)
Auto-reset timer (L and T) EM 1000 series	2 Nos.
Biosettler	2.83 hrs HDT and 1.85 hrs HDT50 cm (O) 150 cm (L)

The characteristics and magnitude of these constituents in the LTC wastewater are presented in Table 5.2. In the LTC process, 0.15 m^3 of waste ammonia liquor (WAL) is produced on carbonizing one tone of coal for production of domestic coke [1]. Prior to the treatment, the wastewater was adjusted to pH 7.0 – 7.5, using phosphoric acid. Diluted (1: 10) LTC wastewater was used during the investigation. The unit was operated for a period of nine months for generation of design data for a full-scale plant.

Pilot Plant for the Treatment of Inorganic Cyanide Containing Wastewater

This Institute had successfully demonstrated that inorganic cyanide could be detoxified in a CMAS, using specific dominant cultures of *Pseudomonas* sps. as a starter seed. The byproduct of cyanide biodegradation is ammonia and this could be stabilized to nitrogen gas by employing a laboratory scale oxidation pond alongwith CMAS.

Based on laboratory scale studies, a pilot plant was designed, fabricated and installed at an alkali-cyanide manufacturing plant as detailed in Table 5.3 and Figure 5.2. The manufacture of alkali cyanide generates wastewater containing high concentration of cyanide [2] (Table 5.4). Therefore 1: 10 diluted wastewater containing cyanide was used during this study. The wastewater was supplemented with phosphate as a nutrient, and peptone as a growth factor. The pilot plant was operated at pH 8.5 in a continuous mode for a period of six months. The performance of the pilot plant was evaluated to generate the design data for a full scale plant. All the parameters were analyzed using analytical grade chemicals and methods of estimation as per Standard Methods [3].

Table 5.2: Physico-chemical Characteristics of Low Temperature Carbonization Wastewater

Parameters	*Concentration*	
	Maximum	*Minimum*
pH	9.97	8.98
Colour (absorbance at 470 mm/platinum scale)	23,000	13,000
Suspended solids (mg l^{-1})	800	500
Specific conductance (mhos)	27.5	22.5
Alkalinity (mg l^{-1})	15000	12000
Phenol:		
(a) Steam volatile as C_6H_5OH (mg l^{-1})	9,500	6,250
(b) Total phenol (mg l^{-1})	17,500	9,250
Ammonia nitrogen (free) (mg l^{-1})	7,500	7,000
Thiocyanates SCN^- (mg l^{-1})	350	117
Cyanide (CN^-) (mg l^{-1})	22	11.4
Sulphate ($SO_4^=$) (mg l^{-1})	75	25
Sulphide (mg l^{-1})	74	67
Iron (Fe) (mg l^{-1})	62	28
COD (mg l^{-1})	43,000	40,000
BOD, 20°C (mg l^{-1})	21,750	15,200
TOC (mg l^{-1})	15,100	13,700
Hexane extractable materials (mg l^{-1})	350	172

Problems Experienced during Operation of the Pilot Plants for Treatment of Phenol and Cyanide Bearing Wastewaters

The siting of the pilot plant for treatment of LTC wastewater with reference to the topography was ideal since there was no water logging in the pilot plant area. The pilot plant was located near Coal Tar Fuel (CTF) holding tank, since waste ammonia liquor reservoir (phenolic waste reservoir) was situated nearby.

However, the pilot plant for treatment of LTC wastewater was situated in the eastern direction to the existing bricketing plant, which produces fine coke particles. During summer season, wind direction was west to east that is towards the pilot plant. Whenever there was stormy or high speed wind condition, coke particles were carried over to the bioreactor and resulted in increased solid concentration of the reactor. Thus, it was felt necessary that the pilot plant should have been covered, which might have prevented such adverse effect on the performance of the bioreactor system.

The wind direction was observed to be south-north for most of the period during the operation of the pilot plant for treatment of wastewater containing cyanide. The prevailing wind direction and speed also adversely effect the performance of the bioreactor due to fallout of air-borne particulates emitted from the nearby amine

Table 5.3: Dimensional Details of the Pilot Plant for Treatment of Cyanide Bearing Wastewater

Units	*Magnitude*
Wastewater Holding Tank	
Capacity	5 m^3
Material of construction	Mild Steel
Shape	Cylindrical
Mechanical gadgets for conveying wastewater	Flow metering pumps
Bioreactor	
Process type	Activated sludge
Flow rate	3 m^3d^{-1}
Capacity	3 m^3 (approx)
Hydraulic detention time	24 h
MLSS	1000 – 1500 mg l^{-1}
Size	0.75 m x 1.4 m x 3 m (liquid depth)
Influent cyanide concentration	100 – 200 mg l^{-1}
Mode of aeration	Diffused aeration
Source of air supply	Factory's central plant
Material of construction	Mild Steel
Settling Tank	
Shape	Circular
Diameter	0.45 m
Total depth	2.4 m
Bottom shape	Conical hopper bottom
Material of construction	Mild Steel
Sludge Pump	
Shape	Circular (n plan)
Diameter	50 cm
Depth	30 cm
Material of construction	Mild Steel
Sludge recirculation	100% (with the help of sludge recirculation pump)
Stabilization Pond	
Shape	Trapezoidal cross section
Top	7.5 m x 4.5 m
Bottom	6 m x 3 m
Capacity	21.42 m^3 (0.95 m liquid depth)
	30.14 m^3 (1.25 m liquid depth)
Material of construction	RCC

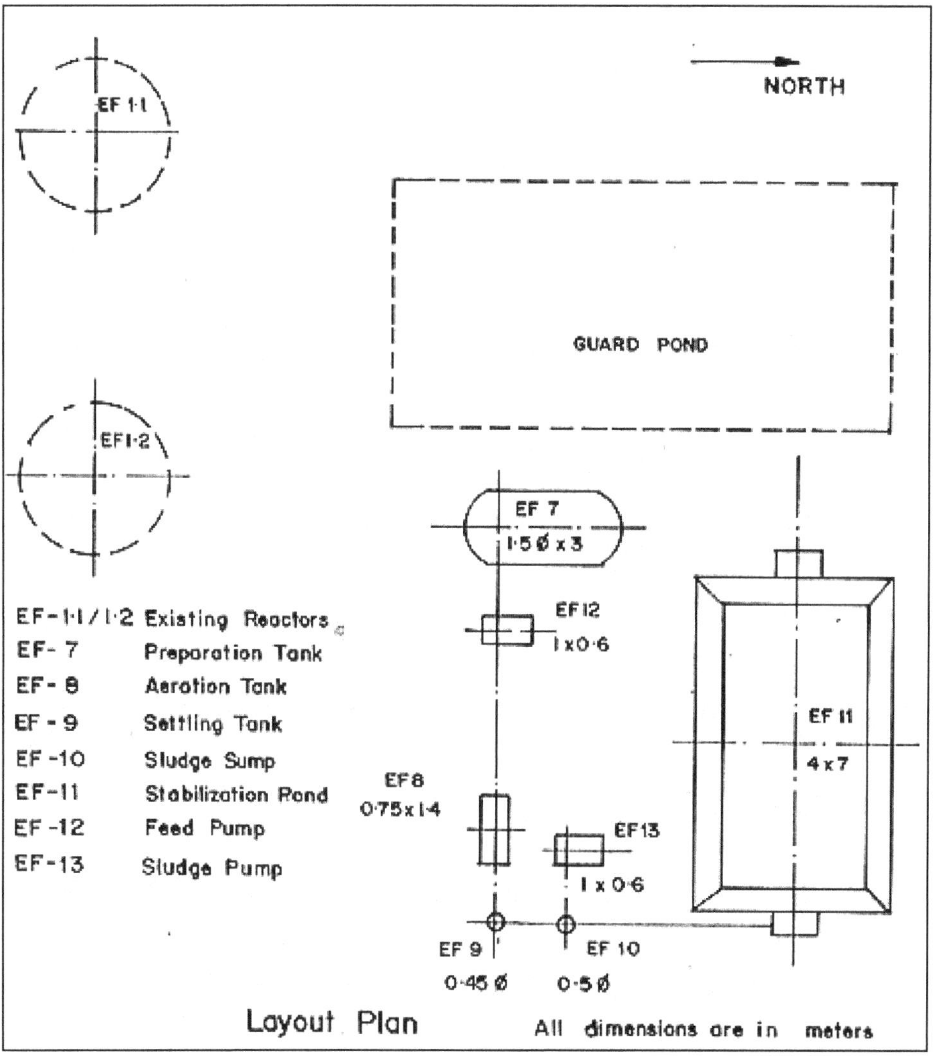

Figure 5.2: Biological Treatment Pilot Plant for Cyanide Containing Wastewater.

manufacturing plant. This particulate matter was getting mixed up with the biomass of activated sludge and also was forming a thin layer on the surface of the oxidation pond.

Table 5.4: Characterization of Cyanide Bearing Wastewater

Parameter	*Concentration*
Cyanide (as CN^-)	2060 mg l^{-1}
COD	1336 mg l^{-1}
Ammonia	Not detectable
pH	10.3

Keeping in view the wind direction and existing cyanide storage facility and cyanide manufacturing chemical reactors, the pilot plant siting was quite safe, but adjacent location of chlorination plant resulted in occasional leakage of chlorine gas, which caused inconvenience to the operators who were looking after the pilot plant operations.

The pilot plant for treatment of LTC wastewater was made up of mild steel and coated with paint from inside and outside. The pilot plant was operated for a period of 9 months continuously. The inner walls of all the units were getting corroded during the concluding phase of the operation of the pilot plant, and this resulted in increased iron concentration in the biomass and also in reactor contents. Similar problem of corrosion in the pilot plant treating cyanide-bearing wastewater was also encountered.

The pilot plant for the treatment of LTC wastewater containing phenolics was commissioned with an initial biomass concentration of 20 mg l^{-1}. The arrangement for propagation of the culture was not available in the initial period. However, a container of 200 litres capacity was procured later on. Thus, the time required for achieving 1000 mg l^{-1} of suspended solid for resuming continuous feeding got delayed. After 20 days only, by fill and draw method, the biomass concentration of the reactor could be raised to 1000 mg l^{-1}.

During the comissioning of pilot plant for treatment of wastewater containing cyanide, cool air was available for aerating the bioreactors to develop the biomass through maintenance of the critical levels of dissolved oxygen in the system. No alternative source for aeration was available in the premises of the industry. Cool air usage resulted in lowering the temperature of the contents of the bioreactor and indirectly affected the quality and quantity of biomass. Installation of heating devices rectified the problem. Technical know-how developed by NEERI requires the usage of sewage as supplemental growth carbon source for the cyanide degrading organisms. Sewage containing other living and non-living entities help in enhancing the settling properties of biomass. The non-availability of required quantity of sewage in the cyanide manufacturing unit affected the biomass build-up of the bioreactor. Peptone addition in place of sewage was a limiting step to increase the microbial population of cyanide degrading organisms. Excess addition of peptone was leading to the shifting of the desired quality of microbial population. Equipment and assemblies required for alternate arrangement were inadequate for batch sludge build-up for commissioning the pilot plant. The other problems regarding sludge build-up were overcome partially by operating the settlers as sludge build-up reactors for the system for a short duration.

Both the pilot plant did not have any cover, and were exposed to the atmosphere. The seasonal variation of temperature at the working place ranged from 10°C to 47°C during the investigation period. The performance of the bioreactor system was affected during the winter season when temperature dropped to 10°C. Similarly, during the summer season when the ambient temperature was more than 45°C, the level of phenolic concentration in the treated effluent was ranging from 50-80 mg l^{-1} (Figures 5.3 and 5.4). When the temperature was above/below the optimal temperature for the

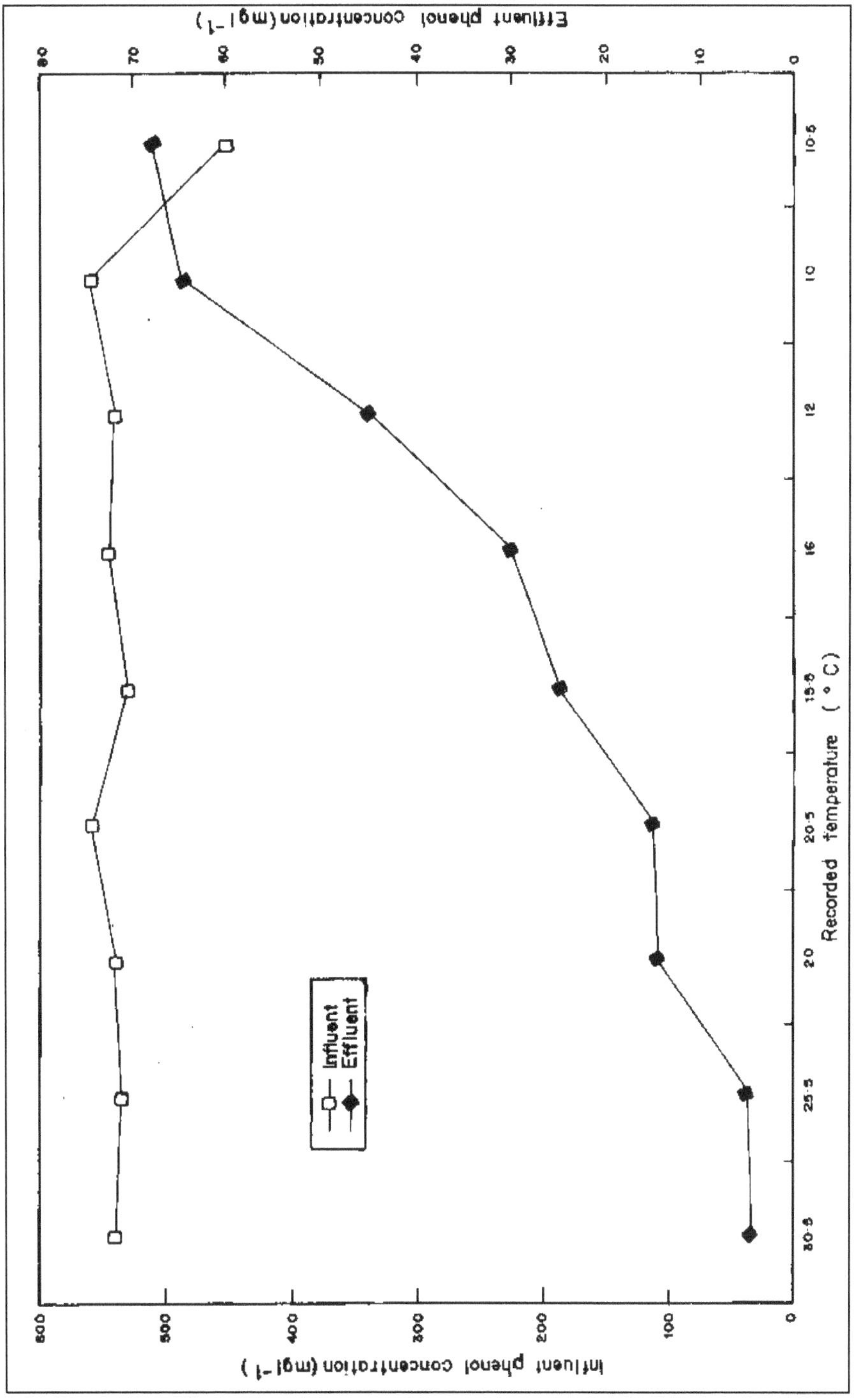

Figure 5.3: Winter Season Variation in the Phenol Concentration in the Treated Effluent Generated during Operation of the Pilot Plant for the Treatment of LTC Wastewater.

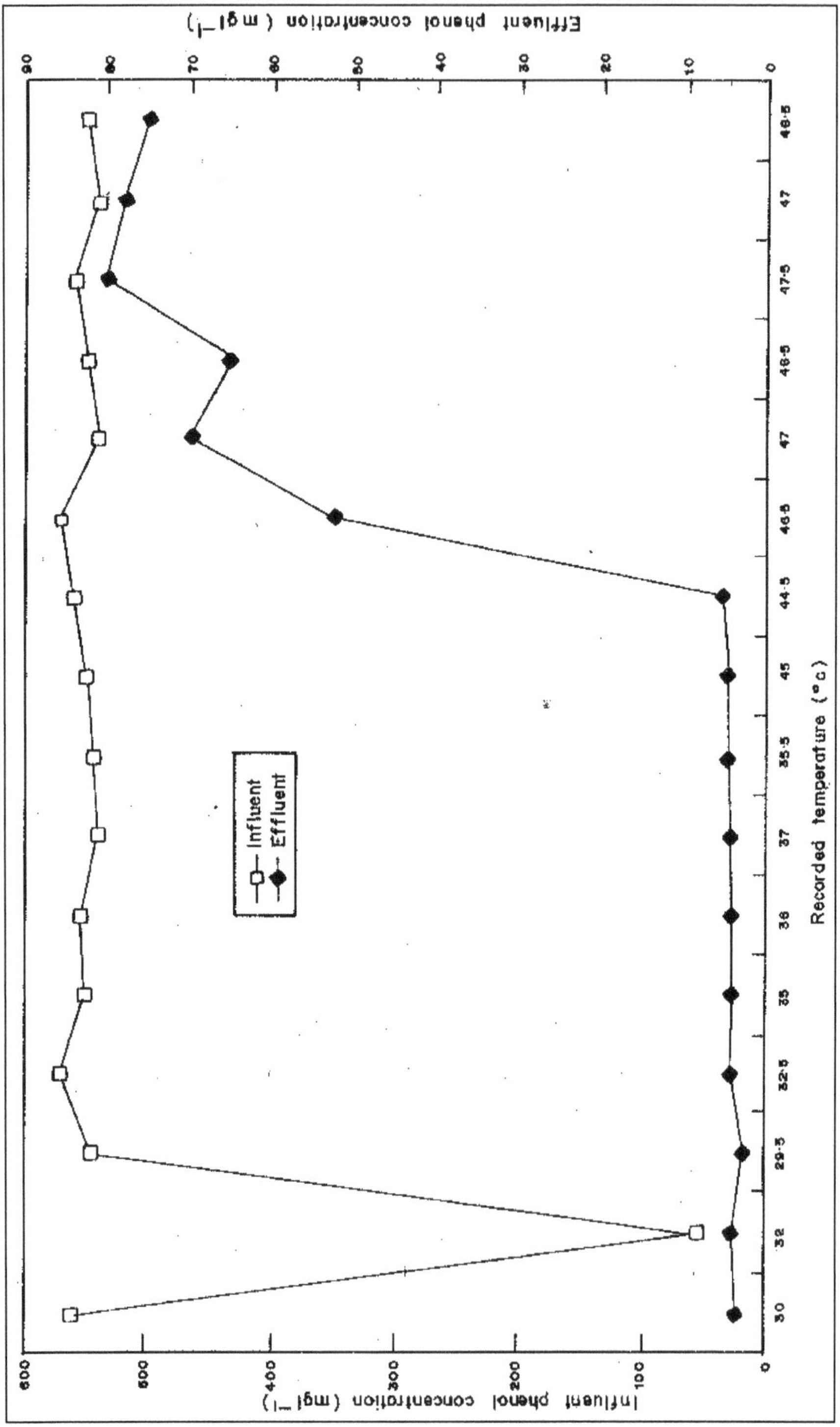

Figure 5.4: Summer Season Variation in the Phenol Concentration in the Treated Effluent Generated during Operation of the Pilot Plant for the Treatment of LTC Wastewater.

microorganisms (optimal temperature, 30-37°C), excessive foaming in the biological reactor was observed.

The diurnal variations of the temperature in the bioreactor for the treatment of cyanide bearing wastewater have been shown in Figure 5.4, which indicate that the temperature of the content of the bioreactor varied from 30°C to 48.5°C. The growth of the microorganisms gets affected in this range of temperature. The inoculated microorganisms in the bioreactor grow efficiently between 30°C and 35°C.

The feed system was a nozzle type structure attached to an overhead tank. The overhead tank was receiving wastewater from equalization basin. The feed rate to the bioreactor was varied by using nozzles of different diameters. The feed of wastewater with lower flow rate incorporated the nozzles with small diameter apertures. The use of such nozzles resulted in choking and spilling of the wastewater from the feed tank.

The settler and sludge recirculation system was always problematic during the pilot plant investigation period. Sometimes, improper operation of recirculation pump with timer resulted in recirculation of water only from the settler to the bioreactor unit, leaving the sludge in the settler. Thus, sometimes, bulking of sludge was also observed in the settler. Manual insertion of a rod upto the bottom of the settler and then agitating at the bottom of settler resulted in proper recirculation of sludge through the pump.

In the CMAS unit of the pilot plant for the treatment of wastewater containing cyanide, height to width and liquid depth in the reactor was requiring high pressure of air to achieve proper mixing in the bioreactor. Baffling of the reactor and making two compartments upto certain distance from top to bottom caused inconvenience in maintaining uniform aeration throughout the bioreactor system. Settler hopper was improperly designed and resulted in the accumulation of biomass at the bottom of settler. The settled biomass was removed manually.

Settler drainage pipe to sludge sump was unsuitable and was giving a back push-up while starting the motor for recycling the sludge and caused disturbance throughout the settler resulting in poor settling.

Operators assigned to run the pilot plant have production oriented mind setting and they were not giving importance to the operation and maintenance of the pilot plant. Operators working in the chemical industry were not aware of the limitations of biological systems and this resulted in improper operation. The biological system was equated with a chemical system and this led to the problems like improper pH adjustment of wastewater, excessive flow rate, inadequate and improper recirculation of sludge to the bioreactor. Cyanide being toxic, proper treatment and disposal requirements are needed for the safe and clean environment. Concepts and importance of clean environment and cyanide waste management through biological system were beyond the comprehension of the operators due to their production-based mind set.

Results and Discussion

Pre-requisites for Biotechnology Based Pilot Plant Investigation for Treatment of the Wastewater

Siting of the Pilot Plants

The topography of the sites selected for the installation of the biological unit must be suitable. During selection of sites for installation of the biological units, the existing surroundings, having various unit operations, should be considered. Sometimes, the units, which are situated nearby the biological units, interfere in the optimum performance as some of the constituents discharged by these units affect the growth and activities of the microorganisms.

Climatic Conditions

Various climatic conditions, *viz.* temperature, wind direction, rainfall, etc., affect directly or indirectly the performance biological treatment units. Temperature profile of the area, where the biological treatment plant has to be installed, must be known. The effect of temperature on biological system must be predetermined and necessary arrangement should be provided in the reactor system for optimal maintenance of the temperature. Dropping down or shooting up of the temperature of the content of the bioreactor affects the activity of the microorganisms. Thus, proper temperature control/ regulating system must be provided while designing the treatment plant, so that sufficient care can be taken to ensure smooth functioning. Similarly, siting of any biological treatment unit must be decided with respect to wind direction. If proper covering is not provided to the biological units, sometimes, they get loaded with inert materials brought in by the wind. The wind blowing in the direction of biological unit carries the unwanted chemical materials from storage yard in the premises of company/nearby industry and can contaminate the biological unit. Thus, the performance of the biological unit gets hampered indirectly by the wind speed/ direction. In an area receiving heavy rainfall, proper leveling/platform is to be constructed for the erection of biological units, as otherwise, the rain water gushes into the treatment units and results in the dilution of the contents of the biological units. This causes loss of biomass from the bioreactor and brings in other contamination from the nearby area. Therefore, a need exists for critical evaluation of the climatic conditions of the area prior to the installation of biological units either within the campus or outside the premises of the industry.

Design/Fabrication/Installation of the Reactor System

Any chemical/biological reactor system designed and upgraded from laboratory to pilot scale is based on specific scale up ratio [4]. Similarly, scale up factor should be optimum from pilot plant to full scale unit. Further, the selection of construction materials required for the fabrication of reactor system must be based on the nature of the wastewater as otherwise the treatment units get corroded and this affects the performance of the bioreactor system.

To meet the eventuality during the shortage/excess of wastewater availability, suitable reservoir must be designed to optimally maintain the biological system.

Proper aeration system for the aerobic bioreactor must be installed prior to the start-up of the plant. All the safety systems/units must be incorporated in designing the pilot plant for the treatment of toxic wastewaters.

Commissioning of the Pilot Plant

The required accessories for commissioning of the pilot plant must be checked with the industry where pilot plant is to be installed and operated. The accessories mainly include power (electricity), water, air (for aerobic system) for the operation of the pilot plant and required chemicals to meet the nutrient needs. The laboratory facilities and availability of trained operators (manpower) must be ensured prior to commissioning of the pilot plant. Further, facilities for storing/propagation of the microbial cultures required for commissioning of the pilot plant must be made available at the site area. Sometimes, it becomes very difficult to arrange such accessories when pilot plant has to be installed in a remote area far away from a city.

All the problems associated with treatment of LTC wastewater were solved and malfunctioning of the units was rectified. The required data for design criteria for a full scale plant was collected and the investigation was completed within 9 months. The pilot plant could remove phenol to the extent of 99.32 per cent, while the removal of COD was 53.4 per cent (Table 5.5).

Table 5.5: Steady State Operational Data for the CMAS for LTC Wastewater at 10 hr HRT

Parameter		No. of Observations	Max. mg l^{-1}	Min. mg l^{-1}	Mean mg l^{-1}	Std. Dvn.	Per cent Reduction
Phenol	Influent	15	570	136	413	114	99.32
	Effluent	9	5.5	1.06	2.8	1.6	
COD	Influent	15	3008	1244	1949	566	53.4
	Effluent	8	1056	649	988	154	
Thiocyanate	Influent	5	251	14.25	20.3	4.1	51.5
	Effluent	5	1.25	7.65	9.87	1.7	
Cyanide	Influent	3	3.61	2.66	3.04	0.50	35.31
	Effluent	3	2.10	1.90	1.9667	0.115	

Dimensional problems associated with pilot plant units for treatment of wastewater containing cyanide could not be rectified while operating the pilot plant. Operational problems of the pilot plant were rectified upto a certain extent by installing temperature control measure and supply of hot air (with the help of heat convector) to the bioreactor, installation of water heater to raise the temperature of the aeration tank as and when required. The diurnal variation of the temperature in the bioreactor ranged from 19°C to 25°C after installation of heating device (Figure 5.5). Insulation of aeration tank was undertaken to minimize the loss of heat from the system and to protect it from the fluctuations of ambient temperature. The problem of proper mixing was solved by providing additional aeration point from top and adjusting it properly between the two baffled chambers of the bioreactor.

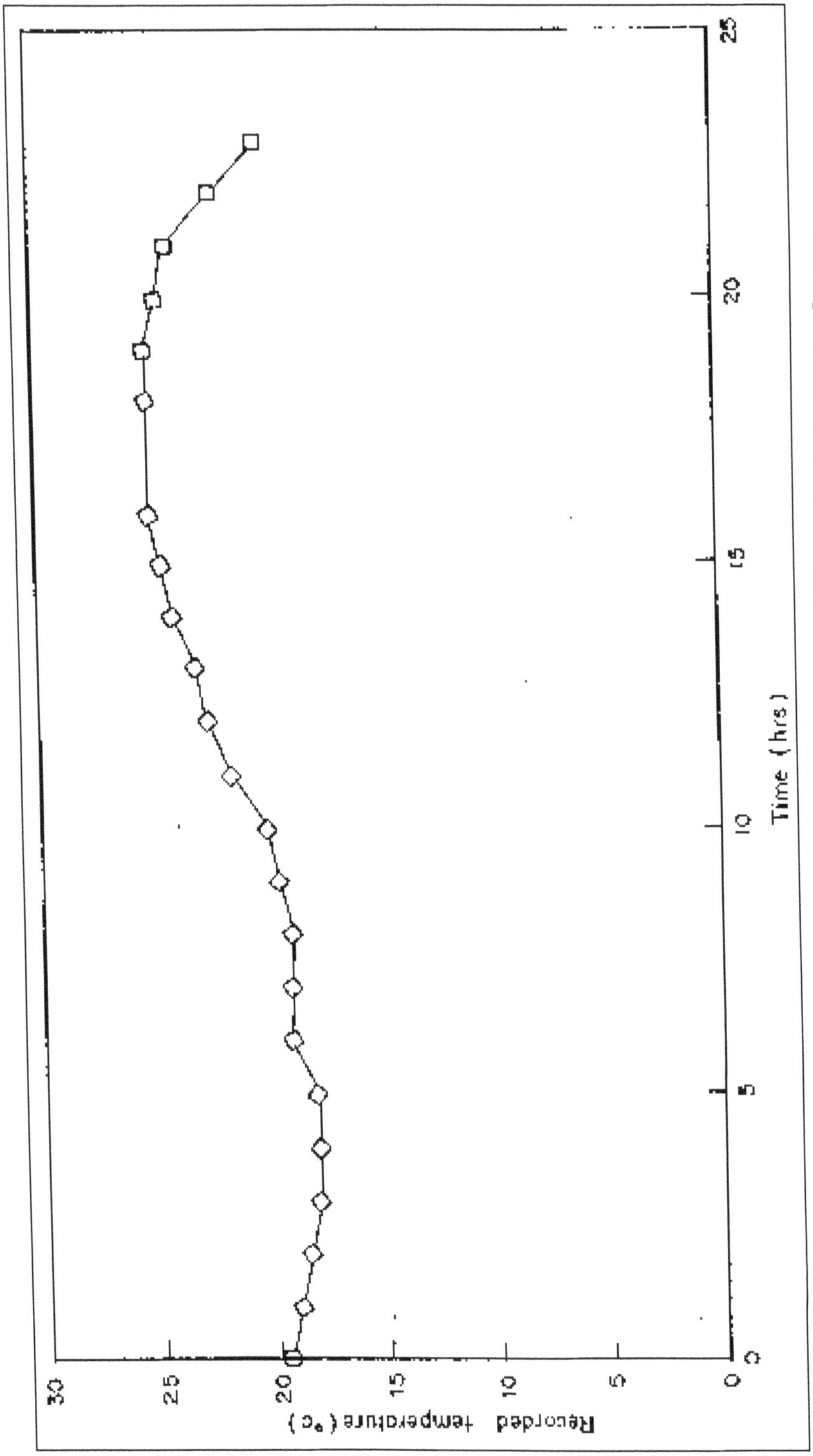

Figure 5.5: Diurnal Variation of Temperature of Content of Bioreactor during Winter Season.

Table 5.6: Steady State Performance of Activated Sludge System for Treatment of Cyanide Bearing Wastewater (HRT of 24 hours)

Cyanide (CN^-) ($mg\ l^{-1}$)		*COD ($mg\ l^{-1}$)*		*Ammonia, N ($mg\ l^{-1}$)*		*pH*		*MLSS ($mg\ l^{-1}$) / MLVSS ($mg\ l^{-1}$)*	*Effluent Suspended Solids ($mg\ l^{-1}$)*
Influent	*Effluent*	*Influent*	*Effluent*	*Influent*	*Effluent*	*Influent*	*Effluent*		
134.00	0.16	224.00	64.00	30.80	101.9	9.2	8.9	1225 / —	110
119.00	0.40	232.00	64.00	28.50	95.2	9.2	8.9	940 / —	120
120.00	0.55	200.00	40.00	21.80	98.0	9.1	8.9	1122 / 920	105
116.00	0.18	196.00	48.00	16.80	95.2	9.2	9.0	1028 / 840	80
105.00	0.80	196.00	40.00	9.50	81.30	8.9	8.8	960 / 760	80
105.00	0.50	196.00	32.00	—	—	8.9	8.9	1007 / 800	80

Adequate sludge recirculation from the settler to bioreactor was also restored by manually putting the sludge from sludge sump into the reactor in the pilot treating cyanide wastewater. The problem arose due to the fabrication material. M.S. used in the fabrication of pilot plant treating cyanide wastewater could not be replaced due to non-availability of substitute reactor or adequate measures of rectification. This resulted in deterioration of sludge quality due to excessively corroded iron particles. Non-reactive and anti-corrosive painting could have been employed before commissioning of the pilot plant, but this could not be done owing to short duration of the studies.

In spite of the abovementioned constraints and the rectifications employed, the biomass concentration of only 1200 mg l^{-1} in the bioreactor could be achieved. These rectifications improved the performance of bioreactor and an influent concentration of 105-120 mg l^{-1} of cyanide could be treated satisfactorily at a hydraulic retention time (HRT) of 24±2 hours (Table 5.6). Thus, it is evident from the above studies that pilot plant investigation requires proper planning for the successful demonstration and repetition of the performance results obtained during the operation of laboratory scale unit for the treatment of the wastewater.

The pilot plant studies based on bench scale experimentation require proper planning and imagination for developing a technology for any system. Scaling up from a bench scale to pilot plant should properly be designed and should be appropriate to encounter the problems.

Conclusion

The necessary modifications were incorporated in both the pilot plants in order to overcome the trouble shooting problems and the investigations were completed. Both the pilot plants were operated and data on design parameters for full-scale plant were generated. The pilot plants for treatment of LTC wastewater could remove phenol to the extent of 99.32 per cent, while the removal of COD was 53.04 per cent. The pilot plant for treatment of cyanide could treat 105-120 mg l^{-1} at hrt of 24 ± 2 hour.

References

1. National Environmental Engineering Research Institute (NEERI), Nagpur and Ministry of Environment and Forests, New Delhi. Detoxification of phenol and cyanide bearing industrial wastes; (1982)
2. Biological treatment of cyanide bearing wastewater: Pilot plant study; Report NEERI, Nagpur (1987).
3. *Standard Methods for the Examination of Water and Wastewater*. 18 edition, Greenberg, A.E., Clescer, L.S., Eaton, A.P. and Frasnson, M.H.A. (eds,) American Public Health Association, Washignton D.C., USA (1992).
4. Shinskay F.G., *Process Control Systems*, McGraw Hill International Editions, New York, USA (1988).

Chapter 6

Sequencing Batch Reactor: An Emerging Technology for Wastewater Treatment

P. Nawghare, A. Srivastava, M. Karthik and S.N. Kaul

CSIR-NEERI, Nagpur

ABSTRACT

Sequencing Batch Reactor is a promising scheme for effluent treatment; the process that can be most reasonably compared is continuous flow systems. It falls in a broad category of unsteady activated sludge systems where the effluent flows from process to the next sequentially in a single reactor. The basic processes involved in an effluent treatment scheme of Fill, React, Settle and Draw is carried in a single reactor. The main difference between this emerging technology and the conventional continuous flow systems is the former provides in time what the latter provides in space. The principal factors that seem responsible for lack of enthusiasm for this technology in and around the developing world is that the process is time based for the different procedures exerted in the effluent treatment scheme. Lack of reliable timers and instrumentation could be attributed to this peculiar situation.

Excellent results for pollutant removals have been observed from using this technology in the advanced nations, as compared to the continuous flow systems. The major advantages in utilizing this technology are the flexibility of operation and processes. The effluent treatment could be adjusted as per their quantity and quality. SBR can be used in all the treatment schemes involving the conventional activated sludge process. SBR is effectively utilized for the treatment of various industrial and domestic effluents. Some of the effluents finding application of SBR are from piggery, brewery, paper mill, tannery, landfill leachates and hazardous wastes

This paper discusses the process variables, operational strategies and design approaches.

Keywords: *Sequential treatment, Batch activated sludge process and Effluent treatment.*

Introduction

Over the past century, awareness of hygiene, as a result of diseases and death resulted in ways and methods to dispose wastes safely. The beginning of 19th century paved a systematic treatment of wastewater treatment led by the discovery of Germ Theory by Koch and Pasteur marked a new era in sanitation. Discovery of the role of microorganisms in the treatment of wastewater followed soon with Arden and Lockett [1, 2] carrying out experiments at Manchester Laboratories. The earlier wastewater treatment schemes were carried for domestic and to a smaller extent industrial effluent. Reactors containing microorganisms were purposefully kept in suspension to oxidise the wastes. After the completion of the reaction, the particles that were allowed to settle were found to be responsible for treatment. It was observed that bacteria grew in a readily settleable floc like structures, which were later called as activated sludge.

The first full-scale activated sludge plant was built at Salford [3,4] based on the work carried out by Arden and Lockett. The wastewater to be treated was introduced in batches following certain periodicity in the treatment process. The contents were then aerated for certain time and allowed to settle till a clear supernatant was drained from the reactor. The cycle was repeated for next batch of wastewater to be treated. The effluent quality was found to be excellent, but operation of the treatment scheme was difficult, as there were no reliable timers to control the duration of the different process time. Simultaneously the continuous flow treatment schemes gathered momentum, as they were comparatively easier to operate and maintain. It is understood that due to lack of automation in controlling the operation led to declining interest in this reactor scheme. This paper brings out a renewed interest in the field of batch treatment for the industries.

The various reactions *viz.* physico-chemical and/or biological are carried out in a reactor. Some of the different types of reactor [5] that are utilized in the wastewater treatment [5] are shown in Table 6.1.

Sequential Bach Reactor for Wastewater Treatment

Sequential Batch Reactor is a treatment scheme where in a tank/sets of tanks operate on a fill, treat and draw on sequence of operations. The tank (s) are filled for a certain period and then operated as a batch reactor. After desired level of treatment is achieved the mixed liquid is allowed to settle for a predetermined time, and the clarified supernatant is discharged form the tank.

The sequence of operations carried for effective treatment involve five phase *viz.* Fill, React, Settle, Draw and Idle as presented in Figure 6.1. A brief discussion of the five phases of the SBR treatment is discussed herein under.

Table 6.1: Various Reactors Used for Wastewater Treatment[5]

Sl.No.	*Reactor Type*	*Description*	*Sketch*
1.	Batch	Flow is neither entering nor leaving the reactor and the liquid contents are completely mixed.	
2.	Plug flow	Wastewater passes through the tank and is discharged in the same sequence in which they enter. The particles retain their identity and remain in the tank for a time equal to theoretical detention time.	
3.	Complete flow stirred tank	Complete mixing occurs when the particles entering the tank are dispersed immediately throughout the tank. The particles leave the tank in proportional to their statistical population	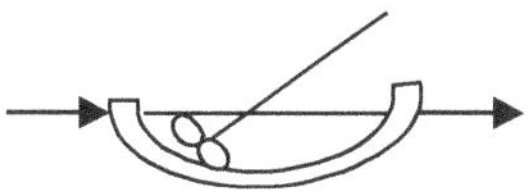
4.	Arbitrary flow	Any degree of partial mixing between plug flow and complete mixing	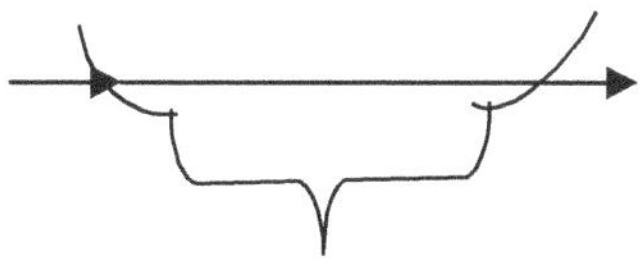
5.	Packed bed	The reactor is filled with some type of packaging medium such as rock, slag, plastic or ceramic solids	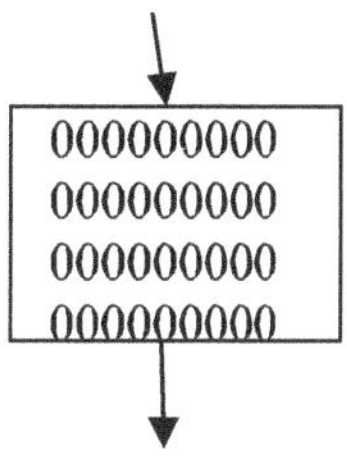
6.	Fluidized bed	This reactor is similar to the packed bed in many aspects but the upward movement of fluid–air or water through the bed, expands the packing medium. Controlling the flow rate of the fluid can vary the porosity of the packing.	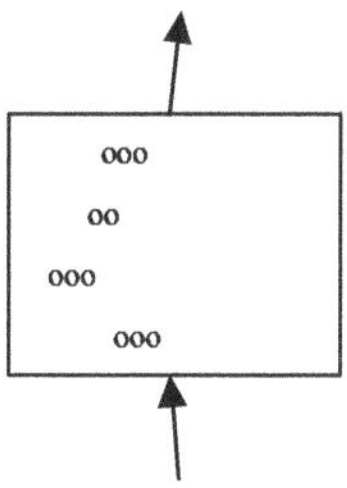

FILL Phase

The first phase in the sequence is filling of the reactor. The influent to the tank may be either raw wastewater or primary effluent. Filling could be achieved by pumping or through gravity. When a number of tanks are operated, the wastewater is added by gravity, devices like an adjustable weir or an automated valve are operated to divert the flow within the tanks. The tanks are generally an earthen or oxidation ditch, a rectangular basin or any other concrete or metal type structure. In the Fill

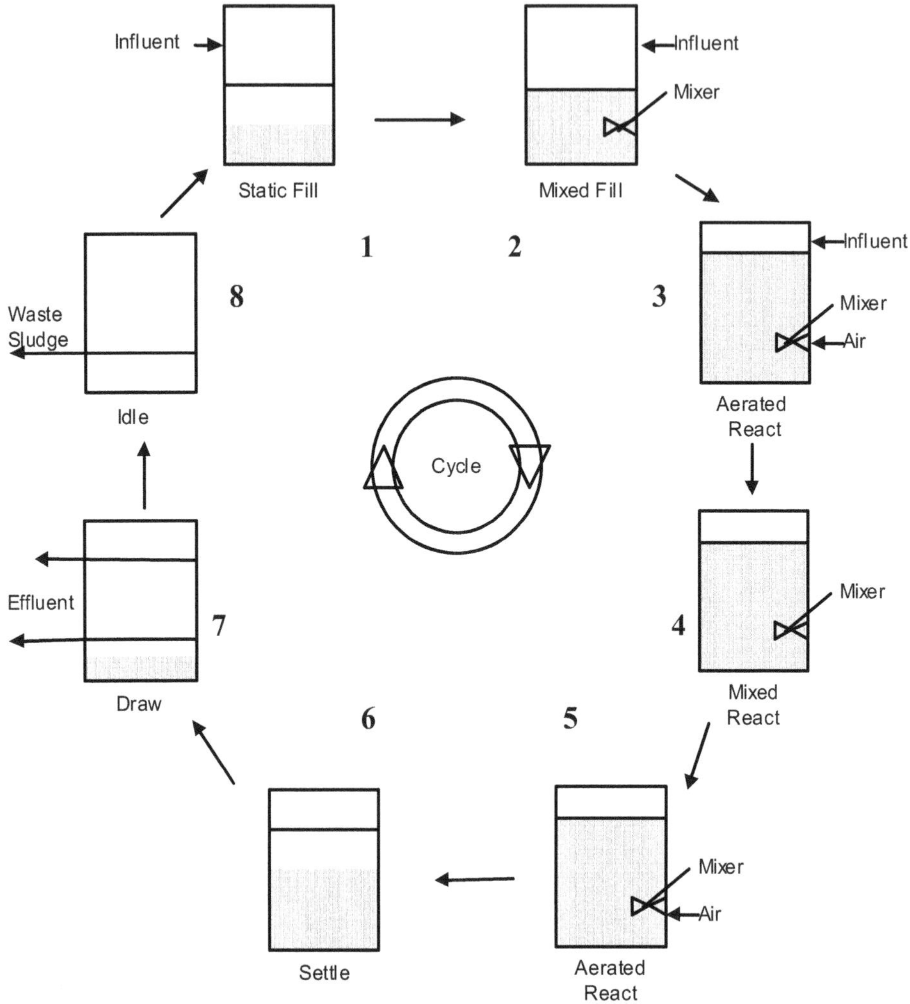

Figure 6.1: SBR Operation for One Cycle Showing Five Discrete Time Periods as FILL, REACT, SETTLE, DRAW and IDLE.

phase, the wastewater is added to the biomass, which is present in the tank from previous cycle. The liquid level in tank increases from the set level to the maximum of 100 per cent. The set volume may be as little as 25 per cent or as great as 70 per cent. The Fill phase can be designed to terminate before the maximum level is reached, by either limiting the FILL time to a predetermined maximum or eliminating IDLE phase.

Aeration is carried during this phase by various diffusing systems such as surface aerator and diffusers. A phase in Fill period, where no mixing or aeration is carried is called Static Fill. This phase utilizes minimum energy input and high substrate concentration at the end of the FILL. Similarly, Mixed Fill imparts mixing of wastewater without aeration. Level sensing devices, timers, or online probes can switch the aerators and/or mixers on and off as desired. Fill is stopped, when the tank is full or diverted into another tank in the SBR treatment scheme.

REACT Phase

Reactions are initiated during Fill, and are continued during this phase. The react period is the time during which the tank receives no flow. Alternating conditions of low dissolved oxygen concentration (no.4 of Figure 6.1) and high dissolved oxygen concentration (no. 3 of Figure 6.1) is adopted as per the requirement of wastewater treatment. The liquid level remains at maximum throughout React. Wasting of sludge could be carried as means of controlling sludge age. The duration of this phase can vary from zero to more than 50 per cent of cycle time depending on the level of treatment achieved. React phase completes the reactions initiated during the Fill phase.

SETTLE Phase

The reaction completed in the React phase is allowed to settle down, wherein the solid liquid separation takes place under quiescent conditions. Generally no inflow or outflow is carried from the tank is practiced. The major advantage in the SBR is clarification process is carried in the same tank. The biomass is retained in the tank until some fraction is decided to be wasted. The time in Settle typically ranges between 0.5 and 1.5 h. Sludge wasted at the end of Settle phase is more concentrated. Longer settling periods are harmful, as sludge may rise to the surface.

DRAW Phase

This phase clearly means removal of clarified effluent. The mechanism for withdrawal is shown below.

1. A pipe fixed at some predetermined level with the flow regulated with an automatic valve
2. A pump, depending upon the hydraulic grade line of the system.
3. An adjustable or floating weir at or just beneath the liquid surface can be used

The period of draw can range from 5 to more than 30 per cent of the total cycle time.

IDLE Phase

The phase between Draw and Fill is termed Idle. Sludge can be wasted effectively during Idle. The frequency of sludge wasting can be designed to range between each cycle to once in every 2 to 3 months. Aeration or mixing can be provided, after sludge wasting. Idle can be avoided depending upon treatment policy by filling the tank, as soon as the tank in Draw reaches the minimum liquid level.

A summary of the periods, using an example that highlights the hydraulics for a reactor is presented in Figure 6.2. As can be seen from the Figure 6.2, the volume of the liquid increases during the fill period from an initial volume to the maximum volume at a rate determined by flow variations unless solids are wasted during the

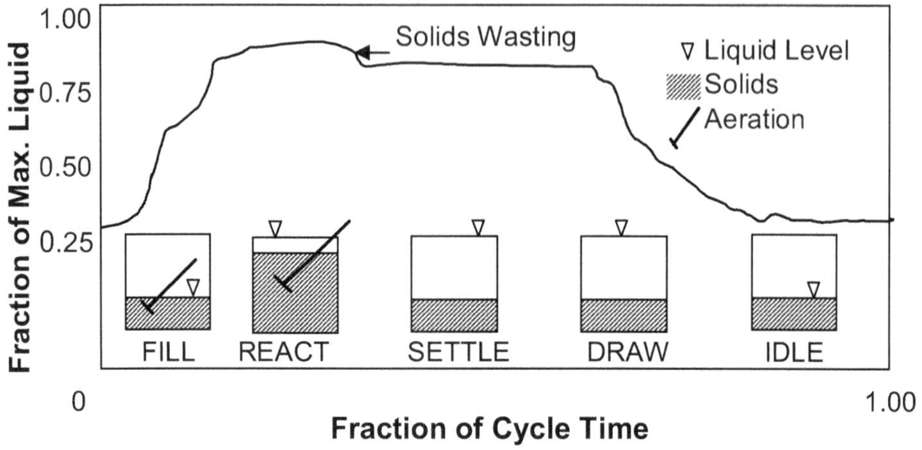

Figure 6.2: Liquid Volume versus Time in SBR Reactor (Hypothetical).

react period, as shown. The volume would remain at the maximum through the settle period and until draw. The volume during the draw period decreases to the minimum and remains there during the idle period. SBR behaves both as a biological reactor and as a clarifier [6].

Devices for Aeration, Mixing and Decanting

Aeration

Aeration is the process by which oxygen is introduced into the system for the oxidation of organic and inorganic wastewater compounds. In addition to the transfer of oxygen into the water, aeration usually provides sufficient mixing and removal of reaction products (gases) from the water. A number of devices/methods are available to aerate. Some of the devices are surface aerators, diffusers and jet aeration. Selection of aeration device, two aspects are of importance: energy efficiency and reliability. An energy efficient aeration device should be selected, as aeration is among the most energy intensive operations in wastewater treatment systems, consuming 50-90 per cent of the total energy cost [7].

Mixing

Mixing of wastewater components (substrate) to the microorganism's needs to be ensured by mixing devices. Some of the mixing devices used is jet mixer, surface rotor and a submerged propeller. Mixer devices should be selected to achieve good mixing with limited introduction of oxygen due to any turbulence at the surface. Figure 6.2 shows a typical effluent volume with respect to various phases of treatment.

Decanting

A variety of decanters are used to remove the treated supernatant from the reactor. Three categories are distinguished. The decanters remove the supernatant from the surface of the reactor by a motor to rotate the trough with the overflow weir, the speed of rotation to lower the weir is used to control the rate of draw. In another method decanter floats on the surface of water and withdrawl rate is determined by the hydrostatic pressure at the inlet of the decanter. In some instances, decanter is installed at a fixed depth in the SBR. When designing a decanter, precautions to account for foaming sludges must be included.

Operational and Control Strategies

Process

SBR system can be operated as plug flow or completely mixed. One or more parallel sets of tanks are operated on a common cycle. Different control strategies could be adopted for meeting effluent standards as the operations are batchwise. Supply of oxygen or other process control parameters for biological reaction are easily supplied, since these parameters can have a considerable effect on overall system operation.

SBR system also could be used for equalization during which Fill and React is not carried out. In areas where large flows occur, sets of tanks are operated simultaneously in different phases of treatment. Normally for small flows or where effluent is only intermittent, one tank system could be used [6, 8, 9, 10]. However, all the phases Fill, React, Settle, Draw and Idle would have to be completed to meet the desired effluent quality.

A general scheme of SBR plant is shown in Figure 6.3.

The scheme consists of primary treatment unit, influent holding tank (optional), SBRs (number depends on the quantity of wastewater to be treated) and an effluent buffer tank (optional). If an influent holding tank is used the conditions in the SBR are separated from the variations of the influent wastewater as soon as the fill phase is ended. The duration of each process phase can either be fixed or can be adjusted to the actual loading conditions and the effluent requirements (variable cycle time).

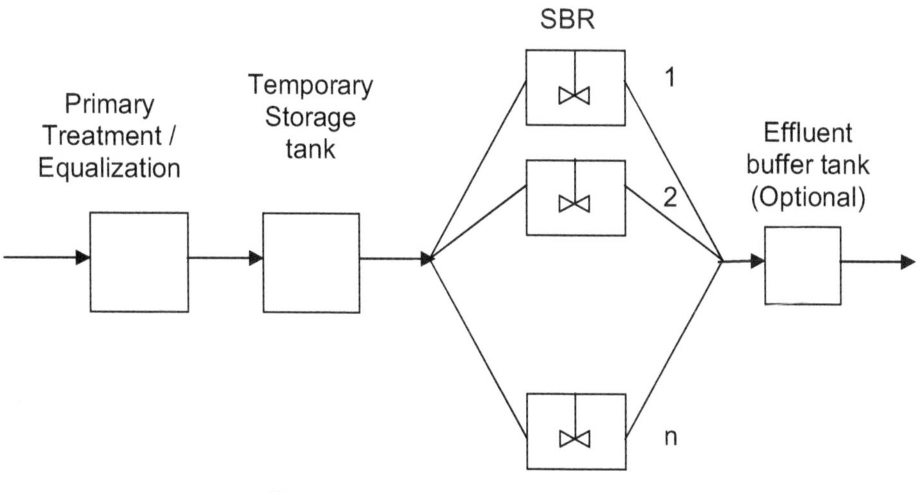

Figure 6.3: Scheme of an SBR Plant.

Without a holding tank the number of available SBRs determines the duration of fill phase and the duration of each cycle (*e.g.* with two SBRs the fill period makes up 50 per cent of the cycle time). An effluent buffer tank can be used to smooth the variations in the effluent flow rate, which may be necessary for large SBR systems discharging into small receiving water bodies.

Instrumentation Control

Instrumentation plays a major role in large-scale SBR operations. The length of each cycle is dependent on length of each phase in cycle. Formulation of operational strategies for smooth working of SBR system is very important. The online control of biological process allows flexibility to operate and decide on strategies to optimize the SBR treatment plants, both in terms of removal efficiencies and in terms of costs. The most used control parameters are ORP (Oxidation – Reduction potential), DO (Dissolved oxygen) and pH, as their probes are easy to use and less expensive. The oxidation – reduction potential represents the basic chemical energy potential of an oxidation – reduction reaction. This parameter has been applied to monitor the overall reactions of biological and physico-chemical processes. Researchers have demonstrated the efficiency of these measurements as control parameters both under aerobic and anoxic conditions. Some authors suggest a control method based on a fixed value of ORP [11, 12], whilst others advice a strategy based on the relative dynamics of ORP and DO [13, 14, 15]. Many researchers explored more efficient and reliable automatic control approaches of biological nitrogen removal by monitoring ORP in SBR system [16, 17, 18 and 19].

The on line monitoring of pH, Oxidation-reduction potential and dissolved oxygen in SBR systems has following advantages:

- ✰ The biological process control
- ✰ The optimization of the aerobic stage (when nitrification is completed, the oxygenation can be stopped).
- ✰ The optimization of the denitrification stage (when denitrification is completed a subsequent anaerobic stage can be avoided.)

Online monitoring can save time and lesser volume without reducing the efficiency. Recent advances in application of automatic control to SBRs for cost reduction and process optimization are some of the main causes of interest in the SBR technology [20]. Since ORP and pH profiles pinpoints the end of nitrification and denitrification, on-line monitoring of ORP and pH is observed as a useful technique in nutrient removal process control [14, 18, 21, 22, 23]. The experimental results of the study [24] showed that application of on-line ORP and pH monitoring to dynamically control a continuous inflow SBR resulted in significant saving in construction cost. The operational policies are related to fixing of duration of each phase in each cycle is mostly dependent on process objective. Figure 6.4 shows concentration of substrate with respect to various phases of SBR treatment.

Table 6.2: Common-operating Strategies to Meet the Treatment Objectives [25]

Sl.No.	*Treatment Objective*	*Fill Policies*	*React Policies*
1.	Organic carbon and suspended solids reduction, minimum energy consumption or sludge production	Static, mixed, then aerated	Aerated
2.	Organic carbon and suspended solids reduction and nitrification	Static, mixed, then aerated	Aerated
3.	Organic carbon and suspended solids reduction and denitrification	Static, mixed, then aerated	Aerated, followed by mixed then aerated
4.	Organic carbon and suspended solids reduction and biological phosphate reduction	Mixed (short period) then aerated	Aerated
5.	Industrial organic wastewater, toxic at high concentration	Mixed (short period) then aerated	Aerated (long period)

Tables 6.2 and 6.3 presents a number of operation schedules; each designed to meet specific treatment objectives. For COD removal from regular municipal wastewater the SBR cycle may consist of a short fill phase followed by an aerated phase, settle, and draw. The same cycle can be used to achieve nitrification by

Table 6.3: Operating Schedules to Meet the Treatment Objectives [25]

			Cycle Time								
(a)	COD Removal	Fill	×								
	(short SRT)	Mix									
		Aerate		×	×	×	×	×	×		
		Settle								×	
		Draw									×
(b)	COD removal nitrification	Fill	×								
	(long SRT)	Mix									
		Aerate		×	×	×	×	×	×		
		Settle								×	
		Draw									×
(c)	COD removal nitrification	Fill	×			×		×			
	denitrification	Mix		×		×		×			
		Aerate			×		×		×		
		Settle								×	
		Draw									×
(d)	COD removal nitrification,	Fill	×		×			×			
	denitrification, enhanced	Mix		×	×			×			
	biological phosphorus	Aerate				×	×		×		
	removal	Settle								×	
		Draw									×

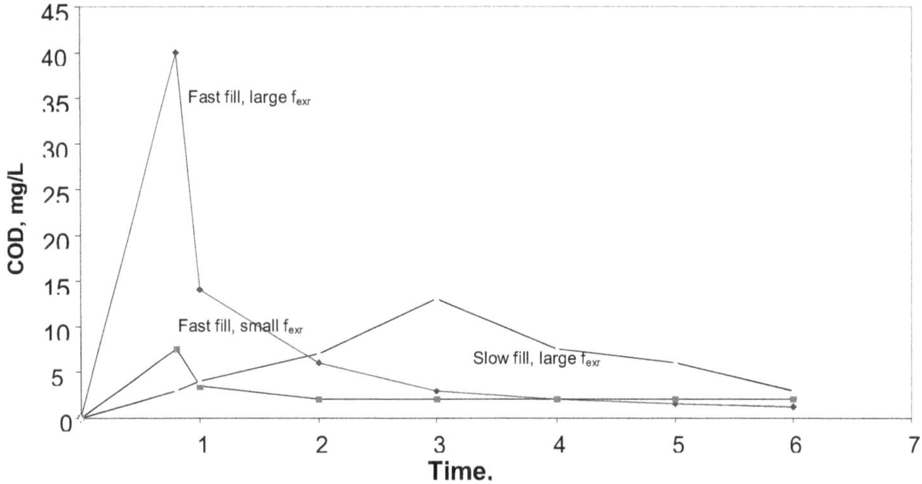

Figure 6.4: Concentration Gradients for Substrate Depending on Volumetric Exchange Ratio (f_{exr}).

increasing the sludge age. Nitrogen removal has been achieved by introducing an anoxic phase in to the cycle. Oles and Wilderer [26] proposed a SBR cycle with three fill phases, where the second and third fill is used to supply electron donors required for denitrification. To achieve enhanced biological phosphate removal an anaerobic phase must be included in the operation. The first fill phase is used to denitrify any nitrate remaining in the reactor after draw. Then the readily biodegradable COD of the second fill phase is made available for p-release. The COD of the third fill phase used as a carbon source for denitrification. From the literature studies it is evident that the operational policies related to different phases and their length in a cycle are mostly dependent on the objectives to be achieved.

Comparison with other Continuous Flow (Aerobic/Anaerobic) Processes

SBR and the conventional continuous flow system (CFS) differ in a way that the former provides in time, what the CFS provides in space [6]. SBR can meet even more stringent effluent criteria that are being promulgated, since they are operated in batches and flexibility to treat on merit of the wastewater. In fact a properly designed semi batch reactor should achieve a higher effluent quality than CFS. In contrast to continuous flow system, both biological reactions and biomass separation take place in the same tank [27]. This is a single tank batch system, sequencing on individual cycle and thus provides both low capital and operating costs. Conventional waste management technologies commonly adopted in tropical climates are not only expensive but also warrant exacting operation and maintenance requirements.

In addition to this one more important parameter associated with SBR and not found in CFS is flexibility of operation. The time of all the phases (fill, react, aeration etc.) can be easily changed, which facilitates the SBR to tolerate variations in organic and hydraulic loading and to sustain the shock loads.

After the react phase, the microbial sludge is allowed to settle in the same tank which has comparatively a larger volume than that of secondary clarifier used for conventional continuous flow activated sludge plant. SBR has the characteristic ability to incorporate aerobic and anoxic phases in a single reactor, if desired.

There are many continuous flow systems in which the air supply has been cycled on and off in an attempt to provide for both nitrification and denitrification without the addition of an external supply of substrate (for example methanol). The balance between air supply and external substrate concentration in a continuous flow system with or without recycle is quite a sophisticated operational problem, which is believed to be unsuccessful in developing countries. A batch reactor on the other hand can be easily operated.

In the SBR process, during each cycle dynamic conditions are prevailing. Control and periodic repetition of the short term unsteady state allow for, in the long run, enhancement of certain effects such as enzymatic activity, accumulation of metabolic products and selection and enrichment of specific groups of microorganisms [(27)]. Similar effects are achieved in continuous flow activated sludge system with a plug flow or cascade flow bioreactor.

The SBR has the advantage of a more flexible operation as stated earlier with true plug flow characteristics. Since the treatment process proceeds over time in one reactor, the process control can be conventionally executed by means of auto-analyzers (*e.g.*, for dissolved oxygen, pH, nitrate) and computers. On the other hand, continuous flow systems are more flexible with respect to recycle of sludge and wastewater. Recycle flows can be directed to any reactor in the system where as the comparable operation is not possible in the SBR. SBR treatment profile functions similarly as in a continuous flow stirred tank reactors scheme, where the concentration of substrate reduces in similar pattern. Figure 6.5 shows the concentration profile of organic substrate in a continuous flow stirred tank reactor (CFSTR) and SBR.

In a similar manner for anaerobic treatment of wastes, anaerobic sequencing batch reactor (AnSBR) is utilized. The commonly used mode for anaerobic digestion is the Upflow Anaerobic Sludge Blanket reactor which is continuously fed in an up-flow mode and the influent and effluent must be fed and withdrawn uniformly across the horizontal plane of the reactor. In contrast, the AnSBR is intermittently fed. An elaborate wastewater feed and gas/effluent collection system is not required for the AnSBR, since AnSBR is batch-fed and drawn, there is no possibility of short circuiting of waste through the reactor and no re-circulation of effluent is needed.

Point wise comparison between the sequencing batch reactors and the conventional continuous flow system is given in the Table 6.4.

In a comparative study [(28)], the performance of full scale SBR and rotating biological contactor (RBC) was utilized for treating sewage at different operational conditions. The results of the study demonstrate that both SBR and RBC units treating sewage are capable of producing good effluent quality with average BOD and SS

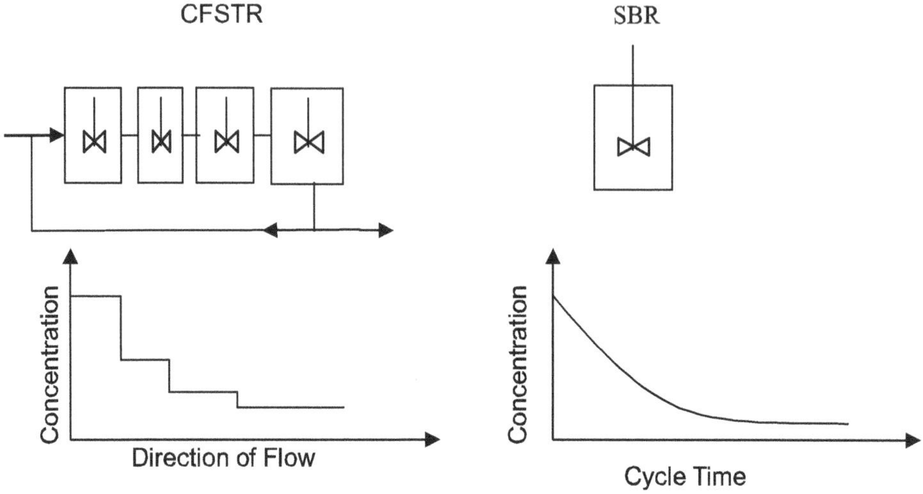

Figure 6.5: Concentration Profile of Organic Substrate in CFSTR and SBR.

valves below 20 and 30 mg/L respectively. However in the long run, performance of the SBR was more consistent in producing better effluent quality (*i.e.* BOD < 10 mg/L and SS < 20 mg/L) than RBC.

Table 6.4: Comparison between SBR and Continuous Flow Reactors

Sl.No.	*Parameter*	*Batch Process*	*Continuous Process*
1.	Concept	Time sequence	Spatial sequence
2.	Inflow to reactor	Periodic	Continuous
3.	Discharge from reactor	Sequenced	Continuous
4.	Aeration	Sequenced	Continuous
5.	Mixed liquor	Always in reactor, no recycle	Recycle through reactor and clarifier
6.	Clarification	Quiescent hydraulics	Hydraulic motion
7.	Flow pattern	Perfect plug	Complete mix or approaching plug
8.	Equalization	Inherent	None
9.	Flexibility	Considerable	Limited
10.	Hydraulic sizing	Variable	Uniform
11.	Capital cost	Low	Medium
12.	O and M Cost	Low	Low
13.	Possible automation	Yes	No
14.	Physical structure	Compact	Larger than SBR
15.	Sludge Yield	Low	Moderate
16.	Effluent quality	Good	Variable

The nutrient removal could be easily carried out in the SBR as sequential phase of oxic and anoxic conditions could be easily demonstrated to carry nitrification and denitrification in the same reactor.

Factors Affecting the Performance of SBR

Effect of Temperature

Temperature is one parameter, which affects the performance of SBR [29]. The performance of SBR is different under identical loading conditions. Higher removal efficiency could be obtained at temperatures ranging from 20–35°C [29, 30]. Large number of filamentous organisms was produced at high temperatures. However higher removal efficiency also could also be obtained at lower temperatures if wastewater is diluted [31]. Figure 6.6 presents the removal efficiency of various parameters.

Biological treatment of wastewater at higher temperature has the advantage of providing effective treatment with minimum energy and operating cost. Chemical compound analysis of feed and various treated effluents showed a better removal of long chain fatty acids at thermophilic conditions (60°C) in comparison to mesophilic conditions (35°C) [32].

To demonstrate the temperature effect Ndon and Dague [33] conducted a laboratory scale anaerobic SBR experiment. The reactors were operated at different HRT' s (48, 24, 16 and 12 hr). Results showed 80-90 per cent soluble COD removal efficiency at various substrate concentrations, temperatures and HRT's. At lowest experimental temperature of 15°C, the AnSBR was able to achieve 80 per cent total COD removal even at the shortest HRT of 12 hr at low substrate concentrations of 600 and 400 mg/

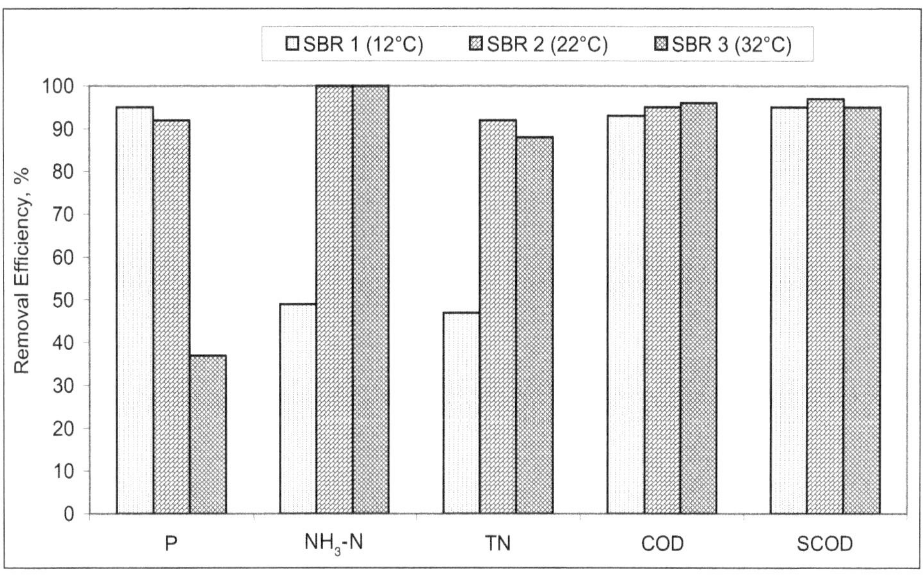

Figure 6.6: Effect of Temperature on the Removal of different Pollution Parameters in SBR.

L COD. The findings showed that the AnSBR could be used for the treatment of low strength wastewater at low temperature, and COD removals were slightly affected by change in temperature.

Effect of Long Idle Periods

Sludge storage can be used as an effective control to adjust plant capacity to large influent variations. The storage of sludge can be done in an external tank or in the mainstream system. Mainstream storage is possible in a treatment plant with parallel treatment trains, where individual trains can be individually switched on and off. The SBR technology is well suited for this temporary sludge storage because reactors can easily switched off individually and operated in an idle mode. A number of researchers have investigated the effect of storing bacteria without external substrate· Morgenroth [34] conducted batch experiments to find the effect of long term idle period on the performance of SBR. SBRs were operated for idle times ranging from 6 to 20 days and was found that after 6 weeks of storage and without aeration the sludge still had above 55 per cent of the initial volumetric ammonia oxidation rate. No adverse effects on sludge settleability were observed with an SVI below 140 mL/g in the lab scale SBRs. In the laboratory scale activated sludge sequencing batch reactor idle periods had no effect on nitrogen removal. For longer idle periods ammonia oxidation was not significantly affected but reactor operation had to be adjusted for nitrite accumulation. It was observed that to prevent nitrite accumulation, time for denitrification was to be increased by supplying an external carbon source. Also with the increase in time of idle more quantity of ammonia was oxidized to nitrate.

Effect of Oxygen Supply

Oxygen supply methods also affected the performance of SBR. Dangcong [35] studied high ammonium nitrification studies in SBRs by two oxygen supply methods *viz.* controlled and uncontrolled oxygen supply method. In the controlled oxygen supply method dissolved oxygen concentration ranged between 2-3mg/L. However, in uncontrolled method, the concentration of DO changes with the oxygen utilization rate. The experimental results showed that Oxygen supply methods have significant effects on SBR performance. High ammonia nitrogen oxidation took place simultaneously with the uncontrolled oxygen supply method. Ammonium oxidation proceeded smoothly during the low DO period of the cycle, however nitrate oxidation was completely inhibited at low DO but recovered during the high DO period of the same cycle. Ammonium oxidation developed an ability to sustain the fluctuations of DO, but nitrate oxidizers did not.

Application of SBR in Wastewater Treatment

Nutrient Removal

SBR is utilized for nutrient removal strategy successfully in many industrial effluents [36,37,38 and 39]. The overall reactions in an intermittent aeration sequencing batch reactors activated sludge process have been studied on full scale by Kazmi and Furumai [36]. Several field investigations were conducted during settling in different seasons. Nitrate Mixed Liquor Suspended Solids, Phosphate, and Total Organic

Carbon profiles were obtained at specified depths. Nitrogen mass balance during settling revealed that significant nitrogen removal (>30 per cent) can be achieved even at lower temperatures by maintaining higher MLSS (2700-3000 mg/L). Residual DO at the commencement of settling has an inhibitory effect on denitrification under low MLSS conditions.

The flexibility of SBRs also suited to biological phosphorous removal as this flexibility allows the development of conditions necessary for biological phosphorus removal. The biological phosphorus removal can also be achieved in SBR [37] with a relatively low COD/TKN ratio. The process produced a highly nitrified effluent and sludge with excellent settling characteristics. The SBR could also be designed to operate as a non-nitrifying system and still achieve biological phosphorus control. However, biological phosphorus removal capability reduces slowly over time when no P-removal strategies have been adopted. Figure 6.7 indicates the reduction of biological phosphorus in effluent sludge with time.

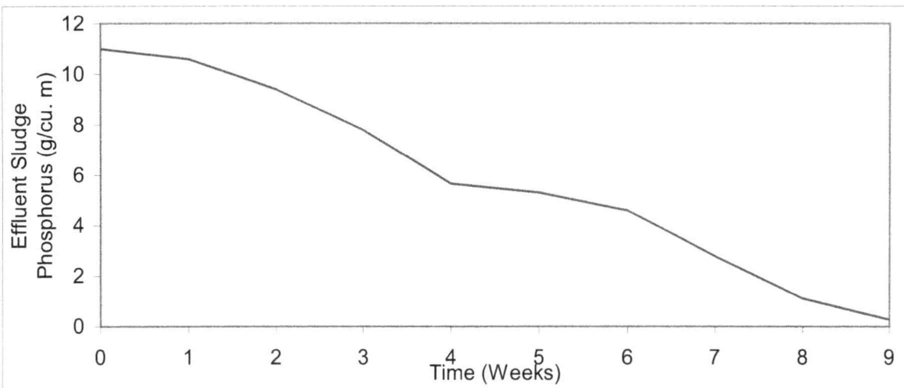

Figure 6.7: Reduction in Biological Phosphorus in SBR Sludge.

Randall *et al.* [38] operated three anaerobic/aerobic sequencing batch reactors resulted in the optimization of enhanced biological phosphorus removal (EBPR).

Carucci *et al.* [39] conducted a lab scale study for biological phosphorus removal with different organic substrates in an anaerobic and aerobic SBR. It was observed that the enhanced biological phosphorus removal was clearly established with a sludge containing G-bacteria. Apart from separate removal of nitrogen and phosphorus from effluents combined removal could be obtained in an SBR [40]. Metabolism of enhanced biological phosphorus removal in an (aerobic/anaerobic) SBR was also investigated [41] using glucose as a sole carbon source and it was observed that high removal was achieved in the process.

In another experiment, denitrifying phosphorus removal was verified by Ng, Ong and Hu [42], the results demonstrated the ability of anaerobic/anoxic strategy to enrich the growth of denitrifying phosphorus bacteria capable of taking up phosphate under anoxic conditions, using nitrate as electron acceptor. Phosphorus removal efficiency ranging from 40-100 per cent could be attained in an A/ASBR. Simultaneous anoxic phosphate uptake and biological denitrification under anaerobic/anoxic condition occurred in the system.

A study conducted by Belia and Smith [(43)] demonstrated addition of pure culture of acinatobactor ewoffii to an activated sludge resulted in the development of enhanced biological phosphorous removal in 14 days. A controlled reactor allowed developing an enhanced culture of poly-phosphate bacteria, reached the same removal rates after 58 days. Both the control and bio-augmented reactors achieved 90 per cent COD and nitrogen removal rates.

Higher enhanced biological phosphorus removal (EBPR), in a sequencing batch reactor could be obtained [(44)]. The experimental results showed that high EBPR can be obtained in a lab-scale anoxic/anaerobic SBR using synthetic wastewater containing acetate, protein and glucose. However, phosphorus removal gradually declines as acetate addition is lowered and compensated with glucose. Further decrease on the removal of phosphorus occurs when the glucose addition is stopped.

Similarly two separate studies conducted by Ketchum and Liuo [(45)] on lime and ferric chloride for phosphorus reduction from synthetic wastewater and municipal secondary effluent showed a significant saving in cost of chemicals (ferric chloride), when tertiary phosphorus reducing system was operated in sequencing batch mode.

Higher removal efficiencies could be obtained in a continous flow systems modified into SBR [(46)]. Modification improved organic and nutrient removal efficiencies without corresponding increase in tank volume and blower capacity. The removal efficiencies were as high as 97 per cent for organic, 99 per cent for nitrogen and 78 per cent for soluble phosphorus. Oxygen requirement is also reduced simultaneously with only piping changes and timer installations a major capital investment. Energy savings estimated to be about 40 to 53 per cent were achieved with an improved effluent quality.

Benefield *et al.* [(47)] investigated the feasibility of establishing removal of enhanced biological phosphorus (EBP) while treating a wastewater containing pentachlorophenol (PCP) in a SBR system under different control strategies. The results of the experiments showed that enhanced biological phosphorous were obtained at shorter cycles of operations. Hence, it could also be understood that the SBR system could be operated in regard to wastewater quality for different cycles.

Hamamato *et al.* [(48)] tried to maximize the removal of nitrogen and phosphorus in SBR from the wastewater using intermittent cyclic process. The intermittent cyclic process repeated aerobic and anaerobic treatments, which corresponded to aeration and mixing. The results of the study showed that with different aeration time ratios, the rate of nitrogen and phosphorus removal rates increased as the aeration time ratio decreased. A high average nitrogen removal of 86 per cent and average phosphorus removal rate of 82 per cent were recorded. These figures were comparatively equivalent to those achieved in laboratory studies.

Hazardous and Toxic Substances Removal

Landfill leachate has been generally known to as high strength hazardous wastewater. This is due to its large variability in organic, inorganic and heavy metals, which strongly depends on the age and type of solid waste of a landfill site. The present day SBR technology can be widely used for the treatment of hazardous and toxic wastes. SBR treatment is used in combination of various other treatment techniques to treat wastes from landfill. The methods employed ranged from combination of electro-Fenton oxidation with SBR (49). Such a combined chemical oxidation and biological treatment was found to be highly efficient and offered an attractive alternative in dealing with the high variability of pollutant content in the landfill leachate. The test results revealed that, final treatment of the leachate effluent by SBR elevates wastewater quality to the standard suitable for direct discharge or for reuse as non-potable water.

Biological treatment of leachate prior to activated carbon filtration has proven successful and cost effective in many areas (50). SBR process was selected for this treatment as it offers the advantages of better stability and greater operational control compared with continuous flow system (51). In an experiment biological treatment of synthetic leachate containing cyanide, complete removal of cyanide was observed from the wastewater (52). Results from the studies conducted by Herzbrun (9) indicated that the SBR could provide substantial saving in energy cost by removing organic compounds found in hazardous waste biologically, rather than with activated carbon. The role of organic contaminant removal had shifted from adsorption on activated carbon to degradation in the SBR. The use of carbon for polishing purposes resulted in significant cost and energy savings. SBR could be used effectively for landfill leachate treatment (53).

Yalmaz and Ozturk (54) conducted lab-scale experiment to investigate the use of the SBR technology in the biological ammonia removal from landfill leachate containing high concentration of ammonia nitrogen. The result of the study concludes that SBR is very effective treatment technology for landfill leachates with high NH_4^+—N content, as 90 per cent removal of total N was achieved by using $(CaH_3CO_2)_2$ as external carbon source in the denitrification phase. Settling properties of the SBR sludge were excellent and characterized with very low SVI.

Phenol is an inhibitory and toxic compound present in wastewater from various industries. This toxic compound in biological wastewater treatment contributes to process instability, and provides warning of impending failure. There are many methods for the treatment of phenolic wastewater. But owing to the nature of the process and advances in process control technology, there is renewed interest in the SBR for the treatment phenolic wastewater. A bench-scale SBR was utilized to degrade phenol at a high loading rate (55) with COD removal efficiency of 97 per cent at a SRT of 4 days and a HRT of 10 hours was achieved.

The biodegradability of high concentration of phenol and o-cresol in wastewater and the impact of these toxicants on sludge settleability was investigated (56). The reactor achieved more then 99.5 per cent removals of phenol and o-cresol at toxicant loading in the ranges of 0.1-0.8 kg phenol/m^3/day and 0.1-0.6 kg O-cresol/m^3/d. The effluent total suspended solids concentration was around 12 mg/L which was conforming the statutory standards for river disposal.

Biodegradation of a synthetic mixture of several phenol compounds typically found in coal conversion waste was investigated in a SBR [57]. The mixture contained phenol, O-, M- and P-cresols, 2,3- and 3,4- di-methylphenols and had a total COD concentration of 7750mg/L. In general COD removal efficiencies were greater than 99 per cent, with effluent COD concentration averaging between 30 and 52 mg/L.

Phenol can also be treated by a biological activated carbon sequencing batch reactor (BAC–SBR). It has been extensively investigated by many researchers [58, 59, 60 and 61] and found that treatment efficiency of more than 90 per cent of COD was obtained at all SRT' s. Carbon addition in the SBR system can promise the improvement of the system performance at low F/M ratio.

Combined anaerobic and aerobic treatment in SBR was also found very efficient in the mineralization of recalcitrant pollutant [62]. A reactor packed with volcanic stones and inoculated with microorganisms from activated sludge processes treating municipal and industrial wastewater was used to treat P-Nitro Phenol (PNP); removal efficiencies as high as 100 per cent were obtained.

Azo dyes are organic compound, which are difficult to both light and microbial attacks. The combined anaerobic and aerobic environment can be used to achieve the mineralization. In the anaerobic stage-the azo dyes can be bio-transformed to amines, which can be completely mineralized in the aerobic step [63, 64 and 65]. Biological treatments of azodyes provide lower operating cost and improved applicability [66, 67 and 68].

A study [68] on the degradation of azodyes (acid and bases) in an aerobic bio-filter operated in a sequencing batch reactor was carried in a effluent containing Acid Red 151 and textile effluent containing base dyes. The degradation efficiencies ranged between 60 and 99 per cent. Mineralization studies showed that 73 per cent (as carbon) of the initial azo dye was transformed to CO_2 by the consortia with removal efficiencies varying from 76 to 97 per cent. In all the cycles, the system presented approximately 80 per cent of the colour removal.

SBR is used in the biodegradation of the herbicide [69] 2,4-Dichlorophenoxy acetic acid (2,4-D) using 3 sequencing batch reactors (SBR). The experiment showed complete removal of phenol and dextrose with more than 99 per cent 2,4-D removal.

Application to Specific Wastewater

Piggery Wastewater

The wastewater from the piggery is generally very concentrated compared to domestic sewage with high levels of nutrients, biological oxygen demand (BOD), fiber and minerals [70]. The SBR process can be a good alternative to other conventional treatment processes. Removal efficiencies as high as 99 per cent for ammonia could be achieved [71].

Brewery Wastewater

Nguyen [72] tried treatment of brewery wastewater, using a feed back control system based on dissolved oxygen concentration in the mixed liquor. This controlled system was modeled after the self-cycling fermentation (SCF) technique. The system was operated at 25 and 35°C and under growth rate limiting condition and total COD and BOD removals were found to be between 83 and 73 per cent and 83 and 92 per cent respectively.

Paper Mill Wastewater

Wastewater from paper mills can also be treated in a SBR. Majorities of the constituents of paper mill wastewater are readily biodegradable [73]. The COD decreases as a result of the biodegradation. The highest overall removal and the best sludge settling properties for the paper mill wastewater was obtained at high sludge age (more than 15 days) and react period of 6 to12 hrs.

Tannery Wastewater

An another new application of SBR is the sequencing batch membrane reactor for tannery wastewater treatment [74], which had high strength organic matter and a heavy metal (chromium). The system contains a hollow fiber micro-filtration unit in a sequencing batch reactor (SBR) for solid liquid separation, which aims to enhance sludge concentration in the reactor for stable removal of organic matter and heavy metals.

Biodegradation of Nitroglycerin

SBR also has its application for the degradation of nitroglycerin [75 and 76]. The microbial transformation commence after reactor inoculation with mixed liquor from a wastewater treatment facility resulted in complete mineralization of nitroglycerin in the SBR.

Advantages of SBR

Extensive lab and pilot scale studies have shown that various types of wastewater could be treated using sequential batch reactor. It also provides a viable alternative to continuous flow systems in nutrient removal as well as in carbon and suspended solids removal. It offers a great deal of operational flexibility, as it allows for easy adjustment of aerobic, anoxic and anaerobic periods through temporal control of aeration and filling. Application of automatic control to SBRs for cost reduction and process optimization are some of the main causes of renewed interest in the SBR technology. Because of the flexibility associated with working in time, the SBR can be operated as either a labour- intensive, low-energy, low–sludge yield system or a minimal labour, high-sludge yield system for the same physical plant. The actual operation policy can be adjusted in accordance with economic condition by modifying the settings of a control mechanism. Labour, energy and sludge yield can also be traded off with initial capital costs.

Design Approaches

The perplexing abundance of possible design choice in SBR makes it an important task to develop unified design basis. SBR design for an industrial wastewater generally requires extensive treatability studies. Design for domestic wastewater on the other hand is relatively simple and can be centered on the selection of the proper tank sizes, the inlet and outlet configuration, the aeration system, and the control mechanism.

Tank sizing and aeration system is central to the proper design of the SBR. The initial volume is determined based on a number of factors including desired loading and detention time and expected settling characteristics of the activated sludge. The time of FILL depends upon the volume of each tank, the number of parallel tanks in operation and the extent of diurnal variation in the wastewater flow rate. The reactor can be designed with similar principles applied for the design of continuous flow activated sludge systems. The parameters needed to be determined during design calculations are as given in further reading.

- **Operating strategy:** The duration and order of the different phases have to be introduced in a short or long fill period, and the SBR cycle may contain one or more fill phases. Fixed cycle times can be defined or the cycle time can be adjusted in real time depending on wastewater composition and concentration, using on-line sensors.
- **Sludge age:** To calculate the effective sludge age, the biomass during settle and draw is assumed to be dormant. Thus, the effective sludge age is smaller than the effective solids retention time (Q_x, effective) is smaller than the solids retention time (Q_x).

  Effective solids retention time

  $$Q_x \text{ effective} = \frac{Q_x (t_c - t_s - t_d)}{t_c}$$

- **Volumetric exchange ratio (f_{exr}) and duration of the cycle (t_c):** Volumetric exchange ratio is the ratio of exchange volume to the sum of the minimal volume and exchange volume. The amount of wastewater to be treated is determined by volumetric exchange ratio and duration of the cycle resulting in a hydraulic retention time (Q_H). To achieve a high initial substrate concentration and a large concentration gradient ones the cycle time a large volume exchange ratio is required as shown in Figure 4. Low substrate concentration can be obtained by slow fill and large f_{exr}, which may be beneficial to avoid substrate inhibition, when the incoming wastewater is highly concentrated.

  $$Q_c = \frac{(t_c)}{f_{exr}}$$

- **Duration of different phases:** The duration of different SBR phases is determined according to the design equations and can be optimized using off-line computer simulation or on-line control.
- **Number and volume of SBRs:** The number of SBRs required for the treatment can be chosen according to the process needs. Both cost and operational flexibility increases with the number of reactors.

Different methods have been discussed in literature to design a SBR process. Some of them are as follows:

Design Based on Continuous Flow Systems

System design of an SBR can be based on available procedures for continuous flow systems. A stepwise design of SBR is presented in Table 6.5. It is assumed that the mass and activity of the activated sludge in the SBR will be the same as in continuous flow system. Higher metabolic activities have sometimes been reported for SBRs compared to continuous flow systems, which may be due to the fact that it is difficult to achieve plug flow conditions in a continuous flow system. The duration and order of the SBR phases will be selected proportional to the order and volume of the reactors or zones in the continuous flow systems [77].

Design Based on Mass Balances

A similar stepwise design procedure directly based on mass balance is presented in Table 6.6. After selection of solid retention time (SRT) the total mass of sludge in all reactors is calculated based on the influent COD removed. Minimum of water volume remaining in the reactor after draw (V_0) is determined by the volume of the settled sludge. The exchanged volume (ΔV), depends on the selected cycle time. The total volume of the SBR (V_R) is the sum of minimal volume and exchange volume, V_o and ΔV.

Table 6.5: Design Procedure Based on Design of Continuous Flow System [77]

1	INPUT: Design information from continuous flow system, operating strategy for the SBR	
2.	Estimate effective fraction of t_c	$\text{effective fraction} = \dfrac{t_c - t_s - t_d}{t_c}$
3.	Effective mass of sludge in the SBR equals the mass of sludge in the continuous flow system	$M_{X,SBR} = M_{X\,Cont} \cdot \dfrac{t_c}{t_c - t_s - t_d}$
4.	Select MLSS in SBR alter fill and, thus, $n.V_R$.	$n.V_R = \dfrac{M_{X,SBR}}{C_X}$
5.	Select V_0 depending on the volume of the settled sludge.	min (n.Vo) = M_{xSBR} · SVI. SF_V or $f_{exr,max} = |-C_x.SVI.SF_V$
6.	Select exchange ratio or cycle time	
7.	Select number of SBRs (n) and calculate missing value (exchange ratio or cycle time)	$f_{exr} = \dfrac{t_c.Q}{n.V_R} < f_{exr,\ max}$
8.	For a system with nitrogen removal select unaerated and aerated cycle times according to unaerated and aerated volumes of the continuous flow system	$\dfrac{t_D}{t_D + t_N} = \dfrac{V_D}{V_D + V_N}$
9.	Check t_s and t_d based on settling velocity and maximal draw flow rate, respectively	

Table 6.6: Design Procedure Based on Mass Balances [77]

1.	INPUT: operating strategy for SBR	
2.	Select effective solids retention time (O_x, effective)	
3.	Estimate effective fraction of tc.	$\text{effective fraction} = \dfrac{t_c - t_s - t_d}{t_c}$
4.	Calculate mass of sludge in all reactors	$M_{x\,SBR} = Y_{H.net}\ Q_{\Delta}.C_s.Q_{x,\ effective}\ \dfrac{t_c}{t_c - t_s - t_d}$
5.	Calculate volume of settled sludge in all reactors	$n.V_o = M_{x,SBR} \cdot SVI \cdot SF_V$
6.	Select exchange ratio or cycle time	
7.	If exchange ratio was selected, then calculate cycle time from	$tc = \dfrac{n.Vo.f_{exr}}{Q.(1\quad f_{exr})}$
8.	Calculate exchange volume of all reactors	$n.\Delta. = Q.t_c$
9.	Calculate total volume of the SBRs	$n.V_R = n.V_O + n \bullet \Delta V$
10.	Decide number of SBR (n)	
11.	Calculate min. aerated fraction of reaction time based on min. aerobic solids retention time for, *e.g.*, nitrification	$\text{min.aerate d fraction} = \dfrac{\text{min. } Q_{x,\ aerobic}}{O_x} \bullet SF_x$
12.	Choose number of fill phases and distribution of the influent on the fill phases depending on the wastewater composition *(e.g.*, TKN/COD ratio)	
13.	For a system with nitrogen removal check nitrification capacity and denitrification potential depending on cycle design	$N_{nit} = TKN\text{-}NH_4, e\ N_{org}.e\ N_{sL}.$ $N_{DP} = (1\text{-}Y_{H,\,net}).\ C_{S,\,in}.{}^1/_{2.86}.f_{DN}.n$
14.	Check reacting time for nitrogen removal using denitrification rates.	$t_d = \dfrac{N_{DP}.n.\Delta V}{r_{DN,\,X}.M_{x\,SBR}}$
15.	Check ts and td based on setting velocity and maximum draw flow rate, respectively	

Based on Hydraulic Considerations

SBRs could be designed as a batch or continuous flow through systems. Continuous flow systems are generally designed on the basis of hydraulic requirements.

Design Guidelines for SBR [77]

Guidelines for different system components of the SBR have been discussed in detail in the following paragraphs.

System Sizing and Reliability

- ☆ More than two tanks should be provided.
- ☆ Facilities for the equalization of flow and organic shock load (both high and low) shall be considered for each basin.
- ☆ The design F: M ratios and MLSS concentrations for SBRs should be similar to other conventional and extended aeration processes. For domestic wastewater the F: M ratio provided to nitrify is typically in the range of 0.05 to 0.10 and not required nitrifying 0.15 to 0.4. Higher values may be accepted, if justified. The selected F: M ratio for each tank should account for the unbalanced loading, which may occur in the absence of proportional flow splitting. The reactor MLVSS and MLSS concentrations should be calculated at the low-water level.
- ☆ Solids management to accommodate basin dewatering shall be considered.
- ☆ The added settling time before a discharge shall be considered for systems with fixed decanters. The vertical travel distance for mobile decanters, specific to the equipment supplied, shall accommodate the required decant volume.
- ☆ The low-water depth should not be less than 10 feet (3m).
- ☆ The water depth of any basin where simultaneous fill and decant may occur shall not be less than 12 feet (3.7 m) at the end of the DECANT phase. This low-water level may be reduced to 10 feet (3 m), provided the SBR is treating less than 150 kg/day of BOD_5 and is followed by a pond of 5-day detention time. Treatment facilities with fixed decanters or any other system where the low-water depth cannot be adjusted quickly by the operator should be designed to end the DECANT phase at a higher water level than other types.
- ☆ Downstream units shall be sized to handle peak discharge rates. Flow equalization downstream of the SBRs may be necessary to dampen the flow.
- ☆ Although the control systems may prevent multiple tanks decant for normal operations; or to minimize the size of subsequent treatment units, system flexibility and outfall piping to allow decant of at least two tanks. Simultaneously for SBR plants with more than two basins may be desirable, if downstream facilities and the receiving stream are not adversely effected.

Decanter Design

- ☆ System reliability and the instantaneous delivery of flow shall be evaluated in the design of decanter weirs and approach velocities. The decant method, sizing and system configuration shall limit velocities at the inlet port or at the edge of submerged weirs throughout the DECANT phase so as not to cause or allow entry of floating material, vortexing or disturbance of settled sludge. The treated effluent from each reactor should be free of scum and have suspended solids concentration no greater than 30 mg/L at any time

- The decanter (s) shall draw treated effluent from below the water surface and exclude sum.
- An adequate zone of separation between the sludge blanket and the decanter (s) shall be maintained at all times during the DECANT phase. Floating decanters or similar mechanical devices which allow the treated effluent to be drawn near the water surface throughout the DECANT phase are recommended for added process control. Fixed decanters should not be used in basins where simultaneous fill and decant may occur, or where the sludge may be difficult to settle.
- A means of excluding solids from entering the decanter during the REACT phase should be provided. Provisions shall be included to flush and recycle any trapped solids back to the SBR.

Tank Details

- Pairs of tanks rather than odd numbers are recommended for ease of operations or efficient use of equipment.
- Flow division control facilities preceding the SBR tanks shall be designed for greater operating and maintenance convenience, flexibility, continuity of maximum effluent quality, and ease of installation of future units.
- For systems with four tanks, flow division control facilities should be provided which may distribute the wastewater flows to atleast two tanks at all times.
- It necessary to consider the means and frequency for removal of grit and other debris from the SBR tanks. Adequate space for equipment access (possibly cranes) should be considered, especially if the tanks are deep.
- Comminution, rather than fine screening may be approved for a SBR plant but it is discouraged in the absence of primary clarification.
- Influent baffling using a baffle wall and adequate physical separation of the influent from the decanter shall be provided for any basin which may operate with a continuous feed during the SETTLE and DECANT phases. The baffling must direct the influent wastewater below the sludge blanket. Average horizontal velocities through each baffle wall opening should not exceed 1 0.305m/sec at design peak hour flows.
- A system separate from the decant piping allowing sludge to be wasted during the DECANT and/or IDLE phases of each process cycle shall be provided.
- The capability to transfer sludge from each basin to the other aeration tank (s) shall be provided. If decant pumps are used for sludge transfer, all solids in decant piping shall be flushed and recycled back to SBR.
- All sludge transfer and wasting pumps should be accessible for maintenance without dewatering the tank.
- Scum removal facilities such as telescoping valves or high water overflow, should be provided.

☆ Each tank shall be provided with an overflow to the other aeration tank (s) or storage.

☆ A means shall be provided to de-water each tank of SBR. Tank bottom should be sloped towards a drain or sump.

☆ Protection from uplift when empty should be provided where high groundwater may occur. Common walls shall be structurally designed for liquid on one side only.

☆ All SBR tanks shall have a freeboard not less than 18 inches (460 mm).

☆ Adequate site lighting shall be provided for the SBR tanks.

Aeration and Mixing Equipment

The aeration equipment shall be suitable for varying water depths and cyclical operation in a sequencing batch reactor based on experience with installations handling similar wastes and loading. The aeration and/or mixing equipment also shall not produce flow patterns within the aeration tank, which interfere with quiescent settling. The discharge pressure of the blowers must be established at the maximum water depth.

☆ Provisions to easily remove aeration diffusers without dewatering the tank are recommended to permit maintenance and repair without intercepting operation of the aeration tank or inhibiting operation of the other aeration equipment. If the aeration equipment cannot be serviced and/or replaced without dewatering the tank, a minimum of four SBR tanks should be provided.

☆ Oxygen transfer rates from the aerators based on average water depth between the low-water level and maximum water level for the critical flow conditions shall be considered to provide a dissolved oxygen residual of 2.0 mg/L during aeration.

☆ The blowers shall be sized to deliver the total oxygen demand in a shorter period of time than allocated to the FILL/REACT and REACT phases of the cycle if oxic and anoxic conditions are required during these phases.

☆ The mixing shall be sized to thoroughly mix the entire basin from a settled condition within 5 minutes or the react phase shall be increased.

Controls, Sampling, and Operations Manuals

☆ The motor control centers should be located for convenient operation of the facility. A location in close proximity to the process with a view of the batch reactors is highly recommended

☆ A programmable logic controller (PLC) programmed to meet the required effluent limitations for the design loading with limited operator adjustment is recommended. All factors affecting the flow and organic loading for each cycle shall be considered. In addition, a fail-safe timer control which cannot be adjusted by the operator allowing not less than 20 minutes between the

REACT and DECANT phases shall be provided. Both automatic and manual controls shall allow independent operation of each tank

- ☆ Uninteruptible powers supply (UPS) with electrical surge protection shall be provided for each PLC to retain programme memory through a power loss.
- ☆ Visual display and recording of instantaneous and totalized flow rates, both to the plant as well as discharge, shall be provided
- ☆ Dissolved oxygen and oxidation reduction potential (ORP) monitoring and control shall be provided as necessary
- ☆ Effluent composite sampling shall be representative and accurate. Adequate provisions for sampling and flow measurement also shall be included which will allow the operations of each basin to be monitored separately. All 24 hour effluent composites for compliance reporting shall be flow paced and include samples collected at very beginning and end of each decant event
- ☆ Tank level sensors, pressure transducers and float may be used. Bubbler systems will not be approved
- ☆ Annunicator panel to indicate the following alarm conditions as a minimum:
  - (*a*) High–and low – water level in each tank
  - (*b*) Valve–failure of all automatically operated valves
  - (*c*) Decanter failure
  - (*d*) Blower – low pressure, high temperature and failure
  - (*e*) Pump – high pressure and failure
  - (*f*) Mixer – failure
  - (*g*) Effluent – high turbidity
- ☆ Visual display in addition to flow measurement is recommended and should consists of:
  - (*a*) Mimic diagram of process showing status of pumps, valves, blowers and mixers
  - (*b*) Process cycle and time remaining
  - (*c*) Tank level gauges
  - (*d*) Sludge pumping duration or volume
  - (*e*) Air-flow rate and totaliser
- ☆ A complete and comprehensive operation and maintenance manual shall be provided for the treatment facility prior to start-up.

Conclusion

SBRs were the first applications of activated sludge technology. They were converted into continuous flow systems, because of lack of reliable timer control.

With the developments in automation technology these problems have been sorted out. SBR promises an effective treatment technology for wide range of effluents. The system because of its periodic nature expands the spectrum of treatment capabilities and is an alternative to continuous flow systems in use now.

The analysis in the review shows that SBRs and continuous flow systems are comparable in many ways. In comparison of volume, the SBR will have larger volume than the corresponding aeration tank in the continuous flow system. However, if the volume of SBR is compared to the total volume of a continuous flow plant (aeration tanks plus secondary clarifier), the SBR requires a lesser volume.

One additional advantage offered by the sequencing batch reactor is that every batch of treated effluent could be monitored easily and adjusted to meet the needed quality. Using on-line measurements, the cycle time of the SBR can be adjusted continuously based on the wastewater composition. Various mechanical devices are available for draw, mix and aerate. Most of these devices are proprietary.

An SBR treatment plant consisting of a number of reactors and holding tanks can be designed in a rather compact manner, resulting in overall cost savings. Due to the simplicity and flexibility of operation, the SBR system can bear shock loads and is well suitable for variable wastewater treatment. Even though the present day SBR is a promising treatment process for almost all kinds of wastewater, a lot of groundwork is required to develop design criteria and optimization of the operational process. The following are some of the areas to improve the SBR technology application:

- With the advancement of automation for SBR technology, it is necessary to develop a default nomenclature
- Development of design criteria for SBR
- Sophisticated and improved control technologies
- Determination of optimal duration of the components of the process cycle (fill, react, settle, draw and idle)
- Feasibility of converting existing small scale package plants and septic tanks to SBRs over various flow ranges
- To develop methods and approaches to assess the microbial community of mixed population in the SBR system
- To enhance the capabilities of SBRs to treat a wide range and greater concentrations of industrial wastewater pollutants
- Field trials are necessary to find an optimum solution with respect to both sludge de-watering and foam control.

Abbreviations

PFR: Plug flow reactor

C_s, C_{sin}, ΔC_s: Substrate concentration, influent substrate, substrate removal

CFSTR: Continuous flow stirred tank reactor

f_{DN}: Fraction of total electron acceptor demand associated with anoxic reactions

f_{exr}: $\Delta V / (V_o + \Delta V)$, volumetric exchange rate in SBR
M_x, SBR: Effective mass of sludge is SBR
MLSS: Mixed liquor suspended solids
n: Number of SBRs
NH_4,e, N_{org}, e N_{st}, N_{DP}: Ammonia and organic nitrogen in the effluent and nitrogen eliminated with the sludge wastage, Nitrogen denitrification potential
Q: Flow rate
$r_{DN, x}$: Denitrification rate
SBR: Sequencing batch reactor
SBBR: Sequencing batch bio-film reactors
SF_v, SF_x: Safety factors
SVI: Sludge volume index
t_c, t_f, t_r, t_D, t_N, t_s, t_d, t_i: Phase duration with c: cycle, f: fill, r: react, D: denitrification (stirred), N: nitrification (aerated), S: settle, d: draw, i: idle
TKNo, N_{nit}: Total kjeldahl nitrogen, Nitrification capacity
V: Volume
V_o, ΔV, V_R: Minimal volume, exchange volume, total volume of the SBR
V_D, V_N: Reactor volume in a continuous flow plant dedicated to denitrification and nitrification
$Y_{H, net}$: Net sludge yield including decay processes and accumulation of inert solids
Q_H: Hydraulic retention time
Q_x: Solids retention time
Q_X, effective: Qx. $(t_c-t_s-t_d)/t_c$, effective solids retention time
Q_X, aerobic: Qx. t_n/t_c, aerobic solids retention time
$M_{x, Conti.}$: Mass of sludge in continuous flow system
BOD: Biochemical oxygen demand
COD: Chemical oxygen demand
TN: Total nitrogen
TP: Total phosphorus
CFS: Continuous flow system
HRT: Hydraulic retention time
SRT: Sludge retention time
C_x: Sludge concentration
DO: Dissolved oxygen
t_n: Anaerobic period

References

1. Arden E. and Lockett. WT. (1917), The Activated Sludge Process of Sewage Purification, *J. Soc. Chem. Ind.*, 36, 822.
2. Arden, E. and Lockett. W.T. (1914a, b), Experiments on the Oxidation of Sewage without the Aid of Filters, *J. Soc. Chem. Ind.*, 33, 536.
3. Duckworth, W. H. (1914), Aeration Experiments with Activated Sludge, Proc. Annu. Meet. Manchester District Branch of the Assoc. Managers of Sewage Disposal Works, 50.
4. Melling, S. E. (1914) Purification of Salford Sewage along the line of the Manchester Experiments, *J. Soc., Chem. Ind.*, 33, 937.
5. Metcalf and Eddy, (1993), *Wastewater Engineering.*
6. Irvine, R.L. and Busch, A.W. (1979), Sequencing Batch Biological Reactors – An Overview, *J. Wat. Poll. Cont. Fed.* 51, 235-243.
7. Wesner, G.M., Culp, G.L., Lineck, T.S. Hindrichts, D.J. (1978), *Energy Conservation in Municipal Wastewater Treatment*, U.S. Environmental Protection Agency Report, EPA 43019-77-011.
8. Irvine, R. L. (1979), Application of Sequencing Batch Reactors for Treatment of Municipal and Industrial Wastewaters, Research Grant, National Science Foundation's Program of Research Applied to National Needs, July 1976 to April 1979.
9. Herzbrum, P. A., *et al.* (1985), Biological treatment of hazardous waste in sequencing batch reactor, *J. Wat. Poll. Contr. Fed.*, 57, 12, 1163-1167.
10. Staszak, C. N. (1985), Full-scale Demonstration of a Sequencing Batch Reactor for a Hazardous Waste Disposal Site, Report 85-21, New York State Energy Research and Development Authority, Albany, NY.
11. Watanabe,S, Baba, K., Nogita, S. (1985), Basic Studies on an ORP/external Carbon Source Control System for the Biological Denitrification Process, Instrumentation and Control of Water and Wastewater.
12. Heduit, A, Leclerc, L.A., Sintes, L.A., Sintes, L. and Thevenot, D.R. (1988), Aspects of the Contral of Nitrification and Denitrification Reactions in Activated Sludge with the aid of the redox potential, *Water Supply*, 6 (3), 275-285.
13. Charpentier, J., Godart,H., Martin, G., Mango, Y. (1989), Oxidation- Reduction Potential (ORP) regulation as a way to optimize aeration and C,N and P removal: experimental basis and various full-scale examples, *Wat. Sci. Tech.*, 21 (10-11), 1209-1223.
14. Peddie, C.C, Mavinie, D.S., Jenkins, C.J. (1990), Use of ORP for Monitoring and Control of Aerobic Sludge Digestion, *J. Env. Eng. ASCE*, 116 (3), 461–471.
15. Warham,D.G., Hall, K.J, Mavinic, D.S. (1993), Real Time Control of Aerobic – Anoxic Sludge Digestion using ORP, *J. Environ. Eng., Div., Am. Soc. Civ. Eng*, 119 (1), 120-136

16. Plisson-Saune, S., Capdeville, B, Maurt, M., Degain, A., Baptiste, P. (1996), Real time control of nitrogen removal using three ORP bending points; Signification, Control Strategy and Results.

17. Nilson, M.K., Onnerth, T.B. (1996), Strategies for handling of on-line information for optimising nutrient removal, *Wat. Sci. Tech.* 33 (1), 211-222.

18. Hao, O.J. and Huang, J. (1996), Alternating Aerobic Anaerobic Process for Nitrogen Removal: Process Evolution, *Wat. Envir. Res.*, 68 (1), 83-93.

19. Yu, R.F. Liaw, S.L. Chang C.N., Lu, H.J., Chang, W.Y. (1979a), The Monitoring and Control using On Line ORP in the Continuous Flow Activated Sludge Batch Reactor System, *Wat. Sci. Tech*, 35 (1), 57-66.

20. Irvine, R.L., Wilderer, P.A., Fleming, H.C. (1997), Controlled Unsteady State Processes and Technologies–An Overview, *Wat. Sci. Tech.* 35, 1-10.

21. Warwham, D.G., Mavinic, D.S., Hall, K. (1994), Sludge Digestion using ORP-Regulated Aerobic-Anoxic Cycles, *Wat. Res.*, 28 (2), 373-384.

22. AL- Ghusain, J.A., and Hao, O.J. (1995), Use of pH as Control Parameter for Aerobic/Anaerobic Sludge Digestion, *J. Environ Eng. ASCF*, 121 (3), 225-235.

23. Yu, R.F., Liaw, S.L., Chang, C. N., Cheng, W.Y. (1998), Applying Real-time Control to Enhance the Performance of Nitrogen Removal in Continuous-flow SBR System, *Wat. Sci. Tech.*, 38 (3), 271–280.

24. Yu, R.F. Liaw, S.L.,Cho, B.C. Yang S.J. (2001), Dynamic Control of a Continuous inflow SBR with time varying Influent Loading, *Wat. Sci. Tech.*, Vol. 43, No 3, 107–114.

25. Ketchum, L.H. Jr. (1997), Design and Physical Features of Sequencing Batch Reactors, *Wat. Sci. Tech.*, Vol 35, No1, 11-18.

26. Oles, J., Wilderer, P.A. (1991), Computer Aided Designe of Sequencing Batch Reactors Based on the IA WPRC Activated Sludge Model, *Water Sci. Tech.*, 23, 1087–1095.

27. Irvine, R. L., Earley, J. P., Cassidy, D. P., Harvey, S. P. (1997), Biodegradation of Sulphur Mustord Hydrolysis in the Sequencing Batch Reactor, *Wat. Sci. Tech.*, Vol. 35, No. 1, 67-74.

28. Aknna, J.C. and Jefferier, C. (2000), Performance of Family Sized SBR and Rotating Biological Contactor units Treating Sewage at various Operation Conditions. *Water Sci. and Tech.*, Vol. 41, No 1, 97–104.

29. Grant, S. R., Landine, R. C., Wilson D. R., Cocci, A. A., McCarthy, P. J., Brown, G. J., Landine, R. C., Schmidthe, N. W. (1994), Effect of temperature on the performance of SBR treating potato processing wastewater, 49th Purdue University, *Indus. Waste Conf. Proceedings*, Lewis, Publishers, Chelsea, Michigan.

30. Dague, R. R., Timothy, G. and Gourange C. Barnik, (1998), Anaerobic SBR treatment of dilute wastewater at psychrophilic temperatures, *Wat. Environ. Res.*, Vol. 70, No. 4, 155-161.

31. Dague, R. R., Mckinney, R. E., Pfeffer, J. T. (1968), Solids retention in anaerobic waste treatment system. *J. Wat. Poll. Contr. Fed*. 42, R29.
32. Tripathi, C. S. and Allen, D. G. (1999), Comparison of Mesophilic and Thermophilic Aerobic Biological Treatment in SBRs Treating Bleached Kraft Pulp Mill Effluent, *Wat. Res*., Vol. 33, 836-846.
33. Ndon, U.J., and Dague, R. R. (1997), Effect of Temperature and HRT on Anaerobic SBR Treatment of Low Strength Wastewater. *Wat. Res*. Vol. 31, 10, 2455-2466.
34. Morgenrath, E., Obermaryer, A., Arnold, E., Bruhl, A., Wagoner, M., Wilderer, P.A. (2000), Effect of long term idle periods on the performance of SBRs, *Wat. Sci. Tech*., Vol. 41, No. 1, 105-113.
35. Dangcong, P., Bermet, N., Delgenes, J. P., Moletta, R. (2000), Effect of oxygen supply methods on the performance of a SBR for high ammonium nitrification, *Wat. Environ. Res*., Vol.72, No. 2, 195-200.
36. Kazmi, A. A. and Furumai, H. (2000), Field investigations on reactive settling in an intermittent aeration SBR activated sludge process, *Wat. Sci. Tech*. Vol. 41, No.1, 127-135.
37. Manning, J. F., Irvine, R. L. (1985), The biological phosphorus removal in a sequencing batch reactor, *J. Wat. Poll. Contr. Fed*., Vol. 57, No.1., 189-197.
38. Randall, A. A., Benefield, L.D., Hill, W. E., Nical, J.P., Boman, G. K., Jing, S. R. (1997), The effect of volatile fatty acids on enhanced biological phosphorus removal and population structure in anaerobic/aerobic sequencing batch reactors, *Wat. Sci. Tech*., Vol. 35, No. 1, 155-160.
39. Carucci, A., Majone, M., Ramdor, R, Rossetti, S. (1997), Biological phosphorus removal with different organic substrates in an anaerobic/aerobic sequencing batch reactor, *Wat. Sci., Tech*., Vol. 35, 161-168.
40. Paturea, D., Helloin, E., Rustrian, E. Bouchez, T. Delgener, J. P., Maletta, R. (2001), Combined phosphate and nitrogen removal in a sequencing batch reactor using the aerobic denitrifier, Microvirgula aeronitrificans. *Wat. Sci*. No. 1, 189-197.
41. Jeon, C. O. and Park, J. M. (2000), Enhanced biological phosphorus removal in a SBR supplied with glucose as a sole carbon source, *Wat. Res*., Vol. 34, No.7, 2160-2170.
42. Ng, W. J., Ong, S. L., Hu, J. Y. (2001), Denitrifying phosphorus removal by anaerobic/anoxic sequencing batch reactor, *Wat. Sci. Tech*., Vol. 43, No. 3, 139-146.
43. Belia, E. and Smith, P. G. (1997), Bioaugmentation of sequencing batch reactor sludges for biological phosphorus removal, *Wat. Sci. Tech*., Vol. 35, No. 1, 19-26.
44. Tasli, R., Artan, N., Orhon, D. (1997), The influence of different substrates on enhanced biological phosphorus removal in a sequencing batch reactor, *Wat. Sci. Tech*. Vol. 35, No. 1, 75-80.

45. Ketchum, Sr., L. H., Liuo, P. C. (1979), Tertiary chemical treatment for phosphorus reduction using sequencing batch reactors, *J. Wat. Pollu. Contr. Fed.*, Vol. 51, No. 2, 289-304.

46. Choi, E., Oa, S. W. Lee, J. J. (1997), Night soil treatment plant converted into a sequencing batch reactor to improve removal of pollutants and nutrients, *Wat. Sci. Tech.*, Vol. 35, No.1, 233-240.

47. Benefield, L., Kim, S., Randall, C. (1989), Effect of Pentachlorophenol (PCP) on Enhanced Biological Phosphorus Removal in SBR, 44th Purdue University, *Industrial Waste Conf. Proc.* Lewis Publishers, Inc., Chelsea, Michigan, 141-147.

48. Hamamato, Y., Tabata, S., Okubo, Y. (1997), Development of the intermittent cyclic process for simultaneous nitrogen and phosphorus removal, *Wat. Sci. Tech.* Vol. 35, No. 1, 145-152.

49. Sheng, H.L. and Chang, L. L. (2000), Treatment of landfill leachate by combined electrofenton oxidation and SBR Method, *Wat. Res.*, Vol. 34, No. 17, 4243-4249.

50. Ying, W., Bonk, R. R., Lloyd V. J., Sojka, S. A, (1984), Biological Treatment of Landfill Leachate in Sequencing Batch Reactors, Presented at *IWPCF Conf.* In October 1984.

51. Wilderer, P.A. and Schroeder, E.D. (1986), An Wending des Sequencing Batch Reactor (SBR) – Verfahrens zur biologischen Abwasserreinigury. Hamb Berichte zur Siedlungswasser wirt schaft, TU Hamburg – Harbuy, 4

52. Daniel, M.W. and Schnabel, W. (1998), Treatment of cyanide waste in sequencing batch biofilm reactor, Vol. 32, No.1, pp 254-157, Wat. Res.

53. Dockhorn, T., chang, L. and Dichtl, N. (1997), Removal of nitrogen from landfill leachate by using SBR-technology, Proc. Sardinia 97, *Sixth International Landfill symposium*, 303-314.

54. Yalmaz, G., Ozturk, I. (2001), Biological ammonia removal from anaerobically pre-tested landfill leachate in sequencing batch reactors, *Wat. Sci. Tech.*, Vol. 43, No. 3, 307-314.

55. Yoong, E. T., Lant, P.A. (2001), Biodegradation of high strength phenolic wastewater using SBR. *Wat. Sci. Tech.* Vol. 43, No. 3, 299-306.

56. Nakhla, G. F., Al-Harazin, I.M., Farooq S. (1993), Organic Loading Effects on Treatment of Phenolic Wastewaters by Sequencing Batch Reactors, *Water Environ. Res.*, Vol. 65, No.5, 686-690.

57. Brenner, A., Chozick, R., Irivine, R. L. (1992), Treatment of High Strength Mixed Phenolic Waste in a Sequencing Batch Reactor, *Wat. Env. Res.*, Vol. 64, No. 2, 128-134.

58. Malmey, S.W., Bancraft, K., pipes, W.O., Suffet, L.H. (1984), Bacterial TOC removal on sand and G.A.C., *J. Encirm. Eng*. 110, 519-533.

59. Chozick, R. and Irivine R.L. (1991), Preliminary studies on the granular activated carbon-SBBR, *Environ. Prog.*, 10 (4), 282-289.

60. Lin, W. and Weber,A.S. (1992), Aerobic biological activated carbon (BAC) treatment of a phenolic wastewater, *Environ. Prog.* 11 (2), 145-154.

61. Ha, S.R., Oishan, L., Vinitnantharat, S. (2000), COD removal of phenolic wastewater by biological activated carbon- sequencing batch reactor in the presence of 2,4, dichlorophenol, *Wat. Sci. Tech.*, Vol. 42, No. 5-6, 171-178.

62. Zitomey, D.H. and speece, R.E. (1993), Sequencing environmental for enhanced biotransofrmation of aqueous contaminants, *Environ. Sci. Tech.*, 27 (2), 227-244.

63. Seshadri, S., Bishop, L.P., Agha, M.A. (1994), Anaerobic aerobic treatment of selected azodyes in wastewater, *Waste Management*, 14 (2), 127-137.

64. Zissi, V. and Lyberators, G. (1996). Azodye biodegradation under anoxic conditions, *Wat. Sci. Tech.*, 34 (5-6), 495-500.

65. Cruz, A. and Buitron, G. (2000), Biotransformation of disperse blue 79 by an anaerobic sequencing batch bio-filter, *Wat. Sci. Tech.*, Vol. 42, 5-6, 317-320.

66. Beydilli, M. I., Parilatahis, S.G., Timcher, W. C. (1998), Decolourization and toxicity screeping of selected reactive azo dyes under methanogenic conditions, *Wat. Sci. Tech.*, 38 (4-5), 225-232.

67. Lourenco,N.D., Novais, J.M., Pinheir, H.M. (2000), Reactive textile dye colour removal in a sequencing batch reactor, *Wat. Sci. Tech.*, vol.42, No 5-6, 321-328.

68. Quezada, M. Linarer, I., Bwitron,G. (2000), Use of sequencing batch bio-filter for degradation of azo dyes, *Wat. Sci. Tech.*, Vol. 43, No 5-6, pp-329-336.

69. Mangat, S.S. and Elefsiniotis, P. (1997), Biodegradation of the herbicide 2,4-dichlorophenoxy acetic acid (2,4-D) in sequencing batch reactor, *Wat. Res.*, Vol.33, pp. 861-869.

70. Bortone, G., Gemelli, S., *et al.* (1992), Nitrification and Dentrification and Biological Phosphorus Removal in SBR Treating Piggery Wastewater, *Wat. Sci. Tech.*, 26 (5-6), 977-985.

71. Edgertor, E. D., Mcnevin, D., Wong, C. H., Menoud, P., Barford, J. P., Mutchell, C. A. (2000), Strategies for Dealing with Piggery Effluent in Australia- the SBR as a Solution, *Wat. Sci. Tech.*, Vol. 41, No. 1, 123126.

72. Nguyen, A. L., Sheldon, J. B., Sheppard, J. D. (2000), Application of feed back control based on dissolved oxygen to a fixed film SBR for treatment of brewery wastewater, *Wat. Environ. Res.*, Vol. 72, No. 1, 75-83.

73. Franta, J. R. and Wilderer, P. A. (1997), Biological Treatment of Wastewater from Paper Mill by SBR Technology to Reduce Residual Organics, *Wat. Sci. Tech.*, Vol. 35, No. 1, 129-136.

74. Yamamoto, K., Win, K. M. (1991), Tannery wastewater treatment using a sequencing batch membrane reactor. *Wat. Sci. Tech.*, Vol. 23, 1639-1648.

75. Accasthian, J. V., Smets, B.F., Kim, B. J. (2000), Aerobic biodegradation of nitroglecerin in a sequencing batch reactor. *Wat. Environ. Res.*, Vol. 72, No. 4, 499-504.

76. Pesari, H. and Grasso, P. (1993), Biodegradation of an inhibitory non-growth substrate (nitroglecerin) in batch reactors, *Biotech. Bioeng.*, 41, 79-89.

77. ATV (1997), Belebungsandlagen in Aufstaubetrieb (Activated sludge variable volume reactors), Merkbaltt ATV-M210, Hennef: Abwassertechnische Vereinigung.

Chapter 7

Kinetics and Hydrrodynamic Aspects of the Removal of Pollutants from Wastewater in Electrochemical Reactors

L. Szpyrkowicz

University of Vience, Italy

Introduction

Research on the use of electrochemical methods for the treatment of industrial wastewater has demonstrated that many pollutants, organic and inorganic, can be effectively removed by anodic oxidation or cathodic reduction. Electrochemical reactors have been successfully used since the early '60s for the recovery of pollutants from metal plating factories, for silver recovery from photographic baths and in other processes of metal recovery (Rajeshwar *et al.*, 1994). Among various applications of electrochemical processes, those in which oxidation of pollutants can be achieved is particularly important, as it can offer an alternative to traditional chemical oxidation. Thus electrochemical processes have been considered recently for other environmental applications, ideally requiring total combustion of organic pollutants to CO_2 and H_2O or modification of their structure, resulting in improved biodegradability (Comninellis, 1992; Simond *et al.*, 1997; Foti *et al.*, 1999). Application of electrochemical reactors proved feasible for the purification of wastewater from tanning industry (Naumczyk *et al.*, 1996), of wastewater containing cyanides (Ho *et al.*, 1989), phenols (Comninellis and Negrini, 1995), chlorophenols (Polcaro and Palmas, 1997), wastewater containing dyes (Szpyrkowicz *et al.*, 2001a) and deriving from dye production (Abdo and Al-Ameeri, 1987) and others (Murphy *et al.*, 1982). It results that the composition of the wastewater, and in particular the type of pre-treatment it had undergone before electro-oxidation, can influence strongly the performance of electrochemical oxidation. For example studies conducted at the University of Venice (Szpyrkowicz *et al.*, 1994, Szpyrkowicz *et al.*, 1995) with graphite, Ti/Pt, Ti/Pt-Ir, Ti/RuO_x, Ti/PbO_2 and other anode materials proved that purifying tannery wastewater by an electrochemical process can be efficient, particularly if the process is applied for the pre-treated effluents. The rate of removal of a single polluting species proved to be a function of the electrocatalytic properties of the anode material.

In fact, the type of the oxidative process involved in the removal of a pollutant is a function of the material of the anode and application of different anodic materials with different electrocatalytic properties can affect reactor treatment efficiencies. For electrochemical processes occurring via a direct reaction of the pollutant at the electrode surface a term *direct electrolysis* is generally used. The term *indirect electrolysis* refers to those processes in which a redox couple, whose oxidised form is being continuously renewed by an anodic heterogeneous reaction, is involved in a homogeneous phase reaction with the pollutant. The reactors which operate by applying direct or indirect electrochemical oxidation (or reduction) are influenced by stirring in a completely different way.

Electrochemical Reactors and Direct Electro-oxidation

Figure 7.1 explains the concept of a direct electrochemical reaction applied for the pollutant abatement. The graph shows the anodic process as an example transfer are linked in series. At low overpotentials transport rates are normally fast compared with the electrochemical reaction kinetics and electron transfer is the rate determining step (Figure 7.2 (Pletcher and Walsh, 1993)). Current density (reaction rate) increases exponentially with overpotential, which leads to the reactant being progressively depleted at the interface and a concentration gradient in the boundary layer is

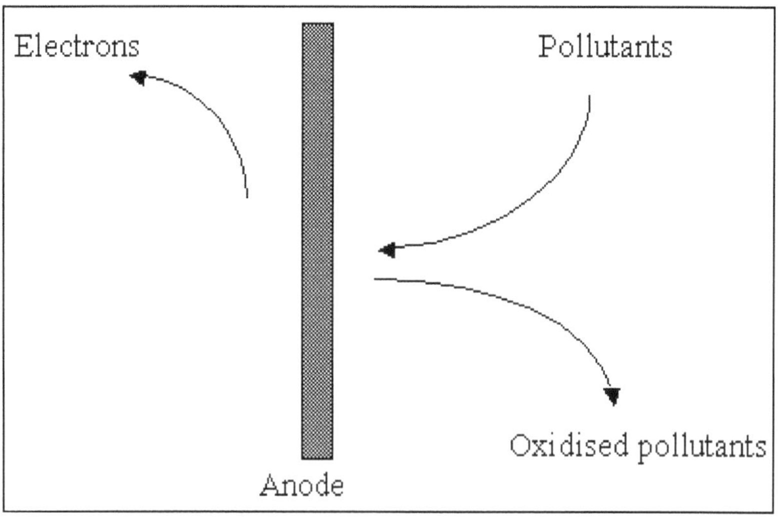

Figure 7.1: The Scheme of the Removal of Pollutants by Direct Electro-oxidation.

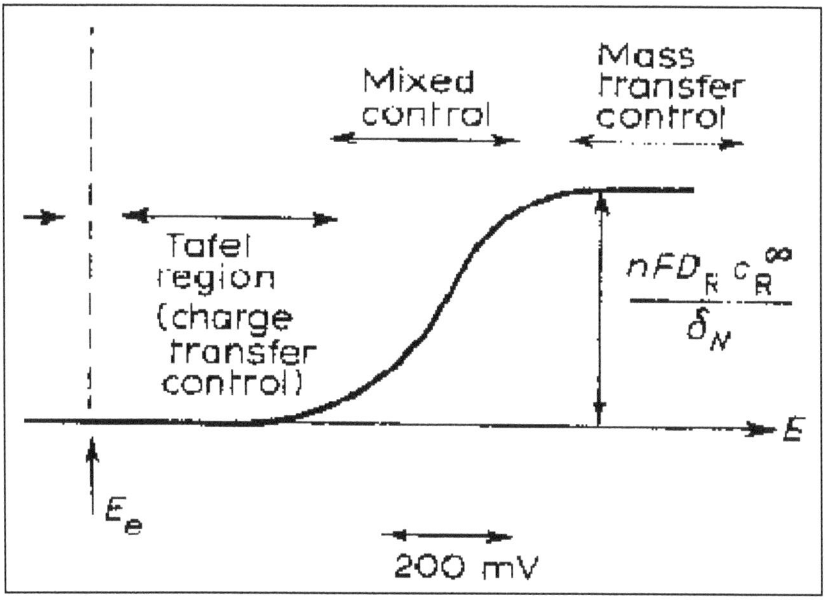

Figure 7.2: A Current-Potential Curve for the Positive Region.

developed. As an overpotential increases, the rates of electron transfer and mass transport are comparable and both processes control the rate of the overall reaction (the region of the mixed control). As the overpotential increases further, the surface concentration of electro-active species tends to zero and the current density becomes independent of the potential; the reaction rate is wholly determined by the mass transport rate.

Under the conditions of the complete mass transport control, assuming that migration of the active species in the electric field is negligible (an assumption valid when the concentration of the base electrolyte is high), the limiting current i_L expresses the duty of the reactor:

$$i_L = k_L\ AnFc^{\infty}$$

where,

i_L–limiting current (A); A–electrode surface area (m^2); n–number of participating electrons; F–Faraday constant (96490 C mol^{-1}); c¥–bulk concentration of the reacting species (mol m^{-3}).

Thus, assuming the reactor being a batch or a plug-flow type reactor, the observed change of the concentration of the reactive species, obtained from the mass balance, is given by:

$$\frac{dc_t}{dt} = i_t / (nFV_R)$$

where,

i_t is the instantaneous current at time t; V_R is the working volume of the reactor.

Under the complete mass transport control $i_t = i_L = k_L AnFc_t$. Substituting, the following expression is obtained:

$$\frac{dc_t}{dt} = k_L\ Ac_t / V_R$$

From the last equation the importance of the mass transfer and of the highly developed electrode surface area in electrochemical reactors operating with direct electrochemical oxidation of pollutants is evident. One way to fulfil this requirement is to operate with three-dimensional electrodes. This solution, however, has also several drawbacks, *e.g.*:

- The difficulty to obtain a uniform potential and current distribution;
- An easy entrapment of gas bubbles (H_2 or O_2), which are likely to evolve simultaneously with the main reactions, while dilute solutions are treated (as is often the case during treatment of wastewater), leading to an increase of the cell resistance and electrical energy consumption;
- The high pressure drop and tendency to clogging.

Consequently often parallel plate-type cells are used to assembly the electrochemical reactor. The preference to use plate electrodes derives from:

- Ease of their manufacture and assembly;
- Their versatility for single-phase or multi-phase processing;
- Adaptability to monopolar or bipolar operation;
- Relative uniformity of current distribution;
- Flexible modularity (additional cells can be easily added to the stack);
- Facility to incorporate membranes, if required.

Current/mass transfer correlations for the reactors with parallel plate electrodes can be defined from the first principles. The general form of this correlation is:

$$Sh = a\mathrm{Re}^b\ Sc^c\ L^d$$

where,

Sh–Sherwood number ($Sh = k_L d_e/D$)

Re–Reynolds number ($Re = vd_e/h$)

Sc–Schmidt number ($Sc = \eta/D$)

L–length of the electrode in direction of flow (m).

The symbols in the dimensionless groups are: d_e–equivalent diameter of flow channel (m); D–diffusion coefficient of electroactive species ($m^2\ s^{-1}$); v–mean linear velocity of electrolyte ($m\ s^{-1}$); η–kinematic viscosity of the electrolyte ($m^2\ s^{-1}$).

The *a, b, c* and *d* are coefficients, whose values differ in function of the hydrodynamic regime and the cell geometry.

It can be shown (Pletcher and Walsh, 1993) that the limiting current and the Sherwood number are related via the mass transfer coefficient, so we have:

$$Sh = i_L d_e / nFADc^{\infty}$$

As a direct consequence of this relation, the duty of the parallel plate reactor is supposed to increase exponentially with turbulence. In particular, for the plug-flow reactor the duty would be a function of the volumetric flow rate of electrolyte, with the exponent equal to *b*.

Table 7.1 reports some values of the coefficients of the mass transport correlation for electrolyte flow past a flat plate in a rectangular channel (Picket, 1979).

Table 7.1: Mass Transport Correlation Coefficients for Electrolyte Flow Past a Flat Plate in a Rectangular Channel

$Sh=aRe^bSc^cLe^d$				*Conditions*
a	*b*	*c*	*d*	
2.54	0.3	0.3	0.3	Fully developed laminar flow; $58<Re<2000$; $2850 < Sc < 51400.17 < L_e<12.5$; S/B 0 0.167 or 0.75
1.85	1/3	1/3	1/3	Theoretically derived for fully developed laminar flow assuming infinitely wide electrodes; $ReSc\ (d_e/L)>10^4$ and $L_e<35$
1.85x	1/3	1/3	1/3	x corrects the correlation for for a finite width between the electrodes. $ReSc\ (d_e/L)>10^4$ and $L_e<35$
0.023	0.8	0.3	–	Fully developed turbulent flow; $2000<Re<20000$ and $L_e>12.5$
0.145	2/3	1/3	1/4	Fully developed turbulent flow.$2000<Re<20000$ and $L_e<7.5$

The following graph (Figure 7.3) gives an example of the performance of the eletrochemical batch reactor with parallel plate anodes operating with a complete recycle under different volumetric flow rate conditions (Ricci, 2002).

Recently an increase of the number of commercial electrochemical reactors based upon the parallel plate geometry is witnessed. Particularly the filterpress arrangement achieves great attention (Picket and Wilson, 1982; Schwager *et al.*, 1980; Stork and Hutin, 1981; Trinidad and Walsh, 1996), due to a possibility of realising high mass transfer rates, enhanced by an implementation of the promoters of turbulence. The enhancement of the mass transport rate can be of a factor of two, in function of the geometry of the promoters (Ralph *et al.*, 1996). However, whilst the expanded plastic mesh turbulence promoters do provide a simple and cost effective method of enhancing mass transport, they increase the pressure drop over the reactor, thereby increasing the pumping costs, and can shield much of the electrode surface producing a non-uniform potential distribution. Furthermore they can also give rise to gas blinding and cause flow segregation. As a consequence the mass transfer rates can even be decreased with respect to those for the unmodified channels, as reported by Wragg and Leontaritis (1997).

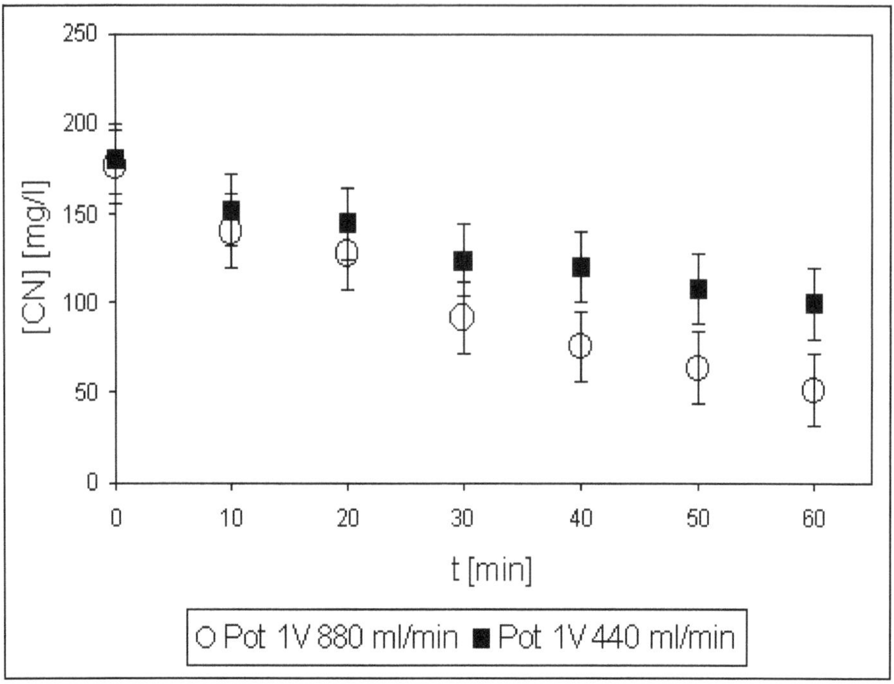

Figure 7.3: Time Trend of Cyanide Concentration during Electrolysis of Copper Complex Cyanide Solution.

If the electrodes working potentials are close to those of water electrolysis, differences in local current/voltage distributions, existing over the electrode surface area, can result in gas evolution (O_2 and H_2). This, in turn, can produce a modification of the hydrodynamic boundary layer at the electrode surface, as the rising bubbles generate radial momentum, which locally destroys the layer and induces mass renewal by bringing fresh solution to the edge of the hydrodynamic boundary layer.

Electrochemical Reactors and Indirect Electro-oxidation

In the process of indirect electro-oxidation of pollutants an electrochemically generated redox reagent is used as a chemical reagent (or catalyst) to convert pollutants to less harmful products, or even to CO_2 and H_2O. The redox reagent acts as an intermediator to transport electrons between the pollutant and the electrode. The general scheme of this reaction is:

$$C \rightarrow C^{+} + e^{-}$$

$$C^{+} + R \rightarrow O + C$$

Figure 7.4 depicts schematically the electrode and bulk reactions involved.

The following conditions should be verified for the indirect electro-oxidation to proceed with a high efficiency:

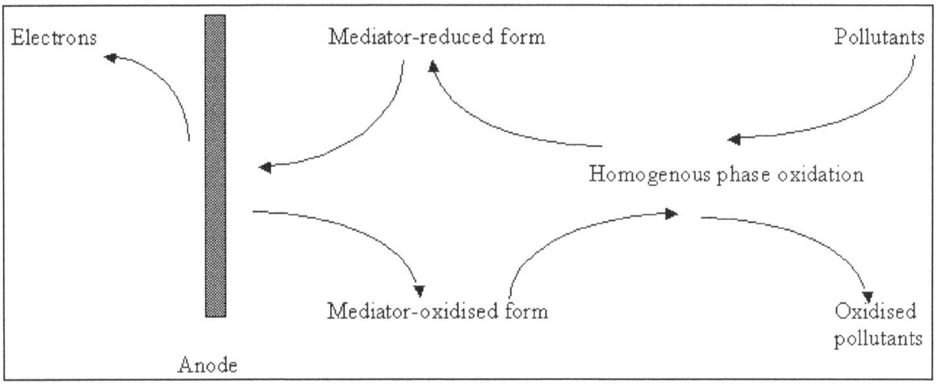

Figure 7.4: The Scheme of the Removal of Pollutants by Indirect Electro-oxidation with an Anodic Regeneration of the Mediator.

- The rate of production of the mediator should be high;
- The rate of the reaction of the mediator with a given pollutant should be higher than the rate with other compounds eventually present.
- The potential at which electrogeneration of the mediator occurs should be distant (lower) from the potential of solvent electrolysis, to prevent lost of energy for an unproductive process;
- The electrode should not be prone to adsorption of the pollutant, to avoid occupation of the active sites, otherwise free for the reaction of the generation of the active form of the mediator.

In fact, indirect electrochemical processes offer a promising alternative solution for destruction of pollutants in industrial wastewaters. They can be particularly attractive if conducted in undivided electrochemical reactors, which can operate at lower cell voltages, and are simpler and cheaper to engineer than their counterparts incorporating ion-permeable membranes, separating anolyte and catholyte. Especially for treatment of low concentrations of pollutants in industrial wastewaters, indirect electrochemical oxidation, mediated in bulk solution by a redox couple with electrochemical regeneration of its oxidised form, is generally preferred to direct electro-oxidation. The most often applied mediators are listed in Table 7.2 (Rajeshwar *et al.*, 1994).

Table 7.2: Redox Mediators Applied in Indirect Electro-oxidation of Pollutants and their Standard Potentials

Mediator Couple	*Standard Reduction Potential (vs SHE)*
Ag (I/II)	1.98
Co (II/III)	1.82
Ce (IV/III)	1.44
Cl (0/-I)	1.36
Fe (II/III)	0.77

Among the redox couples indicated in Table 7.2 the chlorine species are of particular importance, due to the presence of chloride in several types of industrial wastewater and the lack of need for its recovery from the purified effluents. This last advantage makes this mediator particularly competitive with Ag and Co and offers a possibility to conduct the process at low operation costs.

Indirect electro-oxidation of pollutants in wastewater by electrogenerated chlorine has been used widely (Szpyrkowicz et a., 1994; Szpyrkowicz *et al.*, 1995; Do and Yeh, 1995; Comninellis and Nerini, 1995, Bonfanti *et al.*, 2000), because of the low cost of chloride and the relatively high solubility (ca. 0.1 M–Boxall and Kelsall, 1992) and strongly oxidising properties of active chlorine.

The influence of hydrodynamic conditions in a mediated electro-oxidation of pollutants is much more complex than during direct electro-oxidation. The dependence of the reaction rates on stirring is not straightforward and cannot be obtained from the first principles.

Electrochemical reactors, which apply the indirect electro-oxidation via a Cl_2/Cl^- couple, operate at potentials at which two reactions generally can occur at the anode:

$$2H_2O \rightarrow O_2 + 4H^+ + 4e^-$$

$$2Cl^- \rightarrow Cl_2 + 2e^-$$

Whereas aqueous chlorine predominates at pH < *ca.* 3.3, at higher bulk pHs, its diffusion away from the electrode is coupled to its hydrolytic disproportionation to form chloric (I) acid (*HOCl*) at pH < *ca.* 7.5 and chloric (I) ions at pH > *ca.* 7.5:

$$Cl_2 + H_2O \rightarrow HOCl + H^+ + Cl^-$$

$$HOCl \rightarrow H^+ + ClO^-$$

Chlorine, chloric (I) acid and chloric (I) ions are collectively often referred to as "active chlorine".

The cathodic reactions involve hydrogen evolution by:

$$2H^+ + 2e^- \rightarrow H_2$$

and/or

$$2H_2O + 2e^- \rightarrow H_2 + 2OH^-$$

The rate of chlorine evolution reaction may be kinetically or transport controlled, depending on the applied current density, chloride ion concentration, and the prevailing hydrodynamic conditions, which may also depend on the rate of anodic gas evolution; hence, these parameters also control the current efficiency of chlorine evolution. For the given experimental conditions the limiting current density can be estimated from the equation:

$$j_L = nFk_L C_{Cl}$$

where,

n–number of the electrons per Cl^- oxidation (n = 1), F – the Faraday's constant (96 485 C); k_L – mass transfer coefficient for chloride ions (approximately = 10^{-5} m s^{-1}), C_{Cl} – concentration of chloride ions (mol m^{-3}).

Hydrogen evolution generates a high local pH at the cathode, so that chloric (I) ions predominate in the cathode diffusion layer; in an undivided reactor they can be reduced by the (mass transport controlled) cathodic reaction:

$$ClO^- + H_2O + 2e^- \rightarrow Cl^- + 2OH^-$$

This reaction limits the maximum active chlorine concentration that can be achieved in such reactors to *ca.* 0.1 M (Boxal and Kelsall, 1992). In addition, further loss reactions such as chlorate formation may occur in the bulk solution, but can be assumed to have insignificant rates at ambient temperatures, because the steady state chloric (I) concentration is kept low by homogeneous reactions of oxidation of pollutants.

Chlorate may also be formed by oxidation of chloric (I) acid (Boxall and Kelsall, 1992; Czarnecki and Janssen, 1992), mass transport usually controlling the rate, depending on the anode material:

$$6HOCl + 3H_2O \rightarrow 2ClO_3^- + 4Cl^- + 12H^+ + 3/2O_2 + 6e^-$$

As this last reaction and also the oxygen evolution reaction acidify the anode diffusion/reaction layer, a large pH gradient exists across the anode-cathode gap, with possible kinetic consequences for oxidation of pollutants, as the reduction kinetics of active chlorine species decrease with increasing pH.

The speciation of the different oxidising species generated during the electrolysis and their profiles inside the reactor are function of the bulk solution pH, as is shown in Figures 7.5a and b for different pHs of the bulk solution.

The oxygen evolution reaction is charge transfer controlled, as is the chlorine evolution reaction at low current densities and high chloride concentrations, but the chloric (I) loss reactions at the anode and the cathode are mass transfer controlled. Thus, increasing turbulence in the reactor promotes the transport of Cl^- ions to, and $C1_2$ (aq) away from, the anode, but also enhances the rate of chloric (I) reduction at the cathode and its oxidation at the anode.

As shown in Figures 7.5a and b, oxidative decomposition of pollutants in the electrochemical reactor under alkaline and acidic conditions involves active chlorine as mediator, constantly renewed by charge transfer reaction at the anode, involving the mass transfer of the chloride ion from the bulk to the electrode, and the transport of the chloric (I) to the bulk. With several oxidisable pollutants present during the electrolysis, fast homogeneous reactions will affect the concentration profiles in the anode diffusion/reaction layer, enhancing the chloride flux to the anode, whereas slow reactions will occur predominantly in the bulk solution.

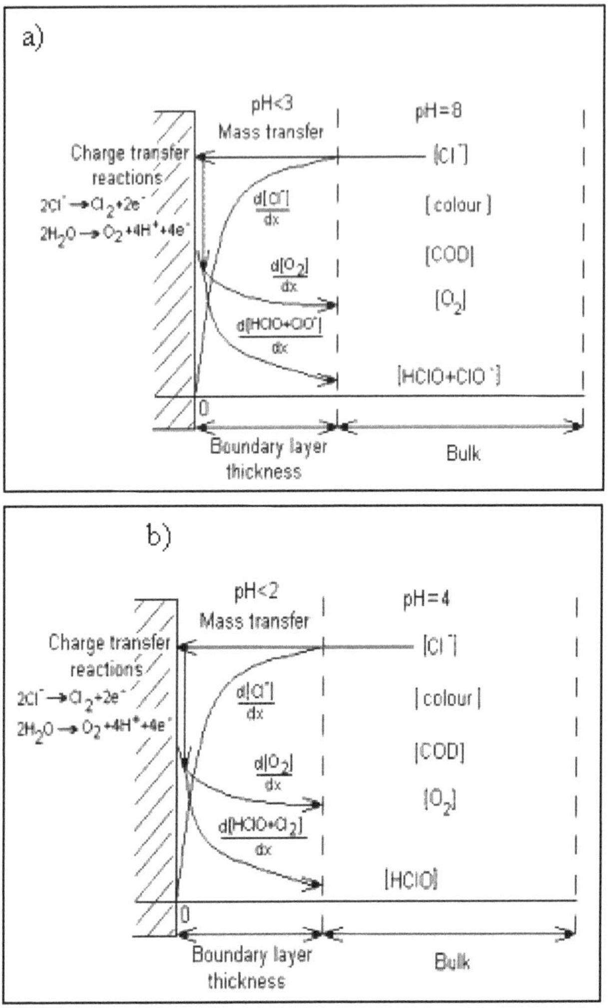

Figure 7.5: Schematic Spatial Distribution of Oxidising Species and their Concentration Profiles Inside the Reactor Under Alkaline (a) and acidic (b) Bulk pH.

A mass balance on active chlorine species, formed in the solution of a batch electrochemical reactor of volume V with current efficiency F_e at anode of area A_a and mass transport rate coefficient $k_{m,a}$, but consumed by further oxidation of chloric (I) acid at the anode, by reduction of chloric (I) ions at the cathode of area A_c and mass transport rate constant $k_{m'c}$ and by chemical reaction with pollutant R and second order rate constant k, can be represented by:

$$V_R \frac{d['Cl_2']}{dt} = \frac{\Phi j_a A_a}{nF} - A_a k_{m,a}[HOCl] - k_{m,c} A_c \left[ClO^-\right] - Vk['Cl_2'][R]$$

The anodic current efficiency F_e depends on the anode current density j_a, $k_{m,a}$ and chloride concentration. Studies conducted in our laboratory (Szpyrkowicz *et al.*, 2001b), showed that in the absence of pollutants at pH 4, current efficiencies were only 0.1 – 0.2, essentially independent of stirring rate, due to the limited solubility of chlorine and hence its desorption from solution, which would be more restricted in the presence of oxidisable pollutants, when homogeneous phase reaction of their oxidation was operative. At the bulk pH = 8, in quasi-quiescent solutions, current efficiencies were *ca.* 0.65, essentially independent of time for the restricted compositional change produced in an hour. When the solution was agitated, the apparent efficiency of chlorine production decreased from the initial value of 0.84 to 0.50 after 60 minutes, due to enhancement of the rates of loss reactions with increasing active chlorine concentration Figures 7.6 and 7.7 show the effect of charge respectively, on the decay of colour and of COD, obtained under isothermal conditions (26 °C). The reactor performance was largely dependent on current density and was influenced by solution stirring and pH; at low current density (1 A dm^{-2}), mixing appeared to have a detrimental effect, due to enhancement of the electrode loss reactions.

In principle, the behaviour of the reactor for such an electrochemical-chemical (EC) process could be modelled, the kinetics of the chemical reaction dictating where its occurs in relation to the electrode surface; fast chemical reactions affect the concentration profiles of mediator species within the electrode diffusion/reaction layer and hence, in this system, the mass transport limited rate of, and current efficiencies for, chloride oxidation (*e.g.* Scott, 1991). Figures 7.6 and 7.7 indicate that at least the initial (< 2 kC) rate of the removal of colour and COD could be described by a pseudo-first order reaction rate. The Hatta number was calculated from:

$$Ha = \frac{\sqrt{k_1 D_{Cl_2}}}{k_L}$$

where,

k_1 – an observed pseudo-first order reaction rate constant (s^{-1}); k_L – mass transport rate coefficient of chlorine (m s^{-1}). The value of D_{Cl2} was assumed equal to 10^{-9} m^2 s^{-1} (Hikita *et al.*, 1975) and that of k_L to 10^{-5} m s^{-1} (Lin and Zhang, 2001). For the specific case of electro-oxidation of disperse dyes (Szpyrkowicz *et al.*, 2001b) *Ha* was equal to *ca.* 0.1. This value of Ha implies that the chemical reaction occurred both in the electrode diffusion layer and in the bulk solution. At low conversions, increased agitation rates enhanced mass transport rates both of the chloride ions to the anode and chlorinated intermediates away from the anode. However, at longer times, increased adsorption densities of those intermediates tended to partially inhibit Cl_2 evolution, as evident from voltammetric tests shown in Figures 7.8a and b. As a consequence, a sharp decrease in the decolorisation rate after a charge ca. 2 kC had been passed was observed, as shown in Figure 7.6. Thus, it can be concluded that under alkaline conditions turbulence initially favours decolorisation, but as intermediates accumulate with time, the benefits of mixing decreases until no difference between quasi-quiescent and agitated conditions is observed.

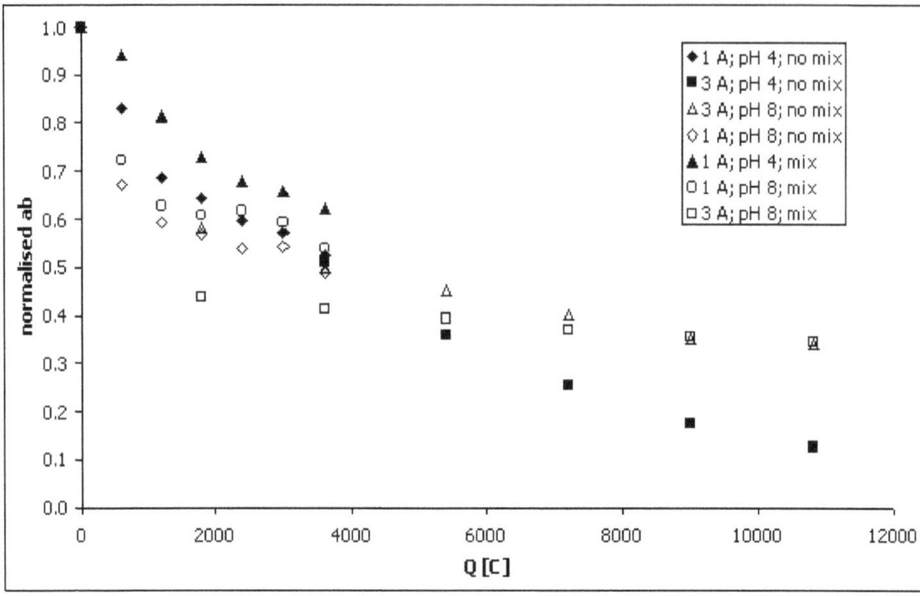

Figure 7.6: Reactor Performance during Removal of Colour under Isothermal Conditions (26ºC).

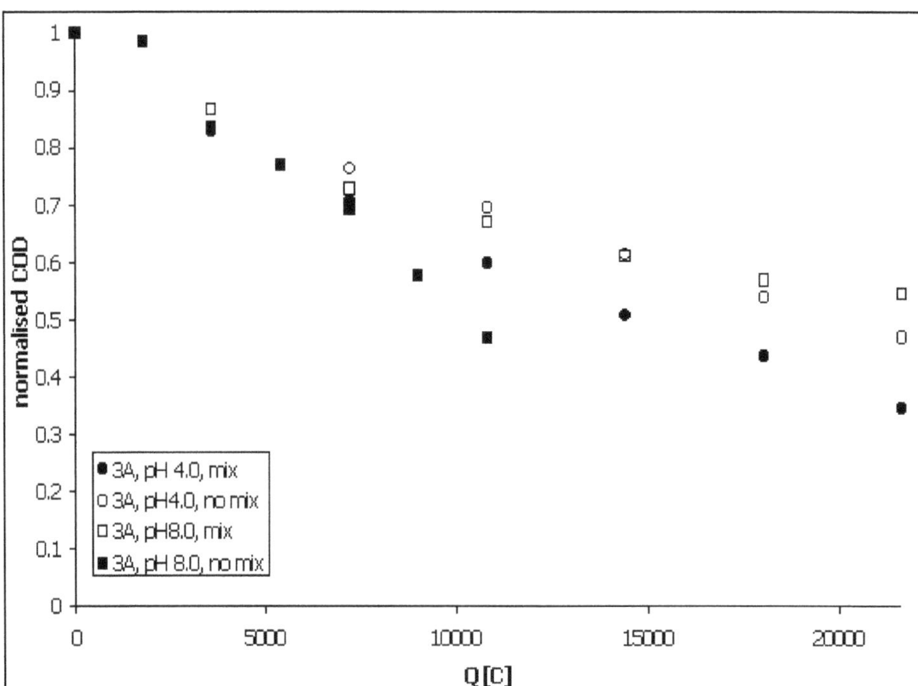

Figure 7.7: Reactor Performance for Removal of COD under Isothermal Conditions (26ºC).

As shown in Figure 7.7 removal of COD showed different behaviour, with the highest removal rate being observed under alkaline, quasi-quiescent conditions. From the performance of the reactor it can be concluded that the degree of stirring influences the selectivity of the reactor towards the removal of colour and COD. This can be explained by promotion of mineralisation of the intermediates by direct anodic oxidation, for which there was some voltammetric evidence (see Figure 7.8). This

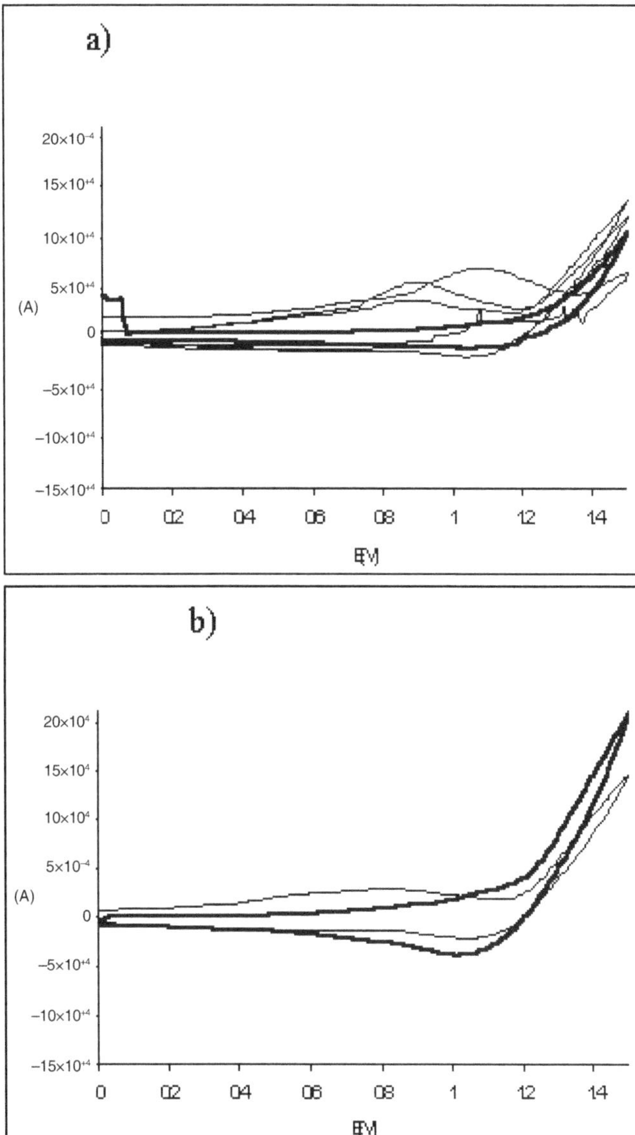

Figure 7.8: Cyclic Voltammograms Recorded at the Ti/RhO$_x$-TiO$_2$ Electrode in a Solution Containing Disperse Dyes; Supporting Electrolyte 0.1 M NaClO$_4$ and 0.1 N NaCl; [a] pH 8; [b] pH 4.0; scan rate 100 mV s^{-1}.

mechanism can explain far better results obtained by electrochemical chlorine mediated oxidation in comparison to homogeneous phase chlorination under neutral or slightly alkaline conditions (Szpyrkowicz *et al.*, 2001a). It can be hypothesised that under acidic conditions, chlorine mediated oxidation of the dyes is acid catalysed, by analogy to other aromatic compounds (Rebenne *et al.*, 1996).

Figures 7.9 and 7.10 illustrate (Szpyrkowicz *et al.*, 2001b) the reactor performance during treatment of dyes-bearing wastewater, in terms of colour and COD removal, respectively, for five agitation rates and in the absence of temperature control, a more common and cheaper practical condition in a batch reactor than isothermal operation. The pH started from about 7, increased to about 9, then stabilised at about 8.5. Quasi-quiescent conditions produced the highest depletion rates of both the colour and COD. The rate of pollutant removal decreased with increasing Reynolds number. Electrolyte temperatures increased from 19° to 26° C at 100 A m^{-2} and 19° to 36° C at 300 A m^{-2}, due to Joule heating, the effect being greater in the absence of additional agitation. The increased temperature caused both the electrode and chemical reaction rates to rise. Hence, when the electrochemical reactor was operated under isoperibolic conditions, with chlorine mediated electro-oxidation being primarily responsible for the destruction of pollutants, it performed better without additional agitation.

Further studies on the influence of stirring during chlorine mediated electro-oxidation of dyes (Moro, 2002), showed that the magnitude of the effect of agitation in an isothermal reactor is comparable with the temperature effects. The pseudo-first order rate constant for electro-oxidation of the Procion Red Rective Dye HEXGL, normalised by the specific surface area of the anode (A/V_R) is shown in Figures 7.11

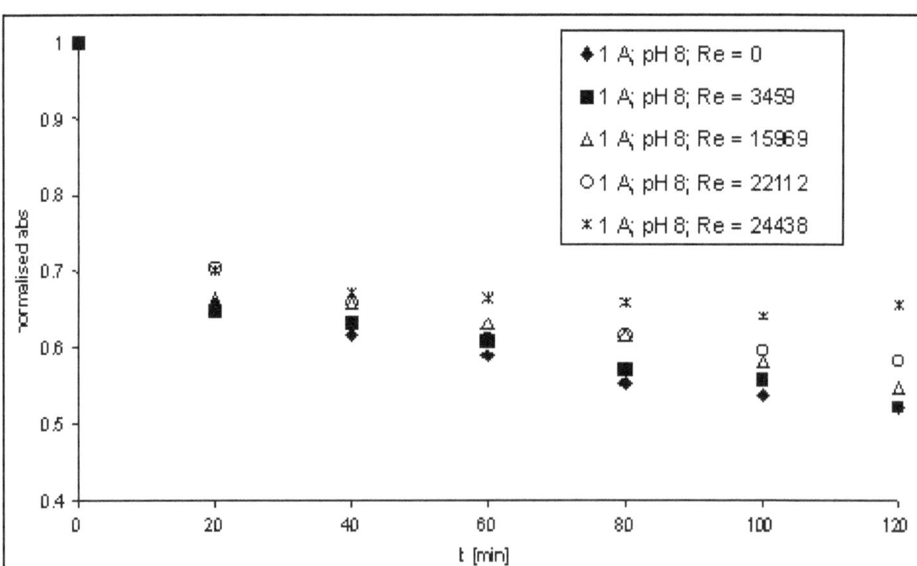

Figure 7.9: Performance of the Reactor under Isoperibolic Conditions during the Removal of Colour.

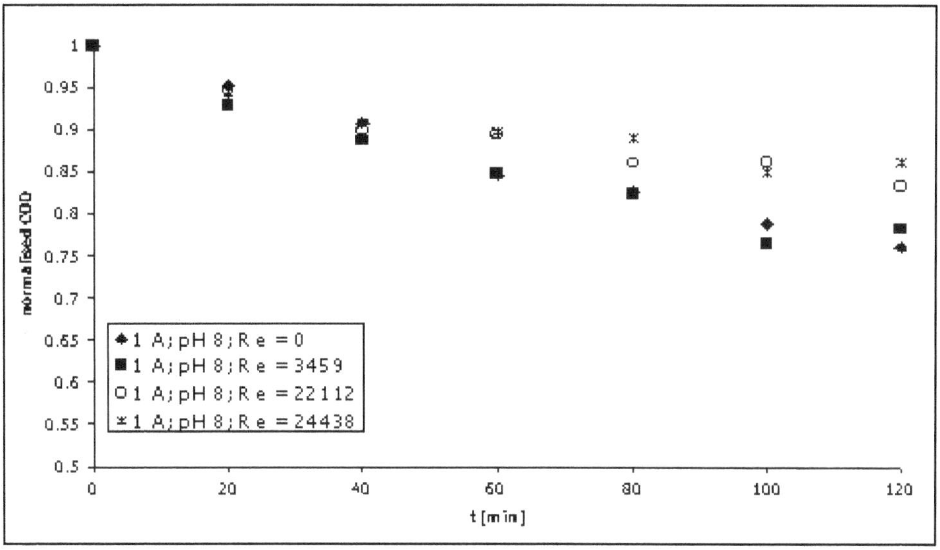

Figure 7.10: Performance of the Reactor under Isoperibolic Conditions during the Removal of COD.

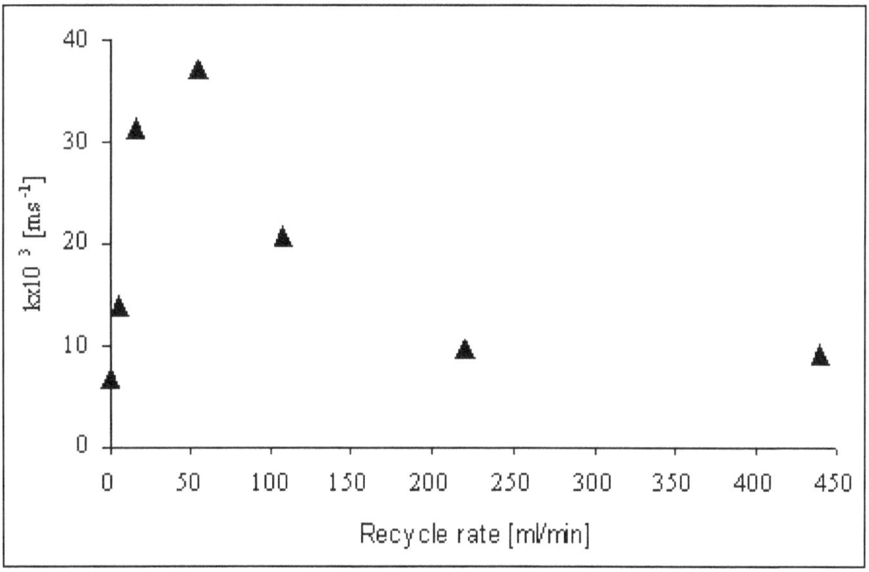

Figure 7.11: Variation of the Kinetic Rate Constant (Normalised by the A/V$_R$ ratio) in Function of the Internal Recycle Rate in the Reactor.

and 7.12, respectively as a function of the agitation (internal recycle) and the temperature of operation.

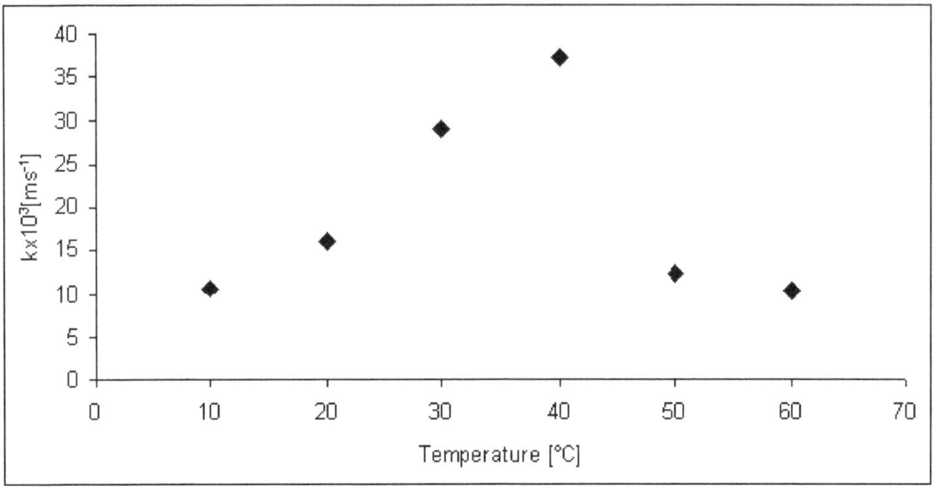

Figure 7.12: Variation of the Kinetic Rate Constant (Normalised for the A/V$_R$ ratio) in Function of the Temperature of Operation of the Reactor.

Conclusion

Direct Electro-oxidation of Pollutants

Stirring results in enhancing the mass transport rates, which increases the pollutants removal rates; this effect can be predicted from the first principles and is expresses by the correlation of the Sherwood–Gilliland type.

Indirect Electro-oxidation of Pollutants

Enhancement of mass transport rates in the reactor by stirring can have either positive or negative effect on the abatement of pollutants, as indirect electro-oxidation involves a complex interaction between electron transfer, mass transport and homogeneous chemical kinetics. In particular the effect of stirring depends whether the electrode reaction of the mediator is under charge or mass transport control. The rate constants for the homogeneous process affect whether the chemical oxidation occurs within the anode reaction layer or in the bulk solution. As a consequence of a complex nature of the involved processes not only the rates of removal of pollutants but also the selectivity of the reactor depends on the hydrodynamic conditions and can be changed by varying the stirring rate.

In particular, the effect of stirring:

- Cannot be predicted from the first principles as the scenario is very complex;
- Depends if the pollutant (or the intermediate formed during its oxidation) can undergo adsorption/oxidation at the anode interface; cyclic voltammetry study is a useful tool to predict the behaviour;
- Depends on the Hatta number (for high values, being the reaction confined to the reaction boundary layer near the electrode, stirring can have a positive effect);

- ✰ Depends if the reactor operates under isothermal, isoperibolic or adiabatic conditions, as agitation affects the temperature profiles and, consequently, the reaction rates in different zones;
- ✰ Its magnitude, in an isothermal reactor, is comparable with the temperature effects.

References

1. Abdo M. S. E. and Al-Ameeri R. S., Anoxic oxidation of a direct dye in an electrochemical reactor. *J. Environ. Sci. Health*, A22, 27-42 (1987).
2. Bonfanti F., De Battisti A., Ferro S., Lodi G. and Osti S., 2000, Anodic mineralization of organic substances in chloride-containing aqueous media, *Electrochimica Acta,* 46, 305–314.
3. Boxall, C. and Kelsall, G.H., 1992, Hypochlorite electrogeneration II. Thermodynamics and kinetic model of anode reaction layer, *Electrochemical Engineering, IChem. E. Symp. Series,* I.Chem.E., Rugby, UK, 127, 59-70.
4. Comninellis, Ch. Electrochemical treatment of wastewater containing phenol. *Trans. IChemE,* 1992, 70, Part B, 219.
5. Comninellis, Ch. and Nerini, A., 1995, Anodic oxidation of phenol in the presence of NaCl for wastewater treatment, *J.Appl.Electrochem.,* 25, 23–28.
6. Czarnetzki, L.R. and Janssen, J.J., 1992, Formation of hypochlorite, chlorate and oxygen during NaCl electrolysis from alkaline solutions at an RuO_2/TiO_2 anode, *J. Appl. Electrochem.,* 22, 315–324.
7. Do, J.-S. and Yeh W.-C., 1995, In situ degradation of formaldehyde with electrogenerated hypochlorite ion, *J.Appl.Electrochem.,* 25, 483–489.
8. Foti, G.; Gandini, D.; Comninellis, Ch.; Perret, A.; Haenni, W. Oxidation of organics by intermediates of water discharge on IrO_2 and synthetic diamond anodes. *Electrochemical and Solid-State Letters* 1999, 2, 228.
9. Hikita H., Asai S., Ishikawa H. and Saito Y., 1975, Kinetics of absorption of chlorine in aqueous acid solutions of ferrous chloride, *Chem.Eng.Sci.,* 30, 607 – 616.
10. Ho S. P. Y., Wang C. and Wan C., 1989, Electrolytic decomposition of cyanide effluent with electrochemical reactor packed with stainless steel fiber, *Wat. Res.,* 24, 1317–1321.
11. Lin C. and Zhang J., 2001, Modeling of hydrazodicarbonamide oxidation by chlorine in a gas-liquid-solid system, Chemical Engineering Science, 56, 667 – 671.
12. Moro S., Electrochemical treatment of textile wastewater bearing the dyes, MS Thesis, 2002, University of Venice (in Italian).
13. Murphy, O., Hitchens, G. D., Kaba L. and Verostko C. E., Direct electrochemical oxidation of organics for wastewaters treatment, *Wat. Res.,* 26, 443-451 (1992).

14. Stork A. and Hutin B., Mass transfer and pressure drop performance of turbulence promoters in electrochemical cells, *Electrochimica Acta,* 1981, 26, 127–137.

15. Szpyrkowicz, L., Naumczyk, J., Zilio-Grandi, F., 1994, Application of electrochemical processes for tannery wastewater treatment *Toxicol. Envirn. Chem.* 44, 189-202.

16. Szpyrkowicz, L., Naumczyk, J., Zilio-Grandi F., 1995, Electrochemical treatment of tannery wastewater using Ti/Pt and Ti/Pt/Ir electrodes *Wat. Res.* 29, 517-524.

17. Szpyrkowicz L., Juzzolino C., Kaul S. N., Daniele S., 2000, Electrochemical oxidation of dyeing baths bearing disperse dyes, *Ind. Eng. Chem. Res.*, 39, 3241–3248.

18. Szpyrkowicz L., Juzzolino C., Kaul S. N., A Comparative study on oxidation of disperse dyes by electrochemical process, ozone, hypochlorite and Fenton reagent, *Water Research,* 2001a, 35, 2129–2136.

19. Szpyrkowicz L., Cherbanski R., Molga E Kelsall G. H., 2001b, Effect of mixing on mediated electrochemical oxidation of disperse dyes in an undivided electrochemical reactor, ICheaP-5 (*Italian Conference on Chemical and Process Engineering*), Florence, 73 – 79.

20. Trinidad P. and Walsh F. C., 1996, Hydrodynamic behaviour of the FM01-LC reactor, *Electrochimica Acta,* 41, 493–502.

21. Wragg A. A. and Leontaritis A. A., Local mass transfer and current distribution in baffled and unbaffled parallel plate electrochemical reactors, *Chemical Engineering Journal,* 66, 1–10 (1997).

Chapter 8

Control of VOCs Using Bio-filters

S.N. Upadhyay

Department of Chemical Engineering and Technology,
Centre of Advanced Study,
Indian Institute of Technology (Banaras Hindu University),
Varanasi

Introduction

Bio-filter is a type of supported film bioreactor used for removal of water soluble biodegradable volatile organic compounds (VOCs), hazardous air pollutants and odorous compounds, present in air emissions at very low to moderate concentrations.

Bio-filteration is more economical and efficient as compared to other contemporary control technologies. It not only removes the pollutants from polluted air but destroys them completely. This is one of the attractive features that provide bio-filteration an edge over other technologies. Unfortunately, this technology has not received the required attention of Indian researchers. Only a few organizations of India – NEERI, Nagpur; IT-BHU; RRL Thiruvananthapuram; and IIT Roorkee are engaged in the R&D activities in this area.

The bio-filteration process occurs in a bio-filter, a kind of supported biofilm reactor. These bioreactors are characterized by microorganisms being attached to a porous solid surface in the form of a thin biofilm. Biofilms are dense layers of microorganisms characterized by their ability to adhere to a solid medium and form a fixed layer of biofilm in which the microorganisms are protected against sloughing off. The VOCs consuming microorganisms immobilized on the solid surface of packing media of the bio-filter from the biofilm. Peat, compost, volcanic rock, wood pieces, other natural solid biomass and some synthetic packing media, such as polystyrene beads and polyurethane foam, have been used as packing materials in bio-filters. Every packing media has some advantages and disadvantages (Upadhyay and Kumar, 2005).

In bio-filteration process, polluted air is purified during its passage through the wet bio-filter media. The pollutant present in the air diffuses to the biofilm where aerobic microorganisms consume it through their metabolic activities and produce innocuous products like H_2O, CO_2, etc. The bio-filter is fed with regular supply of nutrient solution for the growth of microorganisms. This also maintains proper moisture content inside the filter media (Upadhyay and Kumar, 2005).

Selection of suitable packing media is an important step in the bio-filter technology. In view of this, the research work was planned with the objectives of selection and characterization of locally available cheap materials as bio-filter media, evaluation of the performance of the bio-filters using the selected media under varying process parameters, determination of the optimal operating conditions for the bio-filter and study of the reactor stability and durability and mechanical strength of selected media. Here a summary of the results obtained is presented. The detailed results are given elsewhere (Singh, 2007; Singh *et al.*, 2004 a, b; 2006 a,b).

Experimental

Two agro-wastes namely yellow gram stalk and maze cobs and one industrial scrap – polyurethane foam, a locally available cost effective polymeric material, were selected as the bio-filter media. Benzene and toluene were selected as the target pollutants because these are the most common and extensively used VOCs in various industries as raw material or as solvent. These have been also characterized as one of the 188 toxic pollutants according to United State Environmental Protection Agency (US-EPA). These are easily available and have been widely used by other workers hence extensive data are available for comparison.

To judge the suitability of selected materials as bio-filter media, these were characterized and their various physical and chemical properties such as density, porosity, adsorption properties and moisture retention capacity were determined. These materials were also tested against possible hostile conditions such as highly acidic and basic environment, high concentration of pollutants, etc. The results show that the selected media have most of the favourable properties such as high porosity, low bulk density, high water retention capacity and resistant to chemicals.

To know the flow behaviour of packed bed, the residence time distribution (RTD) experiments were conducted and results confirmed plug flow behaviour of the packing. After characterization, continuous biodegradation experiments were performed in three laboratory scale bio-filters, at varying operating conditions to evaluate and optimize the bio-filter performance. Continuous biodegradation experiments were conducted in four different phases. In the first set of experimental runs, yellow gram stalk, maze cobs and polyurethane foam were used as bio-filter media and toluene as the target pollutant. In the last set of runs, polyurethane foam was used as bio-filter media and benzene and toluene as the target pollutants.

Pressure drop across the bed, temperature of bed, moisture content of the inlet stream and pH of leachate collected from the bed were also monitored continuously, which provided valuable information about mechanical strength of packing, biomass growth, excessive bio-oxidation, wet or dry conditions in the bed and type of microorganisms present in the bio-filter.

Biodegradation of toluene vapor was investigated in a laboratory scale bio-filters using cylindrical pieces of yellow-gram *(Cajanus cajan*, a specie leguminosal family) stalk as bio-filter media. It consisted of a plexiglas column of diameter 0.15m and effective height 0.70 m with sampling ports at 0.26, 0.49 and 0.70m from the bottom. Inlet concentrations and volumetric flow rates of toluene were varied from 2.56 to 34.73 g/m^3 and 0.18 to 0.24 m^3/h, respectively.

Using polyurethane foam cubes as bio-filter media, biodegradation of toluene and benzene vapors was investigated separately for 168 and 119 days, respectively. For these runs a laboratory scale bio-filter consisting of five modular sections of size (0.15 x 0.16 x 0 x 0.20m^3) and each filled with polyurethane foam cubes upto a height of 0.15 m was used. Inlet concentrations and flow rates of toluene and benzene loaded air were varied from 0.04 to 2.5g/m^{-3}, 0.69 to 5.45 g/m^{-3} and 0.06 to 0.90 $m^3 h^{-1}$, respectively.

Results and Discussion

Continuous biodegradation experiments showed that steady state was achieved earlier in the units filled with yellow gram stalk and maze cobs (natural media) than in the unit filled with polyurethane foam (synthetic media) because microorganisms find more favourable conditions to adhere and grow on the surface of natural materials than synthetic. Further, the natural materials generally support of a broad spectrum of microorganisms.

For toluene removal, almost constant removal efficiencies of 90, 80 and 90 per cent were achieved upto inlet loadings of 300, 120 and 77 $g\ h^{-1}\ m^3$ using yellow gram stalk, maze cobs and polyurethane foam as bio-filter media. For benzene removal, almost constant removal efficiency of 80 per cent was achieved upto inlet loadings of 68 $g\ h^{-1}\ m^3$ using polyurethane foam as the bio-filter media. Maximum elimination capacities of 342.2, 174.4 and 90.5 $g\ h^{-1}\ m^3$ were achieved at inlet loadings of 580, 292.6 and 126.3 $g\ h^{-1}\ m^3$ using yellow gram stalk, maze cob and polyurethane foam as bio-filter media and toluene as the target pollutant. For benzene, maximum elimination capacity 74.7 $g\ h^{-1}\ m^3$ was achieved at the inlet loading of 134.7 $g\ h^{-1}\ m^3$ using polyurethane foam as bio-filter media. Yellow gram stalk was found to be the most effective bio-filter media for toluene degradation but experiments were conducted at low flow rates only. Toluene was easily biodegradable than benzene. In all cases, bio-filter responded very well to change in operating conditions and could easily recover its performance on re-startup after shutdown. This indicated good stability of the biomass in the bio-filters. Axial variation of pollutant concentration was also investigated in the bio-filters. Results showed that most of the pollutant got eliminated in the upper section of the bio-filter. With the help of this information, optimum height of bio-filter column for given operation can be calculated.

Presence of diffusion and bio-degradation controlling zones were observed in the bio-filters at low and high inlet loadings. Less deterioration and small compaction of bed resulting in low change in pressure drop was observed with polyurethane media (synthetic) than with yellow gram and maze cob (natural). This indicated good mechanical strength and less power required to operate the bio-filter with synthetic media.

Variation of pH in the bio-filters with respect to time not only depends on the type of media and pollutant but also on the type of microorganisms in the bio-filter. All experiments were carried out in the months of March-September at room temperature (29-41°C) and insignificant variation in the performance of the bio-filter was observed. Under steady-state conditions, temperature of bio-filter bed was found 3 to 5°C higher than the ambient temperature. Carbon recovery (per cent) was also estimated in the case of toluene removal using maze cob and polyurethane foam as bio-filter media. Most of the time more than 70 per cent carbon recovery was observed which indicated the effective biodegradation in the bio-filter. More regulated nutrient supply was required in the case of polyurethane foam than yellow gram stalk and maze cobs because the natural media inherently contain most nutrients required for the growth of microorganism.

References

1. Singh, R.S., *Ph.D. Thesis* in Chemical Engineering, Banaras Hindu University, Varanasi, India, (2007).
2. Singh, R.S., Agnihotri, S.S., and Upadhyay, S.N., Removal of Toluene Vapour Using Agro-waste as Bio-filter Media, *Bioresource Technology*, Vol. 27, pp.2296-2301 (2006 a).
3. Singh, R.S., Gunjan Kumar and Upadhyay, S.N., Modelling and Simulation of Bio-filter for the Removal of Volatile Organic Compound from Polluted Gas Stream, *International Conference "BIOCONVERGENCE"*, TIET Patiala, 18-20 November, (2004 a).
4. Singh R.S., Rai, B.N., and Upadhyay, S.N., Environmental Biotechnology and its Role in Air Pollution Control, *International Conference "BIOCONVERGENCE"*, TIET Patiala, 18-20, November, (2004 b).
5. Singh, R.S., Rai, B.N., and Upadhyay, S.N., Performance Evaluation of an Agro Waste Based Bio-filter Treating Toluene Vapours, *Environmental Technology*, Vol.27, pp.349-357 (2006 b).
6. Upadhyay, S.N., and Kumar, V., Biological Control of Air Pollution in "*Concise Encyclopedia of Bioresource Technology*" (Pandey, A., Ed.), The Haworth Press Inc., NY (2005).

Chapter 9

Cleaner Production Practices for Sustainable Development of Pulp and Paper Industry

I.D. Mall, V.C. Srivastava, S. Singh and A. Pachouri

Department of Chemical Engineering,
Indian Institute of Technology Roorkee,
Roorkee

ABSTRACT

Pulp and paper industry is one of the oldest industries and has made continuous strides during recent years. However pulp and paper industries are amongst the top highly polluting industries. Present paper deals with the profile and environmental pollution problems of the pulp and paper industry. This paper deals with environmental impact of pulp and paper production, the technological developments during the last two decades in the area of development of cleaner technologies for the environmental protection and sustainable development of the pulp and paper industries.

Keywords: *Sustainable development, Cleaner technologies, Pulp and paper industry, Paper manufacturing, Fibre recycling.*

Introduction

Cleaner technologies have received considerable interest during last two decades in order to minimise the damage caused to the environment due to emission/discharge of hazardous waste, and various other types of gaseous, liquid and solid wastes. The problem of environmental pollution is increasing consistently in spite of continuous development in the process technology. Fast development and large scale industrialisation have caused depletion of natural life sustaining resources. There is urgent need to minimise further damage to the environment through cleaner technologies using principle of recover, reduction, reuse, recycle and retrofit; use of renewable resources; preventive environmental technologies; and development of environmental friendly microbial system. For ecological sustainable development, cleaner technology will play significant role in future.

History of paper making goes back to over 2000 years, and is one of the oldest core sector industries all over the world. Per capita consumption of paper is sometimes yardstick for measurement of industrialization and economic growth of a country. Paper making like many other industries, has undergone a sea change in the last few decades from being a small scale and totally manual operations to a large scale and substantially mechanized operation. Today, the paper industry accounts for about 2.5 per cent of world's production and 2 per cent of world trade. Global demand for paper and board is forecast to grow from current level of 353 million tonnes to over 410 million tonnes by the year 2010 at an average growth rate of 2.2 per cent per annum [Mittal, 2005]. There are about 500 kraft mills and many thousands of other types of pulp and paper mills all over the world.

Pulp and paper industry is one of the oldest industries in India. With a 20 mills and installed capacity of one lakh tonne per annum at the time of independence, the paper industry has made continuous strides and now there are about 700 mills with a production of 6.8 million tonne of paper, paper board and newsprint. Indian paper industry requires more raw material and energy, generates more waste and has low resource recovery per unit of product as compared to the global paper industry. In this era of globalization, Indian paper industry needs to match global capacities, efficiencies, and quality standards with minimum degradation of environment.

In spite of its significant contribution towards social, cultural as well as economic development of a nation paper industry is amongst the highly polluting industries and is listed in red category. Pulp and paper making consumes huge amount of water and generated large quantity of wastewater containing high BOD, COD, pH, solid content, colour, foam, AOX and furan; air pollutant containing mercaptans, H_2S, SOx, NOx, chlorine; solid waste containing bark, screen rejects, ETP sludge, flyash. A relatively large number of toxic compounds have been identified in the pulp mill effluent, which has been cause of major concern. With discovery of dioxins and furans in pulp mill effluent, there has been wide spread environmental concern over the adverse impact of AOX and environmental pressure has generated for removal of the AOX and other toxic compounds. It is also responsible for emission of gases causing green house effect, acidification, and photochemical smog. The problem of environmental pollution in agro-based paper mills is more serious due to obsolete process technology and lack of chemical recovery and many mills have shifted towards more and more waste paper. Figure 9.1 shows the environmental impact of pulp and paper manufacture on global as well as local environment.

Sustainable development demands attaining the three dimensional balance between economic growth and social and environmental responsibility, therefore

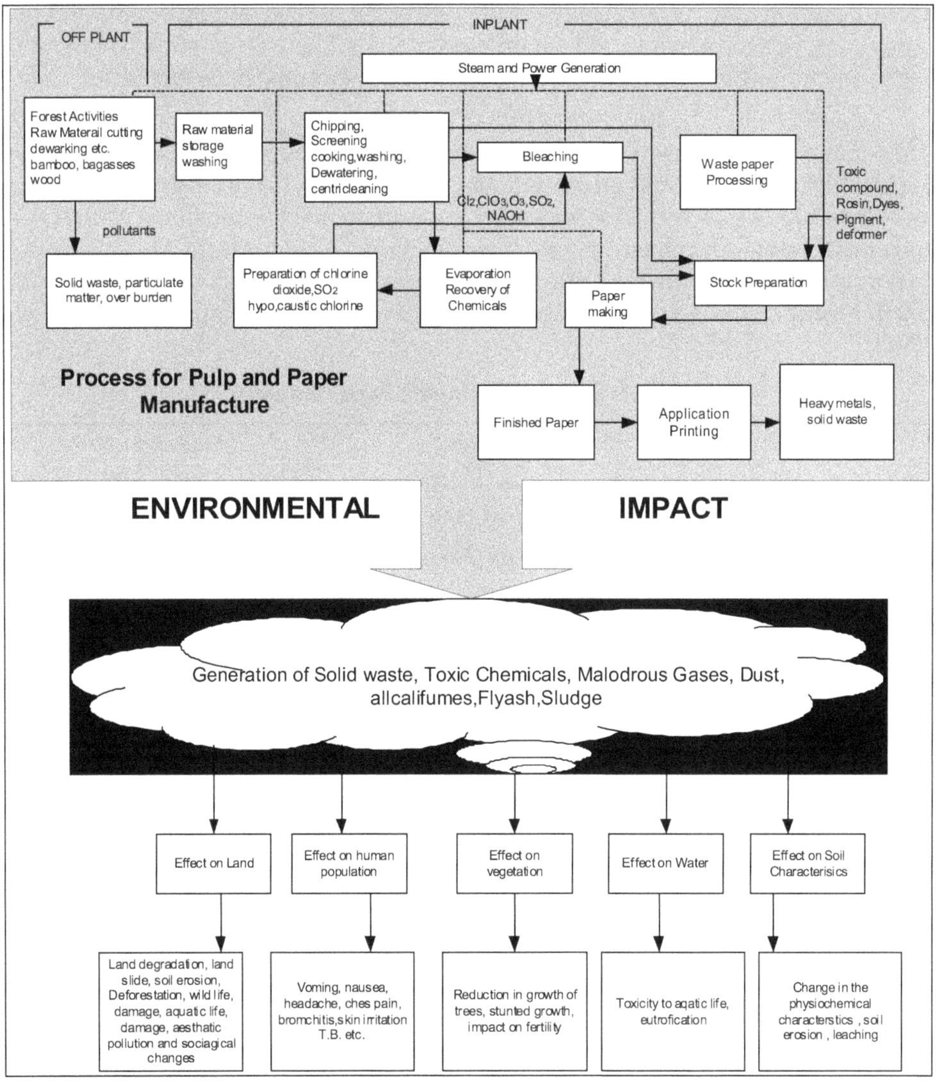

Figure 9.1: Environmental Impact of Pulp and Paper Manufacture.

sustainability entails the continuous fulfillment of the varying needs and some-times conflicting objectives of the multiple stakeholders [Alyousfi, 2004].

The present paper deals with the various aspects of cleaner technologies in pulp and paper mills to meet the gap between present and future for sustainable development.

Present Scenario of Pulp and Paper Industry

Worldwide paper consumption in this century has increased 4 times faster than population. Paper and paperboard worldwide will reach 476 million tones in 2010 and 640 million tones in 2020 with world population of 8000 million people and per capita consumption of paper and paperboard of 80 kg. There are about 500 kraft mills and many thousands of other types of pulp and paper mills in the world.

Table 9.1 shows the structure of Indian paper industry. There are about 700 pulp and paper mills in India with 6.8 million tonnes of paper, paper board and newsprint out of which 5.8 million tons accounts for paper and board. Large integrated (capacity more than 100 TPD) account for 30 per cent of production while medium (50-100TPD) and small paper mills with (capacity less than 50 TPD) accounts for 29 per cent and 35 per cent of total production. The average capacity is as less as 14 TPA. [Mall *et al.*, 2005]. Profile of Indian pulp and paper industry is given in Figure 9.2 [Jain *et al.*, 2007].

Table 9.1: Structure of Indian Paper Industry.

Type of Paper Mill	*Scale of Operation, Tons Per Day*	*No. of Mills*	*Production Share per cent*
Large Integrated Wood based	101-800 (Avg. 300)	33	36
Medium Agro Based	50-100 (Avg. 60)	165	29
Small Waste Paper	5-50 (Avg. 15)	510	35
Total		708	100

Source: Jain *et al.*, 2007.

Challenges to Pulp and Paper Industry

To bridge the gap between the present and the future six major research areas high priority efforts which should be given due consideration are: sustainable forest management, environmental performance, energy performance, improved capital effectiveness recycling and sensors and controls [Copoer, 1997]. Due to environmental pressure and stringent regulatory parameters, and market pressure for eco-friendly product environmental issues has emerged as a major challenge and will continue to be single most important factor influencing the technological change in pulp and paper industry. However, the problem in Indian paper mills is quite complex due to low average capacity of the mills especially the agro-based mills [Mall, 1998].

Although many of the environmental issues in the developed countries have been resolved but situation in Indian paper industry is entirely different due to low capacity plants with poor infrastructure facilities and lack of recovery section in majority of mills using agricultural residues and the environmental problem in Indian paper industry need serious considerations to resolve the issues. Major challenges to Indian Paper industry is given in Figure 9.3 [Rakhilla, 2005]. Major challenges before paper industry are non availability of good quality cellulosic raw materials, emerging environmental issues especially due to presence of dioxin and furan, lack of technology upgradation and also the resources for the technology upgradation [Garg, 2007].

Figure 9.2: Profile of Indian Pulp and Paper Industry.

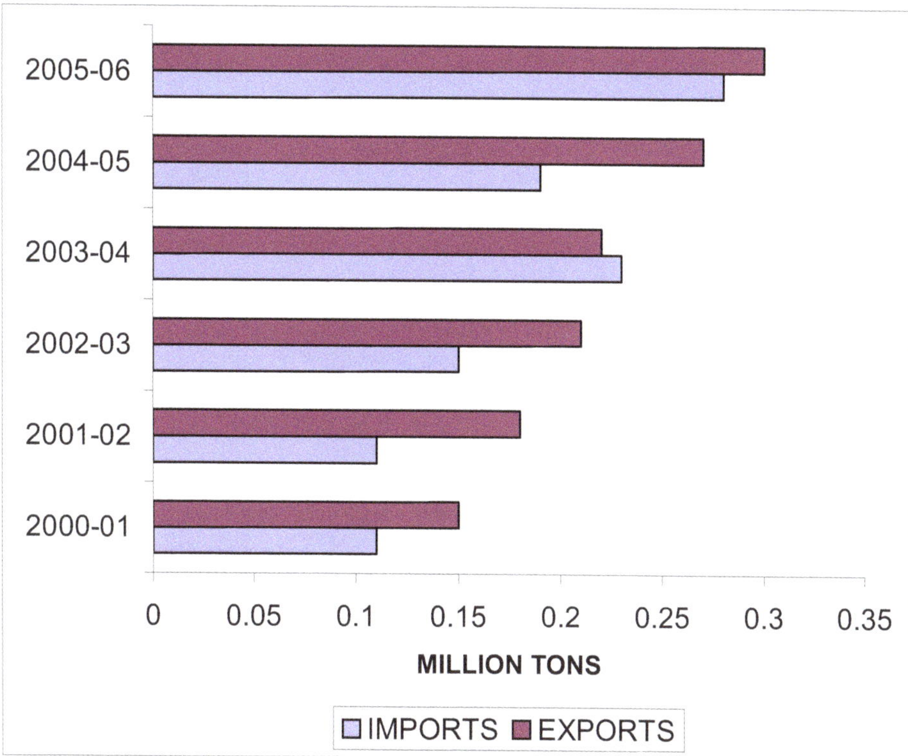

(a) Imports and Exports of various items in India

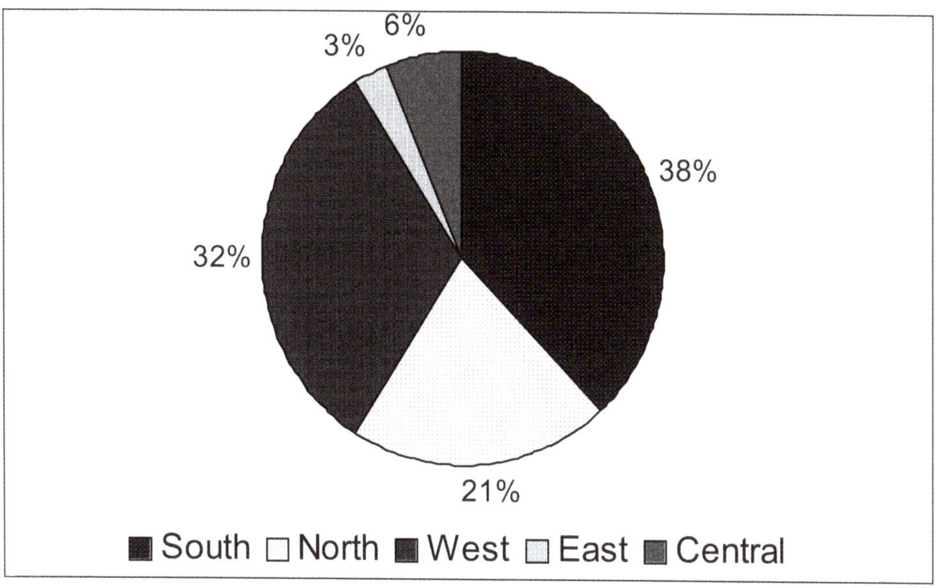

(b) Geographic Spread of Newsprint Capacities in India

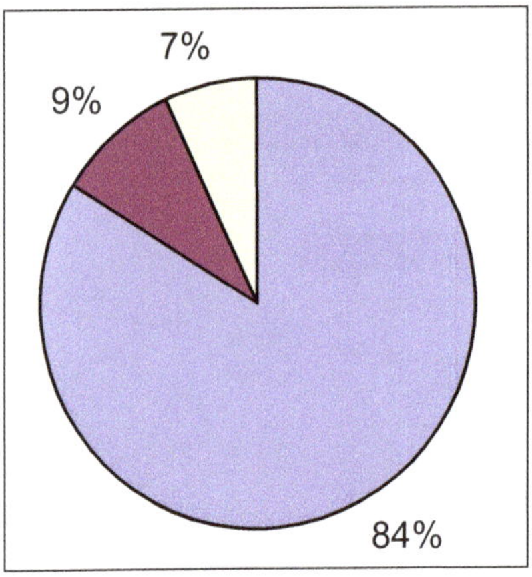

1970 Forest Based

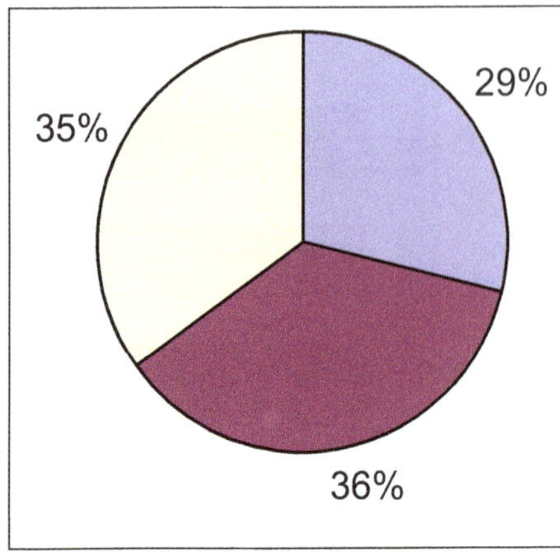

2004 Recycled Paper

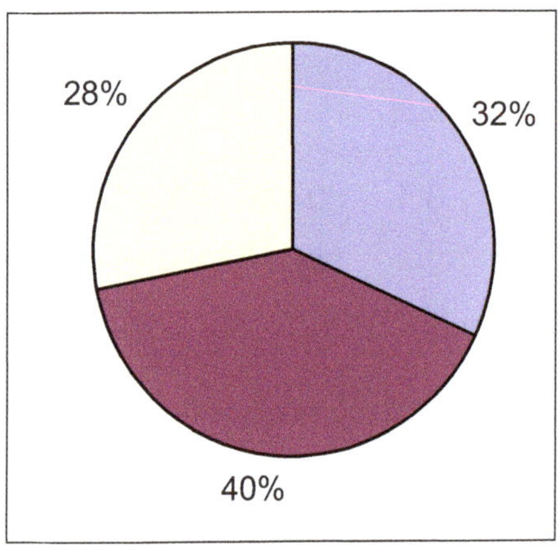

2010 Agro Based

(c) Trends in Use of Raw Material for Pulp and Paper production in India.

Cleaner Production

Cleaner production is the continuous application of an integrated preventive strategy to processes, products and services, to increase eco-efficiency and to reduce risks to humans and the environment [UNEP-IE, April 1998]. Cleaner production is now need of time as it offers lower waste treatment costs and increased production efficiency, lower costs, conservation of raw materials and energy, better access to finance, marketing requirements of environment friendless products, improved environment and better compliance with environmental regulations.

Due to environmental pressure and stringent regulatory parameters, and market pressure for eco-friendly product environmental issues has emerged as a major challenge and will continue to be single most important factor influencing the

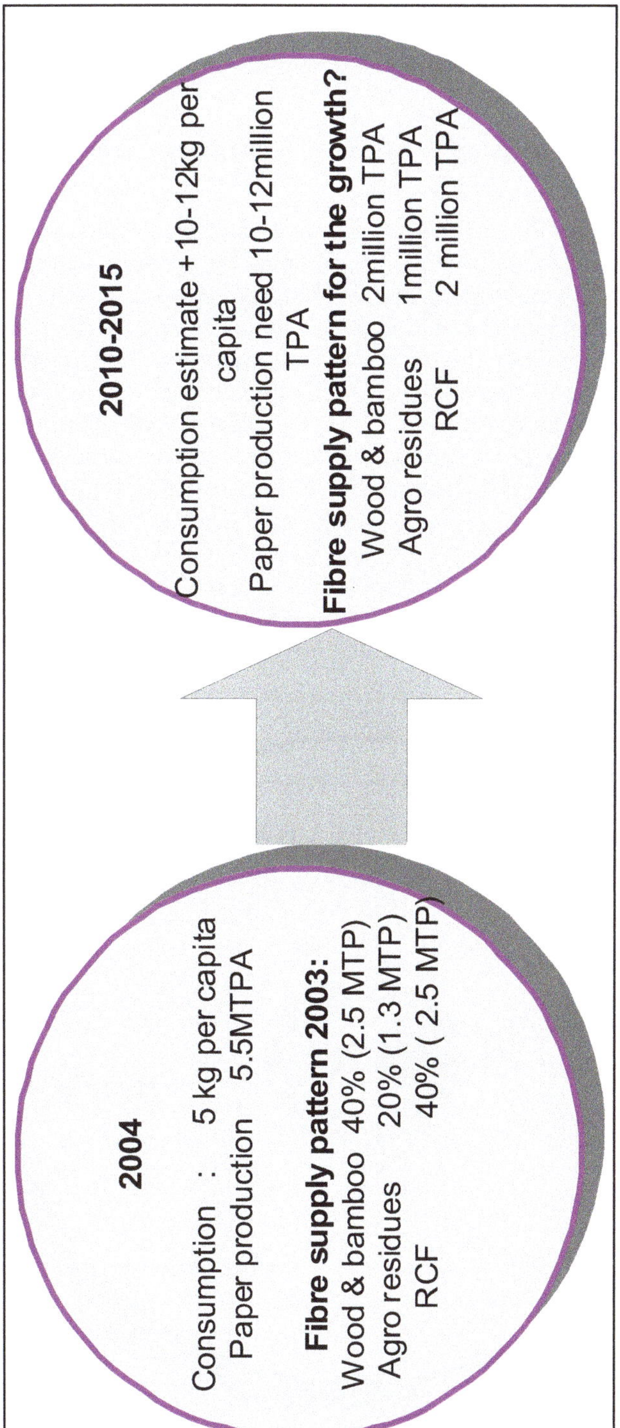

Figure 9.3: Challenges to Indian Pulp and Paper Industry.

technological change in pulp and paper industry. However, the problem in Indian paper mills is quite complex due to low average capacity of the mills especially the agro-based mills. Some of the major developments in various areas of pulp and paper industry are given in Table 9.2.

Innovation in Technology for Waste Management for Cleaner Production

There has been significant innovation and developments in technology in pulp and paper manufacture to improve quality, minimize the energy consumption, reduce the toxicity of bleach plant effluent, and wastewater generation, reduce the color in the effluent, to improve better recycle. Figure 9.4 shows the five R initiatives for cleaner technologies implementation. Figure 9.5 shows cleaner production techniques that may be applied in pulp and paper mills [UNEP Manual, 1999]. Pulping and bleaching has been one of the major areas where major technological development has taken place in the pulp and paper industry. Some of the technological development in pulping and bleaching for reducing toxicity and waste generation are:

Process Modification

Delignification of pulp as much as possible before chlorination stage to decrease residual lignin, improved brown stock washing and screen, improvement in bleaching sequence to reduce or use of eliminate the elemental chlorine, application of bio-pulping and bio-bleaching.

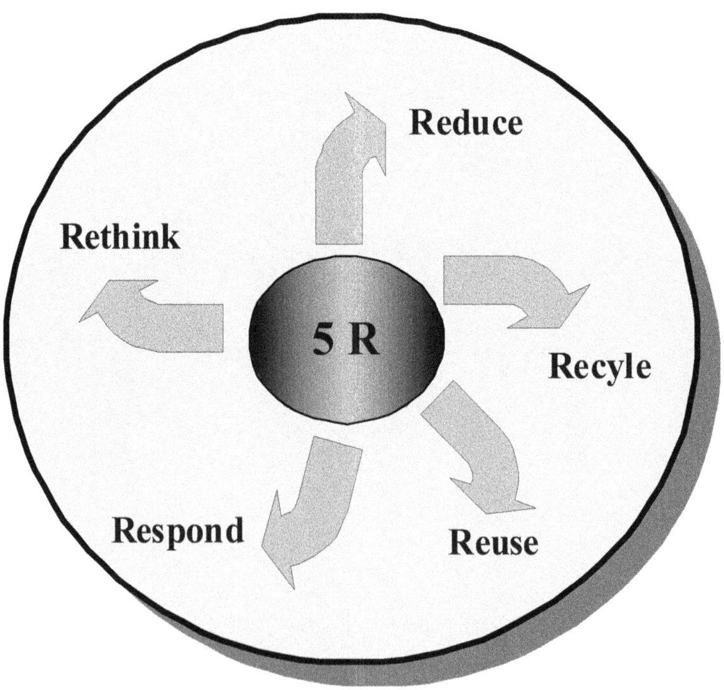

Figure 9.4: Five R Initiatives for Cleaner Technologies Implementation.

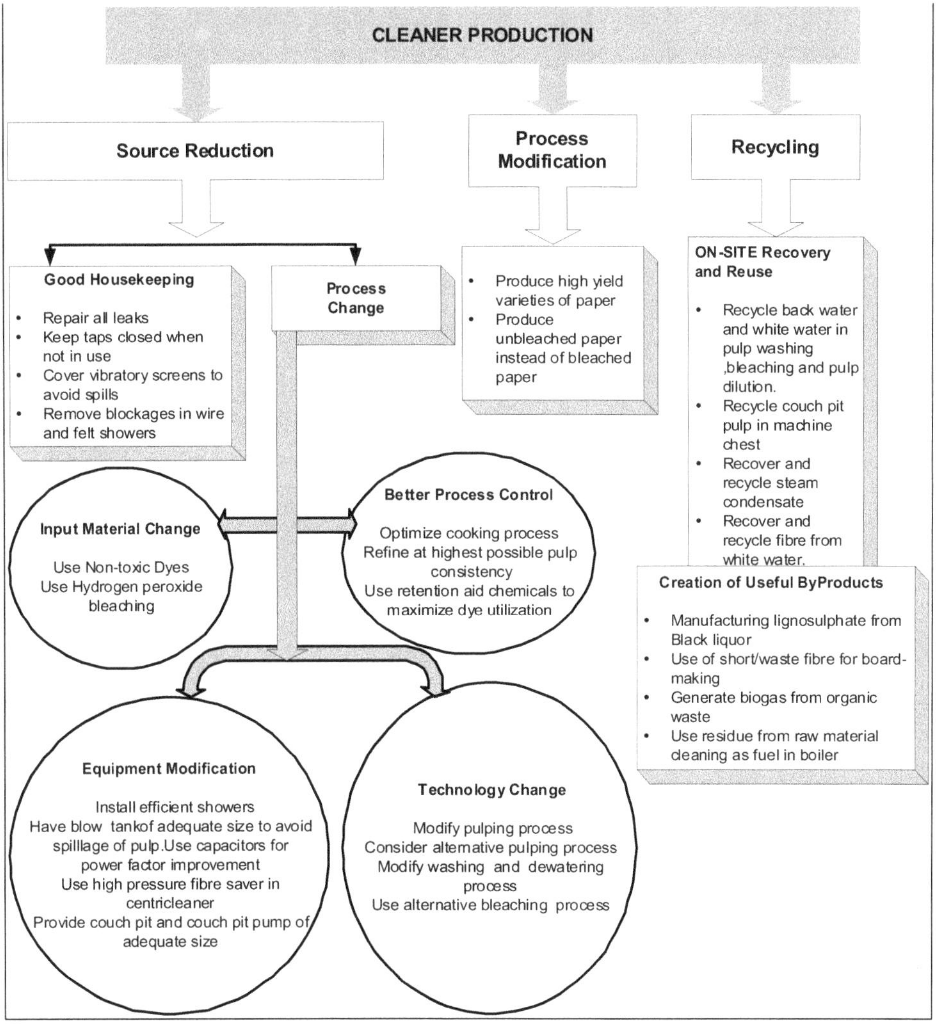

Figure 9.5: Cleaner Production Techniques in Pulp and Paper Mills [UNEP Manual, 1999].

Control of Outside Contamination

Elimination of use of defoamer containing dioxin precursors.

External Control Measures

Stabilization ponds, activated sludge, aerated lagoons, membrane process, adsorption, electrochemical process.

Pulp and paper market forces are also driving the development of pulp bleaching technology and there has been an unprecedented change in the bleaching technology for reducing toxicity. Now the pulp instead of dioxin free pulp today they are classified

Table 9.2: Major Developments in Various Areas of Pulp and Paper Industry.

Area	*Major Development*
Forest management	Sustainable forest productivity with fast growing species with high yield, clonal propagation of soft wood, good harvest management. Main focus areas in forest management are (a) biotechnology, (b) basic physiology of forest productivity, (c) sustainable soil productivity, and (d) remote sensing technologies to improve forest inventory and stand management [Cooper, 1997].
Chipping	Washing system, High capacity chipper, minimizing dust.
Pulping	
Chemical Pulping	Rapid displacement heating (RDH), superbatch, enerbatch, Modified and Etended delignification; MCC with Hi heat washing, isothermal cooking, balck liquor impregnation and modified chips bins, oxidative delignification with oxygen and ozone, anthraquinone pulping, Octahydrodimethy l-anthraquinone (ODIMAQ), Polyoxometalates pulping (POMS), biopulping.
Mechanical pulping	Biopulping: biopulping improves the quality of mechanical pulp and reduces energy consumption. TMP, RTS. RTS shortens the retention time, lowers the temperature of chip pretreatment, which is folled by rapid speed refining through specially configurated refiners plate [Atalla *et al.,* 2004].
Washing and screening	Drum displacer, Pressure diffuser, Displacement presses, combined deknotting and fine screening, high temperature screening before washing, reverse cleaners.
Bleaching	Pulp and paper market forces are driving the development of pulp bleaching technology and there have been unprecedented changes in the bleaching technology for reducing toxicity. Instead of dioxin free pulp, now, pulp are classified as chlorine free (CGF), chlorine chemical free (CCF), Molecular chlorine gas free (MCGF), Non chlorine compound ((NCC), Active chlorine free (ACF), Absolutely chlorine free (ACF), Almost chlorine free (ACF), Elemental chlorine free modified (ECFM), Totally chlorine free (TCF), Totally effluent free (TEF), ECF, TCF, Oxygen bleaching, Ozone bleaching, Mesto Ze Trac™ (high consistency ozone system): ECF-sequence like (Ze)DP or (Ze)DD or a TCF sequence- like ZQ (PO) [Norden *et al.,* 2005].The changes that have experienced a large scale benefits in terms of these reduce the effluent load from bleaching, are to eliminate/reduce chlorine and chlorine compounds in bleaching through introduction of ozone beaching, beaching of pulp in light ECF sequence, like (Ze) DP or (Ze) DD, beaching of pulp in light TCF sequence, like ZQ (PO), introduction of Ozone bleaching and it is also cost effective compared with other ECF bleaching sequences, like D (EOP)D and the application of this methodology can replace large amounts of chlorine dioxide with moderate amount of ozone, bleaching with the sequence (Ze) DP will generate very small amounts of AOX.
Fibre modification	Enhancing the opacity of fibre through pigment particles, better fibre loading, Pressurised sensitive adhesives (PSAs). Fibre loading contributes opacity, smoothness and brightness to high quality paper; this filler also replaces some of the costly bleached kraft fibre.
Paper making	High speed machine with sophisticated instrumentation and control; closed hood and insulation of dry cylinder end; installation of trinip press.

Contd...

Table 9.3–*Contd...*

Area	*Major Development*
Evaporation and recovery	Free flow falling film evaporator, vibrating shear enhanced processing (VSEP™) system [New logic International, www.vsep.com] using ultra fitration and nano-filtration to pre-concentrate solids prior to evaporation; modern recovery boiler with black liquor concentration 85 per cent; installation of lime kiln, recovery system with agro based mills. Low temperature incinerator system (LTIS) for agro based paper mills. Some of the advantage LTIS are a safe process for small medium pulp mills, does not form any smelt, a low cost technology of small/medium pulp and paper mills, a flexible widely applicable technology alkali/neutral sulfite black liquor, an environment friendly clean technology with least problem, minimum maintenance, Can be scaled down to 30-35 TPD pulp mill technology and economics are viable, very low generation time [Gupta and Ramani, 2004].
Recycling of fibre	Improved recycling reduces pressure on wood and often results in lower energy consumption, less chemical usage and lower pollutant production.
Deinking	Enzymatic deinking using celluloses. Celluloses improve de-inking by removing fibrilils from fibre substances and facilitate the removal of tonners particles, increase the freeness and flexibility and reduce the coarseness of dried fibres. Cellulose treatment of recycled fibres reduce disintegration time and increase bleachability of low quality, recyled paper [Atalla *et al.*, 2004].
Effluent treatment system	Membrane separation technology (MST) for treatment of effluent and waste management. Membrane separation technology has been applied in pulp and paper industry in following areas: Kraft lignin and lignosulfonate separation, recovery of inorganic chemicals from kraft black liquor, Recovery of inorganic chemicals and lignin from black liquor from agricultural residue, Concentration of spent kraft and sulfite liquor in pre concentration stage, Wastewater treatment for decolourization and detoxification. Vibrating shear enhanced processing (VSEP™) using ultra filtration and nano-filtration avoiding fouling problem exhibited by conventional membrane system, pressusurized ozone/ultrafiltration membrane systems for removing total solids from paper mill process water by converting it to total suspended solids; Hybractor™ is a hybrid reactor which combine best features of both suspended bed and fixed bed film technologies.

Source: Atalla *et al.*, 2004.

as chlorine free (CGF), Chlorine chemical free (CCF), N, Molecular chlorine gas free (MCGF), Non chlorine compound (NCC), Active chlorine free (ACF), Absolutely chlorine free (ACF), Almost chlorine free (ACF), Elemental chlorine free modified (ECFM), Totally chlorine free (TCF), Totally effluent free (TEF) [Pyre, 1991]. Some of the major approaches for de-colorization and detoxification of bleach plant effluent have been in combination following one or more measures: process modifications, control of outside contamination, external control measures (end pipe treatment). Process modification includes: delignification of pulp as much as possible before chlorination stage to decrease residual lignin pulp, improved brown stock washing and screening, improvement in bleaching sequence to reduce or eliminate use of elemental chlorine.

During the last two decades most of the paper mills have incorporated changes in the process through modified pulping and bleaching technologies and for reducing AOX. Some of the technological developments for reducing AOX are extended delignification, improved washing, oxygen delignification, chlorine dioxide substitution, and oxidative alkali extraction bleaching.

Conclusion

Although Indian paper industry shares with other pulp and paper industries in the world for the emergence of environmental conscientiousness to have clear technologies for better environment and sustainable development, however, still Indian paper industry is far behind the developed countries. Indian pulp and paper industry requires development of further newer technologies, processes and measurements to manufacture pulp and paper products more efficiently and cost-effectively while reducing environmental impacts of operations and maximizing the efficient use and reuse of resources. Sustainable forest management, environmental performance, energy performance, recycling, etc., are the major focus areas for future research efforts. Indian pulp and paper industry will have to plan and act for sustainable developments of paper industry with clear vision and perception of future demand; vision without action and action without vision will be disaster.

Refrences

1. Alyousfi AB. *Practice periodical of hazardous, toxic, and radioactive waste managements*. October 2004, p. 265.
2. Atalla RH, Reiner RS, Houtman CJ. *New technology in pulping and bleaching*. Springer, E.L. Elsevier Ltd. 2004.
3. Cooper JS. The basics of Agenda 2020. *TAPPI Journal*, February 1997, p. 32-35.
4. Garg NK. *IPPTA Journal*, 19 (4), October-Dec. 2007, p. 2.
5. Gupta D, Ramani S. Low temperature incinerator System for agrobased paper mills" *IPPTA J*. Vol 16 No.4 Oct2004, p. 47.
6. Jain RK, Singh K, Roy TK. Indian Paper Industry Raw material Scenario, Growth Prospects and Path ways. *IPPTA Journal*, 3, July-Sep 2007, p. 129.

7. Mall ID, Srivastava VC, Singh B. Environmental protection in pulp and paper manufacture through cleaner technologies. *XXI National Convention of Chemical Engineers*, Indian Chemical Industry Vision 2020, Udaipur, INDIA, September 2-4, 2005.

8. Mall ID. Emerging Scenario in Indian paper Industry: Responding to the Challenges of Globalisation. *IPPTA Convention* issue, Dec. 1998, p. 91.

9. Mittal SK. Strategic moves of pulp and paper industry. Towards achieving sustainable development. *IPPTA Convention* issue Vol. 17, No 2, ii, Jan 2005

10. New logic International www.vsep.com

11. Norden, S., Carre, G., Lundahl, Y., Jain, N.K, "New bleaching processes and fibre line with state-of-the-art technology" *IPPTA Convention* Issue, Vol. 17, No 2, Jan, 2005, p. 29-34.

12. Pryke, D.C., in Chapter 2.1 of "*Environmental solution for the pulp and paper industry*" Edited by Kelly, Ferguson, Miller Freeman Inc. US, 1991, p. 67.

13. Rakkhila, P. A global view on chemical pulping industry and it's recent developments. *IPPTA Convention* Vol. 17, No. 2, Jan, 2005, p. viii-xii.

14. UNEP Manual. Cleaner Production in the Pulp and Paper Industry: Technology Fact Sheets. *UNEP-IE*, 1999.

15. UNEP-IE manual. Cleaner Production in the Pulp and Paper Mills. A training resource package. *UNEP-IE*, April 1998.

Chapter 10

Advances in Biological Nitrogen Removal (BNR) Processes in Wastewater Treatment

R. Biswas and T. Nandy

CSIR–NEERI, Nagpur

Introduction

Wastewater discharges containing nitrogen can be toxic to aquatic life, cause oxygen depletion and eutrophication in receiving water, and affect chlorine disinfection efficiency (U. S. EPA, 1993). Hence reducing nitrogen levels from the discharges is necessary. Nitrogen compounds can be removed from wastewater by a variety of physicochemical and biological processes. Because biological nitrogen removal is more effective and relatively inexpensive, it has been widely adopted in favour of the physicochemical processes. In most modern wastewater treatment plants (WWTP) nitrogen, which is generally in the form of ammonium or organic nitrogen, is removed by biological nitrification/denitrification (reaction (3)). The first step ammonium is converted to nitrate (nitrification, reaction (1)) which is then, in a second step, converted to nitrogen gas (denitrification, reaction (2)). Benefits of the process are the high potential removal efficiency, high process stability and reliability, relatively easy process control, low area requirement and moderate cost.

$$NH_4^+ + 2O_2 + 2HCO_3^- \rightarrow NO_3^- + 3H_2O + 2CO_2 \quad (1)$$

$$5C + 4NO_3^- + 2H_2O \rightarrow CO_2 + 4HCO_3^- + 2N_2 \quad (2)$$

$$4NH_4^+ + 8O_2 + 5C + 4HCO_3 \rightarrow 2N_2 + 10H_2O + 9CO_2 \quad (3)$$

Generally, the conventional biological nitrogen removal process is used for treating wastewaters with relatively low nitrogen concentrations (total nitrogen concentration less than 100 $mgNL^{-1}$). Some wastewater streams such as anaerobic digester effluents, landfill leachate and industrial wastewaters contain high concentrations of nitrogen.

The rise in the demands for environmental protection has forced to develop more efficient nitrogen removing technologies in wastewater treatment. The new technologies that have developed are efficient than the conventional nitrification-denitrification processes in terms of time and space. Based on the knowledge available on nitrogen removing microorganisms, a number of new cost effective technologies have been developed for efficient removal of ammonia from wastewaters (Figure 10.2). The state-of-the-art of these newly developed processes is presented along with their current status on field applications.

Conventional Nitrogen Removal Process

Conventionally, autotrophic nitrification followed by heterotrophic denitrification is a well-established process of removing high concentration of ammonia as well as nitrate from wastewaters.

Nitrification is the term used to describe the two-step biological process in which ammonia (NH_4-N) is oxidized to nitrite (NO_2-N) and nitrite is oxidized to nitrate (NO_3-N). The first step is carried out a group of autotrophic ammonia oxidizing bacteria belonging to genera *Nitrosomonas, Nitrosococcus, Nitrosospira, Nitrosolobus,* and *Nitrosovibrio*. The second step is carried out by autotrophic nitrite oxidizing bacteria belonging to genera *Nitrobacter, Nitrococcus, Nitrospira, Nitrospina,* and *Nitroeystis.*

Generally, nitrification can be accomplished in both suspended growth and attached growth biological process (Metcalf and Eddy, 2003). For suspended growth process, more common approach is to achieve nitrification along with BOD removal in a same single-sludge process, while in case of complex wastewater where there is significant potential of presence of toxic compounds, a two-sludge suspended growth system is considered (Figure 10.2).

Stoichiometrically, for oxidizing each g of ammonia to nitrate, 4.25 g of O_2 and 7.07 g of alkalinity as $CaCO_3$ is required. Besides, DO and alkalinity, nitrification in wastewater process must consider various issues like wastewater nitrogen concentration, BOD concentration, alkalinity, temperature, toxic compounds in wastewater etc. Nitrification is pH sensitive and the optimal pH range is between 7.5-8.0. The rates decline significantly near pH values 5.8-6.0. At activated sludge process, neutral pH of 7.0-7.2 is usually maintained. In single-sludge systems nitrification can take place at reasonable rate at neutral pH, but can suffer competition from fast growing heterotrophs, that have their broad optimal pH range at 6.5-8.5. And also nitrification consumes alkalinity, sufficient alkalinity is essential to buffer the pH drop during nitrification. Nitrifying organisms are also sensitive to wide range of toxic compounds commonly present in wastewater at concentrations that are much lower than the concentrations that would affect other aerobic heterotrophic bacteria. Hence in many cases, nitrifiers continue to grow and nitrify in presence of these toxic compounds but at rates much lower than their optimal rate. Metals are

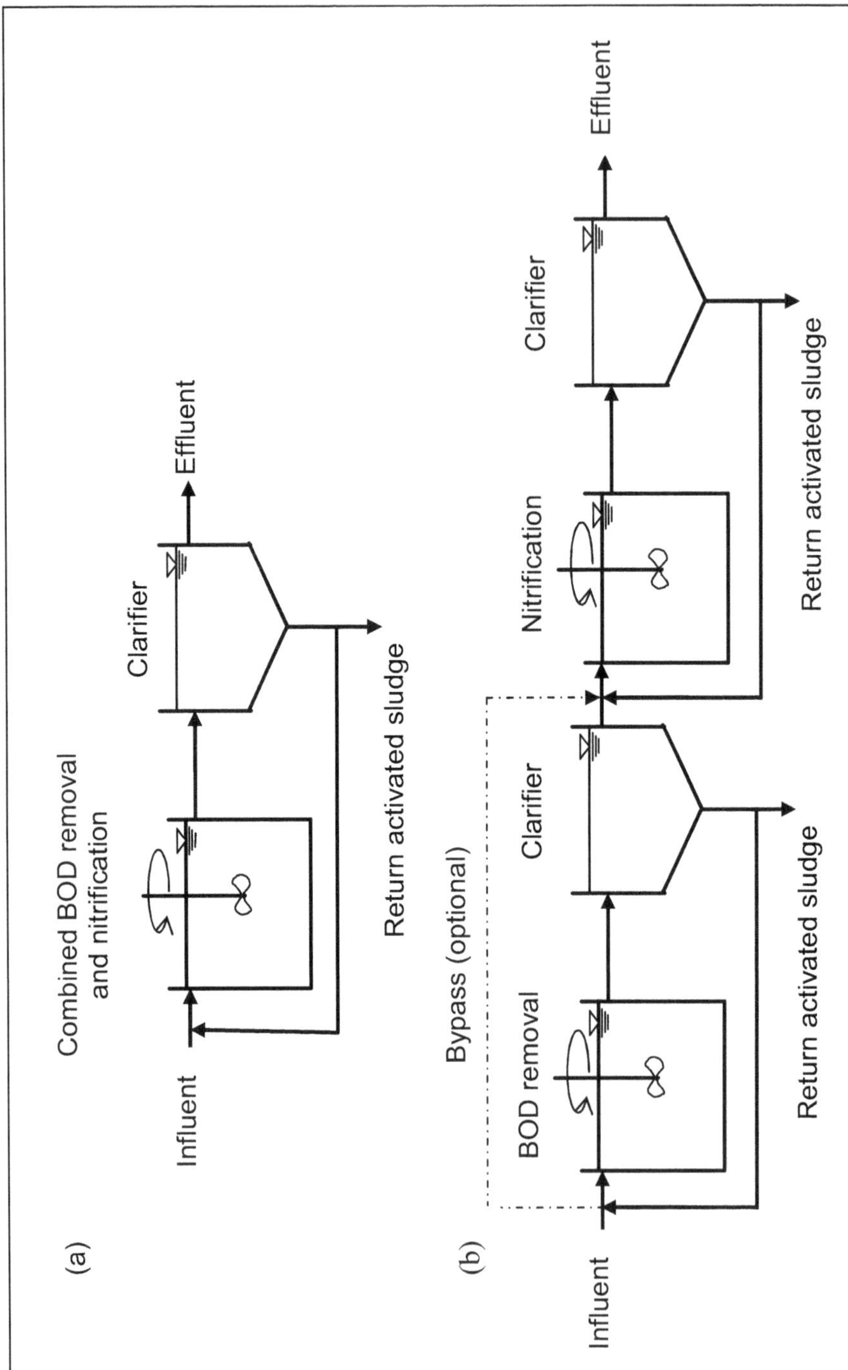

Figure 10.1: Process Configuration Used for Biological Nitrification: (a) Single Sludge Suspended Growth System, (b) Two-sludge Suspended Growth System (Adapted from Metcalf and Eddy, 2003).

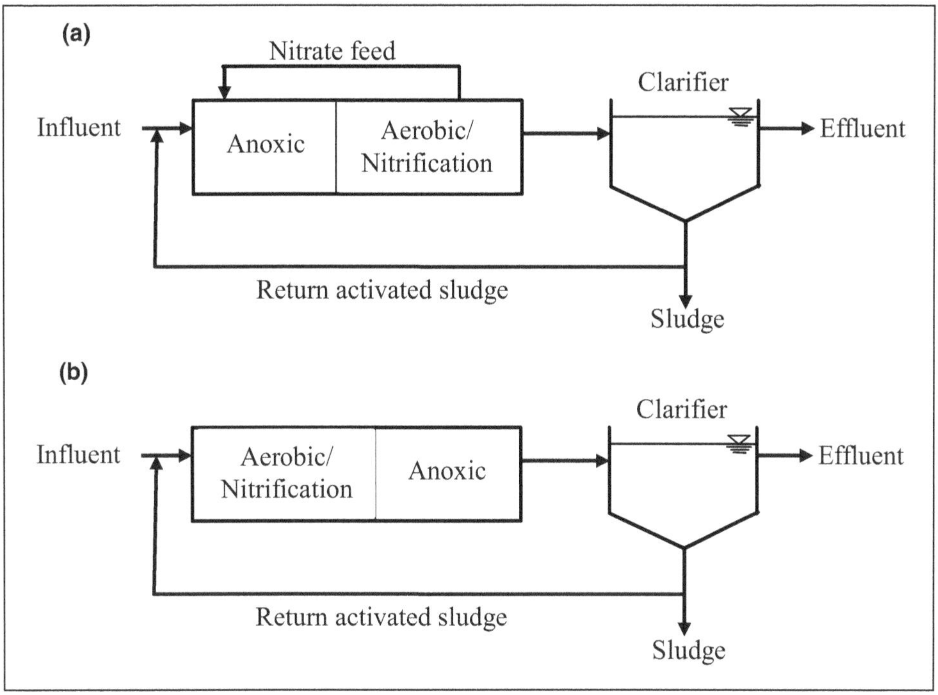

Figure 10.2: Denitrification Process (a) Preanoxic Denitrification, (b) Postanoxic Denitrification (Adapted from Metcalf and Eddy, 2003).

also a concern for nitrification. Nickel (0.25 mg/l), chromium (0.25 mg/l) and copper (0.1 mg/l) are also known to inhibit nitrification completely. Nitrification is also affected by free ammonia (NH_3) and free nitric acid (HNO_2). The inhibition effects are depended upon the concentration of the nitrogen species, pH and temperature. At 20°C and pH 7.0, the NH_4-N and NO_2-N concentrations of 100 mg/l and 20 mg/l may initiate inhibition of ammonia and nitrite oxidation respectively (U.S. EPA, 1993).

Denitrification is biological reduction of nitrate to nitric oxide, nitrous oxide and nitrogen gas under anaerobic conditions. It is an integral part of biological nitrogen removal process (Figure 10.2). Denitrification takes place after ammonia is autotrophically nitrified to nitrate in presence of external carbon source like methanol. A variety of electron donors and carbon sources such as, methanol, acetate, glucose, ethanol, acetic acid, lactic acid and at times wastewaters have been used for denitrification. Methanol (CH_3OH) being relatively inexpensive, has gained widespread application. For complete denitrification, 2.47 g of methanol is required per gram of nitrate nitrogen (McCarty *et al.*, 1969). Carbon source can also be derived from the wastewater. In this arrangement, denitrification (anoxic) tank is followed by a nitrification (aeration) tank. Nitrate produced in the aeration tank is recycled back to anoxic tank. This process is called substrate driven or pre-anoxic denitrification. In another arrangement, denitrification occurs after nitrification and the electron donor source is from endogenous decay. This arrangement is termed as post anoxic denitrification and here the denitrification depends on endogenous respiration for energy and sometimes external carbon source is added. Postanoxic processes include both suspended and attached growth systems.

The differences between autotrophic nitrifiers and the heterotrophic denitrifiers underscore the complexity of design and operation of nitrification/denitrification systems. The autotrophic nitrifiers are slow growing, sensitive to organic carbon concentrations and require high dissolved oxygen concentration. While on the other hand, heterotrophic denitrifiers are fast growing bacteria, resilient to inhibitory compounds and require high organic carbon concentrations with low to absent dissolved oxygen level. Providing heterogeneous environmental conditions to nitrifying and denitrifying organisms is the greatest challenge in successful biological nitrogen removal (Bernet *et al.*, 2001). Moreover, the nitrification reaction consumes a high oxygen concentration, requiring 4.2 g of oxygen for each gram of ammonium nitrogen nitrified (Gujer and Jenkins, 1975). Theoretical oxygen consumption for *Nitrosomonas* and *Nitrobacter* are 3.16 mg O_2 per mg NH_4-N oxidized and 1.11 mg O_2 per mg NO_2-N oxidized, respectively. Moreover, 7.07 mg alkalinity as $CaCO_3$ is required per mg ammonia nitrogen oxidized (Khin and Annachhatre, 2004). Severe pH depletion can occur through acid production during the nitrification process, when the alkalinity in the wastewater is low. Therefore, in cases where the alkalinity of the wastewater is depleted by the acid produced by nitrification, the proper alkalinity must be supplemented by chemical addition, such as lime, for denitrification to occur. Also, during denitrification, the requirement of organic carbon is significant. A variety of electron donors and carbon sources such as, methanol, acetate, glucose, ethanol, acetic acid, lactic acid and at times wastewaters have been used for denitrification. Methanol (CH_3OH) being relatively inexpensive, has gained widespread application. For complete denitrification, 2.47 g of methanol is required per gram of nitrate nitrogen (McCarty *et al.*, 1969).

Nitrogen removal from high COD wastewaters through conventional nitrification-denitrification has many inherent operational problems. The autotrophic nature of nitrifiers makes them vulnerable to vivacious competition by fast growing heterotrophic bacteria. Besides, stiff competition by heterotrophs, autotrophic ammonia oxidizing bacteria are inhibited by many compounds such as cyanide, thiocyanate and acetyline that are commonly present in wastewaters. Still, nitrifying activity has always been reported in conventional activated sludge systems. Such activity is mostly due to the presence of heterotrophic nitrifiers such as *Bacillus* sp., *Pseudomonas stutzeri, Thiosphaera pantotropha Alcaligenes faecalis* and partly due to some autotrophic ammonia oxidizing bacteria (Gupta, 1997; Kim *et al.*, 2005; Lin *et al.*, 2006). Nitrification by heterotrophic bacteria is an alternative mechanism under stressful condition and not a normal energy generating pathway. Moreover heterotrophic nitrification is much slower (generally 10^4-10^5 times) than autotrophic nitrification and in wastewaters with high C/N ratio nitrifying activity of heterotrophs is usually not reported. In chemostat process treating nitrogenous wastewater, *Thiosphaera pantotropha* could achieve nitrification rate equivalent to *N. europaea* only when the heterotrophs outnumber the autotrophs by 250: 1 ratio (van Niel *et al.*, 1993). Thus, the major ammonia oxidizing activity is undertaken by the autotrophic bacteria.

To avoid inhibition of nitrifiers in presence of heterotrophs and organic compounds, few modifications in the operational process have been developed. These include extended aeration system, anoxic followed by aerobic treatment, etc. The advantages in 'anoxic followed by aerobic treatment' process are utilization of available COD in wastewaters for denitrification, and inhibition of denitrification process can be avoided by excessive pH drop by the preceding nitrification step. However, denitrification in the first place could deplete the wastewater from dissolved oxygen. Hence, cost of aeration for nitrification may be more than in conventional process. Furthermore, in wastewaters containing ammonia, denitrification as the first step is not practically possible due to absence of nitrate as electron acceptor. Hence, under such conditions nitrification is an essential prerequisite, at least partial, if not complete. The effluent from first stage partial nitrification can be combined with raw effluent and then be subjected to denitrification unit. Such process calls for third stage nitrification to complete the nitrification of untreated ammonia. No doubt the process becomes more complex to accommodate strikingly different groups of microorganisms for complete removal of ammonia in the wastewater.

ANAMMOX

Generally unusual observations made during biological processes lead to the investigation for understanding the underlying microbial phenomenon. Conversely, sometimes ideas on certain biological processes come first followed by actual hunt on their existence in nature. This happened when an Austrian chemist, Boarda (1977) wondered about the existence of certain nitrifying microorganisms which could catalyze a hitherto unknown reaction based on thermodynamic calculations.

$$NH_4^+ + NO_2^- \rightarrow N_2 + 2\,H_2O \qquad (4)$$

Unaware of such predictions, Mulder detected unusual disappearance of nitrogen in his denitrifying pilot plant in 1999 at Gist Brocades Fermentation Company. The reactor received two effluent streams, one from the methanogenic pilot plant containing ammonium, sulfide and organic compounds and another from a nitrifying plant. Unexpectedly after 100 days of operations, ammonia was decreasing with no built up of NO_2^-or NO_3^-. He coined the term ANAMMOX for anaerobic ammonium oxidation in presence of NO_3^-. However, Mulder was not sure of the biological nature of the reaction. Research carried out in collaboration with J.G. Kuenen and his group, comprising Astrid van de Graaf, Mike Jetten and Marc Strous, not only established the biological nature of the reaction and its mechanism, but also established the identity of the responsible microorganism that belonged to the group *Planctomycetes*. The first ANAMMOX bacteria identified was *Canditatus Brocardia, ANAMMOXidans* (Strous *et al.*, 1999). Further research through physiological and genomic analysis revealed many unexpected facts. CO_2 can be used as carbon source to produce biomass and NO_2^- functions not only as electron acceptor for ammonium oxidation, but also an electron donor for reduction of CO_2.

$$NH_4^+ + 1.32NO_2 + 0.066HCO_3^- + 0.13H^+ \rightarrow 1.02N_2 + 0.26NO_3^- + 0.066\,CH_2O_{0.5}N_{0.15} + 2.03H_2O \qquad (5)$$

Besides NO_2^-, NO_3^- can also act as electron acceptor. The autotrophic ANAMMOX bacteria can oxidize formate, acetate and propionate to CO_2 through dissimilatory pathway. Assimilatory pathway for utilization of organic carbon source does not exist. The organism can generate NO_2^- and NH_3 form NO_3^- using formate, acetate and propionate as electron donor; and thereby capable of producing its electron donors and acceptors.

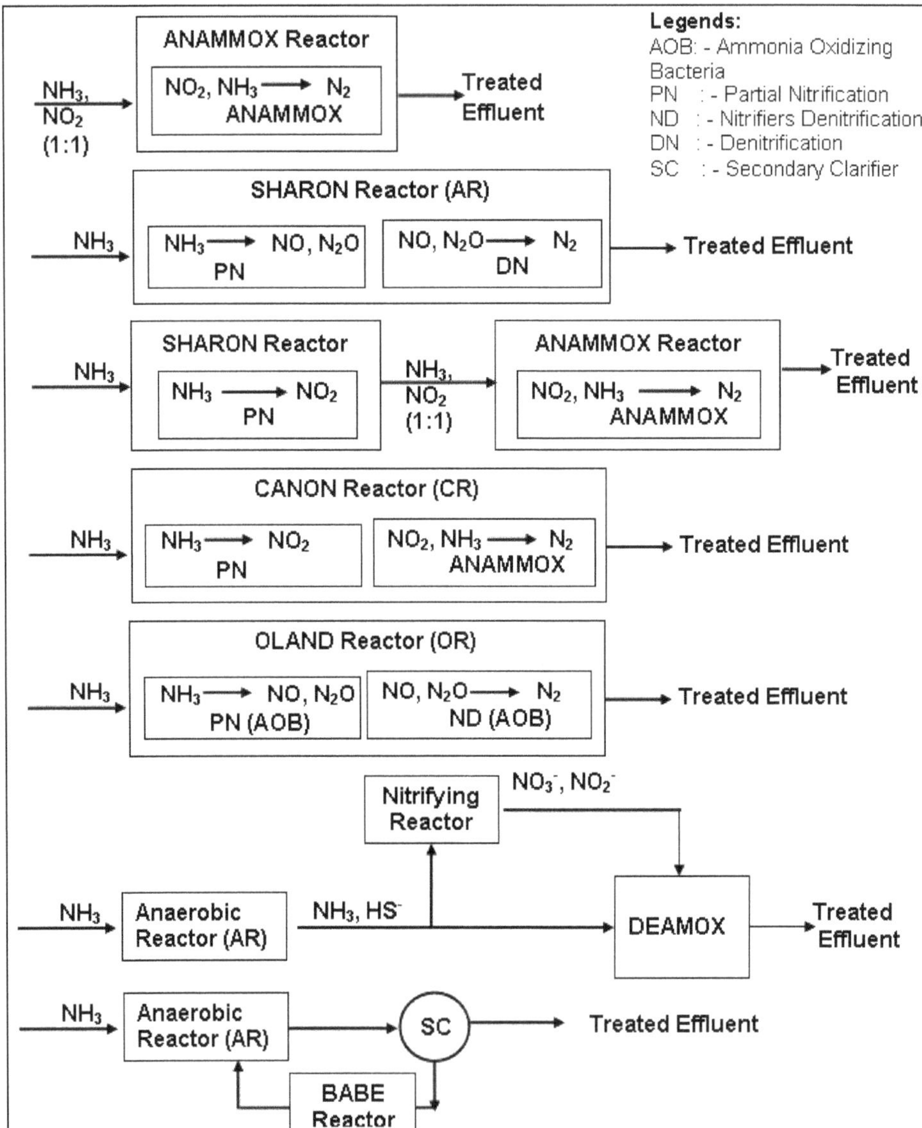

Figure 10.3: Schematic Figures of (i) ANAMMOX Process, (ii) SHARON Process, (iii) SHARON-ANAMMOX Process, (iv) CANON Process, (v) OLAND Process, (vi) DEAMOX Process, and (vii) BABE® Process (Adapted from Kalyuzhnyi and Gladchenko, 2009; Paredes *et al.*, 2007, Bagchi *et al.*, 2011).

The ANAMMOX conversion is an elegant shortcut in the natural nitrogen cycle. The reaction needs NH_4^+/NO_2^- in a ratio of 1: 1.3. Generally, wastewaters always do not have both ammonia and nitrite in the required ratio of 1: 1.3. In combination with nitrification ANAMMOX bacteria convert ammonium (NH_4^+) directly into dinitrogen gas (Figure 10.3). Hence ANAMMOX process has been always applied in association with other low cost technologies that could partially convert ammonia in the wastewaters into nitrite.

The ANAMMOX® process is an innovative biological process that represents a major breakthrough in nitrogen removal. It is a cost effective, robust and sustainable way of removing ammonium from wastewater and from waste gas. Compared to conventional nitrifation/denitrification, operational costs are reduced by up to 90 per cent as are CO2 emission levels. This brings the plant's carbon footprint down to a minimum.

Working closely with Delft University of Technology, Paques developed the patented ANAMMOX® process for commercial purposes. The first full-scale ANAMMOX® plant started up in the Netherlands in 2002. Ten years later there are 11 full-scale ANAMMOX® reactors are operational. ANAMMOX process is widely accepted as it achieved 60 per cent savings in aeration cost and 100 per cent savings in organic carbon dosing (Figure 10.4).

SHARON

The Single reactor High activity Ammonium Removal Over Nitrite (SHARON) process was originally developed for the removal of ammonia via the so called "nitrite route' as represented in Figure 4 (Hellinga *et al.*, 1997; Jetten *et al.*, 1997). In principle, this process involves removal of ammonia to nitrite stage through partial nitrification followed by denitritation of nitrite to nitrogen. Both autotrophic nitrification and heterotrophic denitritation take place under high temperature (35°C) and pH above 7.0 with no sludge retention (Hellinga *et al.*, 1997). Partial nitrification of ammonium to nitrite reduces the requirement of aeration upto 25 per cent and denitrification of nitrite to nitrogen also reduces the cost of addition of electron donor by 40 per cent. Basically the SHARON process involves a single stage completely mixed tank, which is intermittently aerated to accommodate sequential nitrification and denitrification. The process can be carried out in a chemostat such as continuous stirred tank reactor or in a SBR with suspended biomass. Alternatively, two separate tanks, one for nitrification and another for denitrification, has also been used to lower the requirement of aerator capacity. Depending upon the site specific conditions at the wastewater treatment plant (WWTP), the selection of proper configuration is made.

For achieving stable performance, the SHARON process requires partial nitrification of ammonia at nitrite level. Stable partial nitrification can be achieved by eliminating the growth of nitrite oxidizing bacteria (NOB) without harming the activity of ammonia oxidizing bacteria (AOB) in the nitrifying culture (Bagchi *et al.*, 2009). NOB can be washed out of a nitrifying culture through controlling temperature, pH, hydraulic retention time, substrate concentration and dissolved oxygen (DO). The mechanisms to obtain stable partial nitrification are discussed elsewhere. In SHARON

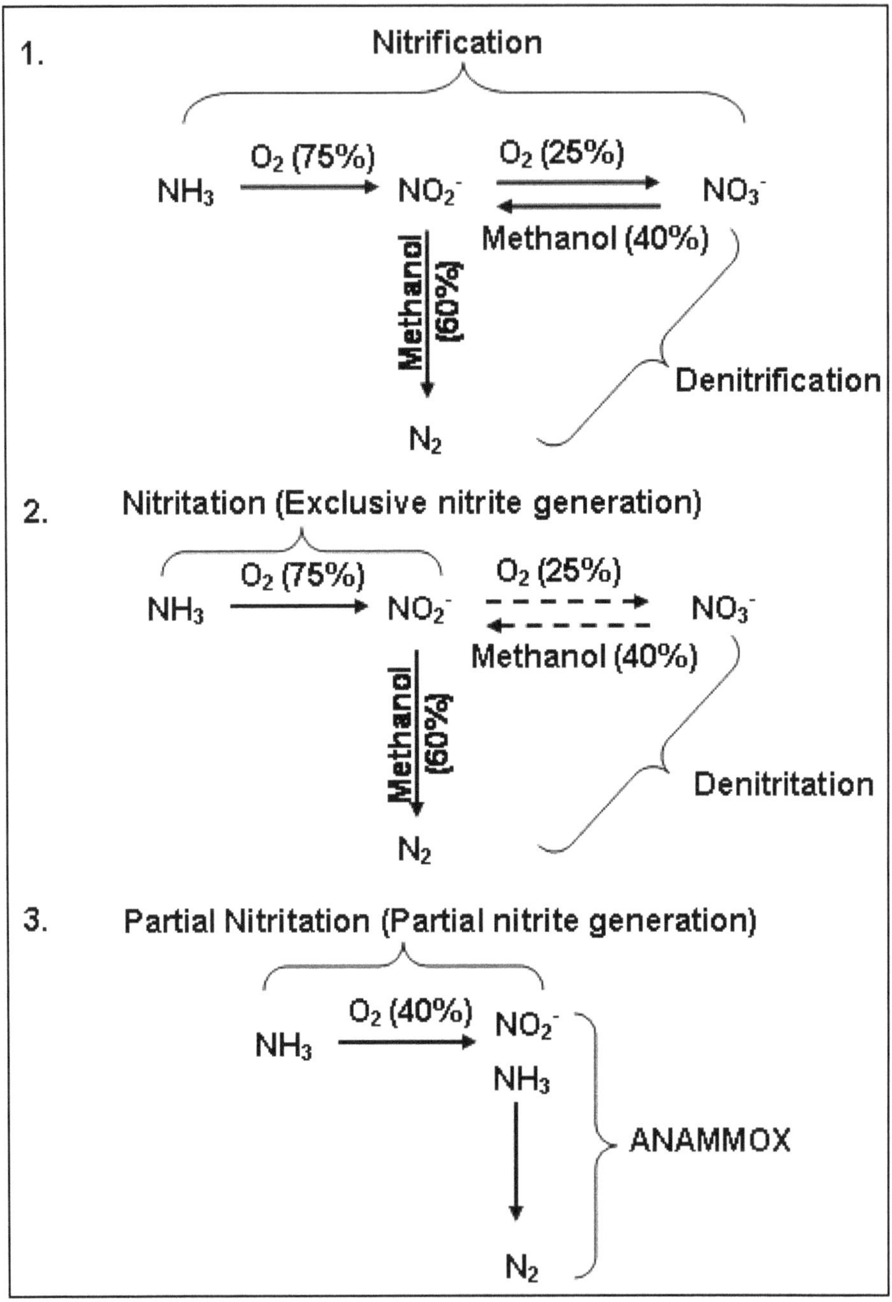

Figure 10.4: Savings in Terms of Cost of Aeration and Cost of Electron Addition in the Three Processes. 1. Conventional Nitrification/Denitrification, 2. The SAHRON Process (through the 'Nitrite Route'), 3. The Partial Nitritation/ANAMMOX. (Adapted from Bagchi *et al.*, 2011).

process, NOB are washed out from the system by maintaining higher operating temperature (35°C) and lower sludge retention time (SRT) or to be more specific, lower aeration retention time (ART) of 1.5 to 2 days (Hellinga *et al.*, 1997). Temperature and ART forms the two major process parameters in SHARON process. In a single stage configuration, cyclic phases of 'aeration' and 'no aeration' are provided for nitrification and denitrification to occur intermittently. The accumulation of nitrite, which is toxic to AOB, is prevented by the denitrification step. Denitrification also restores alkalinity consumed during nitrification process. For denitrification to proceed, organic carbon source has to be additionally supplied. Besides, temperature (between 30-40°C) and ART, inlet ammonia concentration, system pH, DO concentrations and system nitrite concentrations are also to be critically controlled for successful operation of SHARON process.

In wastewater treatment, SHARON process can substantially reduce high ammonia concentration along with organics. The first full-scale demonstration of SHARON® process was carried out at WWTPs of Utrecht (NL) and Rotterdam Dokhaven (NL) (Mulder *et al.*, 2001; van Kempen *et al.*, 2001). At present six full scale plants have been developed at the Netherlands and one at New York, USA. The applied nitrogen loads in these plants are in the range 400-2500 kg N/d (Table 10.1). Because of high temperature it has a high potential for application in tropical regions also.

SHARON-ANAMMOX® Process

In the SHARON-ANAMMOX® process, partial nitrification in SHARON process is used to deliver feed having NH_4^+/NO_2^- ratio close to 1 for the ANAMMOX process. The combination of SHARON and ANAMMOX processes (SHARON-ANAMMOX® process) offers several advantages over the conventional nitrification-denitrification process namely 60 per cent lower oxygen requirement, no need for organic carbon and low production of the excess biomass. This process have been successfully demonstrated at laboratory scale on various types of wastewaters such as anaerobic sludge digesters (Vazquez-Padin *et al.*, 2009), sludge-rejected water from domestic wastewater treatment plant, food and agro industry, coke oven effluent and animal wastewater such as slaughter house wastewater and piggery wastewater. The first full scale application of this process was demonstrated in Rotterdam, the Netherlands, in 2002 (van der Star *et al.*, 2007).

CANON

The Completely Autotrophic Nitrogen Removal Over Nitrite (CANON) process is a combination of partial nitrification and ANAMMOX in a single aerated reactor, employing two groups of bacteria namely, *Nitrosomonas*-like aerobic microorganisms and Planctomycetes-like anaerobic bacteria (Sliekers *et al.*, 2002). This process is critically controlled under limited oxygen condition. Ammonia oxidizing bacteria consumes all available oxygen while partially oxidizing ammonia to nitrite. This creates oxygen free environment needed for the ANAMMOX process. Thus, the cooperation and co-existence of aerobic and anaerobic ammonia oxidizing bacteria is necessary for successful operation of this process (Figure 10.2). The Fluorescent in

Table 10.1: Full Scale SHARON Plants at the Netherlands (Adapted from Bagchi *et al.*, 2011)

Site	*Since in operation*	*System Configuration*	*Wastewater*	*Load (kg N/d)*	*Efficiency (per cent)*
Utrecht	1997	Two (3000m^3 and 1500 m^3)	Sludge dewatering	900	90-95
Rotterdam-Dokhaven	1999	Single (1800 m^3)	Sludge dewatering	850	85-98
Zwolle	2003	Two (900m^3 and 450 m^3)	Sludge dewatering	410	85-95
Beverwijk	2003	Two (1500m^3 and 750 m^3)	Sludge dewatering	1200	85-95
The Hague-Houtrust	2005	Single (2000m^3)	Sludge dewatering	1300	85-98
Groningen-Garmerwolde	2005	Two (4900m^3 and 2450 m^3)	Sludge dewatering	2400	$\geq$95

Situ Hybridization (FISH) assay indicated 40 per cent of AOB and 60 per cent of ANAMMOX bacteria in CANON biomass with absence of aerobic nitrite oxidizing bacteria (NOB) (Sliekers *et al.*, 2002). However, prolonged exposures to ammonia limited condition result in development of NOB in CANON biomass.

Virtually, CANON in an integration of SHARON-ANAMMOX process into one single reactor. Although similar operational controls are needed for both processes, CANON provides more sensitive operational controls in terms of dissolved oxygen, nitrogen-surface load, biofilm thickness, temperature, etc. The ANAMMOX bacteria are reversible inhibited by low concentration of oxygen, above 0.5 per cent air saturation. So an appropriate DO concentration enables the consumption of oxygen to an extent in which DO concentration is not over the toxic threshold for ANAMMOX bacteria and inadequate for the growth of nitrite oxidizing bacteria. The oxygen-limited condition (< 0.5 per cent air saturation) provides an adequate environment for stable interaction between *Nitrosomonas*-like aerobic microorganisms and *Planctomycete*-like anaerobic bacteria (Schmidt *et al.*, 2002).

The CANON process has been tested in different lab-scale reactors. The sequential batch reactor (SBR) has been considered as an effective reactor system for CANON due to efficient biomass retention (Zhang *et al.*, 2008). However, in a gas-lift reactor, higher ammonia removal rate of 1.5 Kg N/m^3.d has also been achieved as against SBR (0.3 kg N/m^3.d).

Few pilot scale and full scale applications of CANON process are reported in literature (Joss *et al.*, 2009; Vazquez-Padin *et al.*, 2009). In a separate study, urea was used as feed for CANON system instead of conventional wastewater. The system adopted very quickly and reached to full capacity within two week of adaptation. This has widened the scope of application of this process to urine waste along with other low C/N ratio wastewaters.

OLAND

The Oxygen-Limited Autotrophic Nitrification Denitrification (OLAND) process refers to the removal of ammonium-N under strict oxygen limitation by autotrophic organisms (Kuai and Verstraete, 1998). Initially, the process was considered to be dependent mainly on aerobic ammonium oxidizing bacteria (AerAOB), but recently it has been reported that under conditions of low oxygen supply and absence of organic electron donors, both AerAOB and anaerobic ammonium oxidising bacteria (AnAOB) co-exist in the OLAND process. The β-subclass of Proteobacteria includes most of the AerAOB, with Nitroso-species (*Nitrosomonas, Nitrosococcus, Nitrosospira,* etc.) as the principal members, while the AnAOB belong to a deep-branching lineage within the order of the *Planctomycetales*. Thus, in principle the process mechanism of OLAND is similar to CANON. Table 10.2 shows comparative performances of OLAND and CANON process. An application of OLAND process on pilot scale is reported for treatment of digested black water using single-stage SBR (Vlaeminck *et al.*, 2009). Similar to CANON, another process known as the deammoniafication process or DEMON has been successfully implemented at WWTP Strass (Austria) since 2006 (Wett, 2007).

Table 10.2: Comparison of Maximum Nitrogen Loading Rate, Nitrogen Removal Rate different CANON and OLAND Processes (Adapted from Bagchi *et al.*, 2011)

Process	*Reactor Type*	*NLR_{max} (kg N/m³.d)*	*NRR_{max} (kg N/m³.d)*
CANON	SBR	0.131	0.075
	SBR	NR	0.08
	SBR	0.22	0.12
	SBR	0.46	0.36
	SBR	NR	0.5
	MABR	0.87	0.77
	Gas lift	5.5	1.5
OLAND	SBR	0.13	0.05
	RBC	0.08307*	0.083*
	RBC	0.716	0.7

* Values are in Kg N/m².d, NLRmax: Maximum Nitrogen Loading Rate, NRR_{max}: Maximum Nitrogen Removal Rate, SBR; Sequential Batch Reactor, MABR; Membrane Aerated Bio-Raector, RBC; Rotator Biological Contractor, CSTR; Continuous Stirred Tank Reactor.

DEAMOX

The Denitrifying AMmonia OXidation (DEAMOX) process was proposed by Mulder in 2004 for the ANAMMOX reaction occurring in the autotrophic denitrifying plant (Kalyuzhnyi *et al.*, 2006). DEAMOX eliminates the requirement of partial nitritation by introducing denitritation of nitrate using sulphide as an electron donor (Equation 6).

$$NO_3^- + 0.25\ HS^- \rightarrow NO_2^- + 0.25\ SO_4^- + 0.25\ H^+ \qquad (6)$$

The process is applicable for sulphur bearing nitrogenous wastewater such as Baker's yeast effluent. In treatment of Baker's yeast effluent, the whole process operates in combination of three reactor system; anaerobic reactor, DEAMOX reactor and nitrifying reactor. In the anaerobic reactor ammonia and sulphide are generated which is partially fed to nitrifying reactor and the remaining is fed to the DEAMOX reactor (Figure 10.2). In the nitrifying reactor, ammonia is nitrified into nitrate and nitrite. The third reactor, *i.e.* the DEAMOX reactor, receives effluent from both the anaerobic and nitrifying reactors. The denitrification of nitrate takes place in presence of sulphide (equation 6) and the ANAMMOX bacteria coexisting in the DEAMOX reactor along with the autotrophic denitrifying bacteria oxidize ammonia using nitrite as electron acceptor (equation 5). Basically the scheme in the DEAMOX process is similar to the autotrophic denitrifying reactor plant in which ANAMMOX was first detected and then discovered by Mulder (1992).

Since, the standard application of this process is restricted to sulphur-bearing wastewaters, the incorporation of heterotrophic denitratation using volatile fatty acids (VFA) as a more widespread electron donor has been investigated to implement the process for commonly found wastewaters. However, the feasibility study showed a severe competition for nitrite between autotrophic ANAMMOX and heterotrophic denitritation resulting in poor removal efficiency.

The typical sulphide driven DEAMOX processes have maximum nitrogen loading rate of 1 kg N/m^3.d with more than 90 per cent removal efficiency (Kalyuzhnyi *et al.*, 2006). Organics-driven DEAMOX process has similar loading capacity (1.24 1 kg N/m^3.d) with relatively lower ammonia removal (40 per cent), mainly due to the competition of heterotrophic denitrifiers. Although the process is 10 times lower than ANAMMOX process, simple process operation is claimed to favour the DEAMOX process over ANAMMOX process. So far, this process has been tested only in bench scale reactors.

SNAD

Low-cost biological nitrogen removing technologies such as ANAMMOX, SHARON, SHARON-ANAMMOX, CANON, OLAND, and DEAMOX have proved to be very effective in removing ammonia from industrial effluents. However, most ammonium rich wastewaters are produced with certain concentration range of organics (such as old landfill leachates) and presence of organics could upset the autotrophic bacteria which are mostly responsible in all low-cost biological nitrogen removing technologies except in the SHARON process. The application of the combined SHARON-ANAMMOX process has been also limited to wastewater with low total organic carbon to nitrogen ratio (C/N) d" 0.15. Increase of C/N ratio to more than 0.3 can lead to competition between heterotrophic bacteria and autotrophic bacteria in the SHARON process (Wang and Kang, 2005).

Studies have demonstrated that ANAMMOX and denitrification process can coexist in the same environment (Chen *et al.*, 2009). Recently partial nitrification, ANAMMOX and denitrification have been simultaneously investigated on bench scale in a single-stage non-woven rotating biological contractor (NRBC) bioreactor (Chen *et al.*, 2009). The Simultaneous partial Nitrification, ANAMMOX and Denitrification (SNAD) process was found to potentially remove ammonium and organics from wastewater in a single NRBC reactor with total nitrogen removal efficiency of 70 per cent and COD removal efficiency of 94 per cent at nitrogen and COD loading rates of 0.69 kg N/m^3.d and 0.34 kg/m^3.d, respectively. Recently, the SNAD process has been demonstrated at full scale for treatment of landfill leachate (Wang *et al.*, 2010; Xu *et al.*, 2010).

AS/PN-ANAMMOX

The AS/PN-ANAMMOX is a two stage process comprising combined activated sludge process-partial nitrification (AS/PN) followed by the ANAMMOX process. The concept promotes coexistence of heterotrophic and autotrophic microorganisms in conventional extended activated sludge process for organic carbon removal and nitrification. Lamsam *et al.* (2008) have demonstrated that partial nitrification can be achieved in activated sludge process (ASP) by controlling three key parameters, namely, DO (0.3-0.5 mg/l), pH (around 8) and temperature (35°C). It differs from the SHARON process by inclusion of an aeration tank and a settler with sludge recirculation. For accomplishing partial nitrification in ASP, sludge retention time

(SRT) of 12-15 days was required (Lamsam *et al.*, 2008). The AS/PN is reported to have several advantages over the conventional extended ASP for organic removal along with nitrification. First, the process consumes 66 per cent less oxygen, as calculated from the theoretical equations (8-10) based on Henze *et al.* (2000) assumption of $C_{18}H_{19}O_9N$ as chemical formula for organic matter. Second, the effluent from AS/PN contains effluent ammonia and nitrite in the ratio of 1: 1, which is desirable for the downstream ANAMMOX process. In addition, a conventional ASP can be up-graded into AS/PN by controlling pH, temperature and DO (Lamsam *et al.*, 2008).

ASP with Nitrification

$$C_{18}H_{19}O_9N + 19.5O_2 \rightarrow 18CO_2 + 9H_2O + H^+ + NO^-_3 \quad (7)$$

ASP without Nitrification

$$C_{18}H_{19}O_9N + 17.5O_2 + H^+ \rightarrow 18CO_2 + 8H_2O + NH^+_4 \quad (8)$$

AS/PN

$$C_{18}H_{19}O_9N + 18O_2 \rightarrow 18CO_2 + 8H_2O + H^+ + 0.5\,NH^+_4 + 0.5NO^-_2 \quad (9)$$

Operating conditions of AS/PN process for coupling with ANAMMOX process were identified as pH: 7.7–8.2 and DO: 0.5–0.9 mg/L to achieve over 85 per cent COD removal as well as partial nitrification. Maximum nitrogen removal rate for the ANAMMOX process was found to be 0.6 kg N/m^3.d. The process was applied for treating sea food processing wastewater at bench scale.

BABE® Process

The Bio-Augmentation Batch Enhanced (BABE)® process was developed to bio-augment nitrification in activated sludge process that operated at sub-optimal solid retention times (Salem *et al.*, 2004). The principle was to implement a nitrification reactor (BABE reactor) in the sludge return line (Figure 10.2). The BABE reactor would be fed with an internal N-rich flow (such as effluent from the sludge treatment). Nitrifiers grown in the BABE reactor would ultimately reach the activated sludge reactor. Hence the nitrification capacity of the activated sludge process (ASP) can be augmented with nitrifiers grown in the BABE reactor. Thus the BABE reactor functions basically as a culture tank supplementing the ASP process with the specially cultivated microorganisms in the culture tank. The BABE technology has been tested in full scale at the WWTP Garmerwolde in Groningen (NL).

The concept of bio-augmentation may be of special interest in a Partial nitritation-ANAMMOX system for bio-augmenting both the cultures in a similar manner and for seeding into the main reactor.

Future Perspective

These applications indicate that the partial nitrification and ANAMMOX based processes are more effective in removing ammonia present in wastewater.

Nitrite is a common intermediate in at least three different oxidative or reductive biochemical pathways that occur in nature (nitrification, denitrification and dissimilatory or assimilatory nitrate reduction). Nitrite accumulation or partial nitrification has been reported in literature for decades. In engineered systems, partial nitrification is of interest as it offers cost savings in aeration as well as in the form of lesser need for addition of organic carbon as compared to the conventional denitrification. However, nitrite accumulation is not a rule of nature. Nitrite is an unstable intermediate of nitrification process. Hence it is very essential for process engineers to understand the process control for achieving successful nitrite accumulation in a reactor.

The Annamox reaction, on the other hand is limited by slow growth rate of amannox bacteria. The practical application of the Anammox process is still limited by its long start-up periods (up to 1 year) due to the very low growth rate and low cell yield of Anammox organisms. Loss of a fraction of the sludge due to wash-out with the effluent could further augment the start-up period. Hence, it is essential to use a reactor with high biomass retention. So far a large range of bioreactor types have been evaluated for the enrichment of Anammox bacteria: fixed bed reactors, fluidized-bed-reactors, UASB-reactors, SBR, gas-lift reactors. Among them, the SBR was accepted for Anammox enrichment for its simplicity, efficient biomass retention, homogeneity of mixture in the reactor, stability and reliability for a long period of operation, stability under substrate-limiting conditions and high nitrogen conversions.

Thus the recent BNR technologies have potential to remove high concentrations of ammonia in wastewaters. In spite of their high removal efficiency, the quantum of full scale applications of these processes is far from trivial. The issues that create bottleneck are to be dealt in details before these processes can claim commercial success.

References

1. Bagchi, S., Biswas, R., Roychoudhury, K., and Nandy, T. (2009). Stable partial nitrification in an up-flow fixed bed bioreactor under oxygen limiting environment. *Environ. Eng. Sci.*, 26 (8), 1309-1318.
2. Bernet, N., Dangcong, P., Delgenes, J., and Molletta, R. (2001). Nitrification at low oxygen concentration in biofilm reactor. *J. Environ. Eng.*, 127, 266-271.
3. Broda, E. (1977). Two kinds of lithotrophs missing in nature. *Z. allg. Mikrobiol.*, 17, 491-493.
4. Chen, H.H., Liu, S.T., Yang, F.L., Yuan, X., and Wang, T. (2009). The development of simultaneous partial nitrification, ANAMMOX and denitrification (SNAD) process in a single reactor for nitrogen removal. *Bioresour. Technol.*, 100, 1548–1554.
5. Gupta, A.B. (1997). *Thiosphaera pantotropha*: a sulphur bacterium capable of simultaneous heterotrophic nitrification and aerobic denitrification. *Enzyme Microb. Technol.*, 21, 589-595.
6. Gujer, W., and Jenkins, D. (1975). The contact stabilization activated sludge process-oxygen utilization, sludge production and efficiency. Water Res., 9, 553-560.

7. Hellinga, C., van Loosdrecht, M.C.M., and Heijnen, J.J. (1997). The Sharon process for nitrogen removal in ammonium rich wastewater. Mededelingen Faculteit Landbouwwetenschappen, Universiteit Gent, 62 (4b), pp. 1743-1750.

8. Henze, M., Gujer, W., Matsuo, T. and van Loosdrecht, M. (2000). Activated Sludge Models ASM1, ASM2, ASM2d and ASM3. *Scientific and Technical* Reports. IWA Publishing, London, UK.

9. Jetten, M.S.M., Horn, S.J., and van Loosdrecht, M.C.M. (1997). Towards a more sustainable municipal wastewater treatment system. *Water Sci. Technol.*, 35, 171–180.

10. Kalyuzhnyi, S.V., Gladchenko, M., Mulder, A., and Versprille, B. (2006). DEAMOX–New biological nitrogen removal process based on anaerobic ammonia oxidation coupled to sulphide-driven conversion of nitrate into nitrite. *Water Res.*, 40 (19), 3637-3645.

11. Khin, T., and Annachhatre, A.P. (2004). Novel microbial nitrogen removal processes. *Biotechnol. Adv.*, 22, 519-532.

12. Kim, K. J., Park, K. J., Cho, K. S., Nam, S. W., Park, T. J., and Bajpai, R. (2005). Aerobic nitrification–denitrification by heterotrophic *Bacillus* strains. *Bioresour. Technol.*, 96, 1897–1906.

13. Kuai, L.P., and Verstraete, W. (1998). Ammonium removal by the oxygenlimited autotrophic nitrification–denitrification system. *Appl. Environ. Microbiol.*, 64, 4500–4506.

14. Lamsam, A., Laohaprapanon, S., and Annachhatre, A.P. (2008). Combined activated sludge with partial nitrification (AS/PN) and ANAMMOX processes for treatment of seafood processing wastewater. *J. Environ. Sci. Health. Part-A Tox. Hazard. Subst. Environ. Eng.*, 43 (10),1198-208.

15. Lin, Y., Tanaka, S., and Kong, H. (2006). Characterization of a Newly Isolated Heterotrophic Nitrifying Bacterium. *Water Practi. Technol.*, 1 (3), doi10.2166/wpt.2006.0052.

16. McCarty, P.L., Beck, L., and St. Amant, P. (1969). Biological denitrification of wastewaters by addition of organic materials: *Proc. 24th Industrial Waste Conference*, West Lafayette, USA, pp-1271–1285.

17. Mulder, J.W., van Loosdrecht, M.C.M., Hellinga, C., and van Kempen, R. (2001). Full scale application of the SHARON process for treatment of rejection water of digested sludge dewatering. *Water Sci. Technol.*, 43 (11), 127–134.

18. Salem, S., Berends, D.H.J.G., van der Roest, H.F., van der Kuij, R.J., and van Loosdrecht, M.C.M. (2004). Full scale application of BABE® technology. Water Sci. Technol., 50 (7), 87-96.

19. Schmidt, I., Sliekers, O., Schmid, M., Cirpus, I., Strous, M., Bock, E., Kuenen, J.G., and Jetten, M.S.M. (2002). Aerobic and anaerobic ammonia oxidizing bacteria–competitors or natural partners? *FEMS Microbiol. Ecol.*, 39,175–181.

20. Sliekers, A.O., Derwort, N., Gomez, J.L., Strous, M., Kuenen, J.G., and Jetten, M.S.M. (2002). Completely autotrophic nitrogen removal over nitrite in one single reactor. *Water Res.*, 36, 2475–2482.

21. Strous, M., Kuenen, J.G., and Jetten, M.S.M. (1999). Key physiology of anaerobic ammonium oxidation. *Appl. Environ. Microbiol.*, 65, 3248– 3250.

22. van der Star, W.R.L., Wiebe, R., Abma, W.R., Blommers, D., Mulder, J.W., Tokutomi, T., Strous, M., Picioreanu, C., and van Loosdrech, M.C.M. (2007). Startup of reactors for anoxic ammonium oxidation: Experiences from the first full-scale ANAMMOX reactor in Rotterdam. *Water Res.*, 41 (18), 4149-4163.

23. van Kempen, R., Mulder, J.W., Uijterlinde, C.A., and van Loosdrecht, M.C.M. (2001). Overview : Full scale experience of the SHARON® process for treatment of rejection water of digested sludge dewatering. *Water Sci. Technol.*, 44 (1), 145-152.

24. van Niel, E.W.J., Arts, P.A.M., Wesselink, B.J., Robertson, and L.A., Kuenen, J.G. (1993). Competition between heterotrophioc and autotrophic nitrifiers for ammonia in chemostat cultures. *FEMS Microbiol. Ecol.*, 102, 109 -118.

25. Vazquez-Padin, J.R., Pozo, M.J., Jarpa, M. Figueroa, M., Franco, A., Mosquera-Corral, A., Campos, J.L., and Mendez, R. (2009). Treatment of anaerobic sludge digester effluents by the CANON process in an air pulsing SBR. *J. Hazard. Mater.*, 166, 336–341.

26. Vlaeminck, S.E., Terada, A., Smets, B.F, van der Linden, D., Boon, N., Verstraete, W., and Carballa, M. (2009). Nitrogen removal from digested black water by one-stage partial nitritation and ANAMMOX. *Environ. Sci. Technol.*, 43, 5035–5041.

27. Wang, C.C., Lee, P.H., Kumar, M., Huang, Y.T., Sung, S., and Lin, J.G. (2010). Simultaneous partial nitrification, anaerobic ammonium oxidation and denitrification (SNAD) in a full-scale landfill-leachate treatment plant. *J. Hazard. Mater.*, 175, 622–628.

28. Wett, B. (2007). Development and implementation of a robust deammonification process. *Water Sci. Technol.*, 56 (7), 81–88.

29. Xu, Z.Y., Zeng, G.M., Yang, Z.H., Xiao, Y., Cao, M., Sun, H.S., Ji, L.L., and Chen, Y. (2010). Biological treatment of landfill leachate with the integration of partial nitrification, anaerobic ammonium oxidation and heterotrophic denitrification. *Bioresour. Technol.*, 101, 79–86.

30. Zhang, L., Zheng, P., Tang, C-J., and Jin, R-C. (2008). Anaerobic ammonium oxidation for treatment of ammonium-rich wastewaters *J. Zhejiang Univ. Sci B.*, 9 (5), 416-426.

Chapter 11

Membrane Bioreactor for the Treatment of Wastewater

***R.S. Singh*[1], *B.N. Rai*[1] *and S.N. Kaul*[2]**

[1]*ITBHU, Varanasi*
[2]*MIT, Pune*

ABSTRACT

With continuing depletion of freshwater resources, focus has been shifted more towards water recovery, reuse and recycling which require an extension of conventional wastewater treatment technologies. Stricter compliance requirements for wastewater discharge, rising treatment costs and spatial constraints necessitate renewed investigation of alternative technologies other than to traditional activated sludge process.

Membrane technology coupled with biological treatment processes has gained considerable attention over the last few years due to its wide range of applicability and high performance of membrane systems. The membrane bioreactor (MBR) is based on the combination of a suspended biomass reactor along with a separation step on porous membrane filtration. When compared to a conventional activated sludge process, the MBR is operated in similar but without the need for secondary clarifier. It has several advantages in terms of water resource protection such as good quality of the treated water that can be reused for irrigation, cleaning or cooling water on industrial site, domestic purposes. Moreover, the quality of water, notably the absence of solids and turbidity make the MBR as a perfect tool for the wastewater treatment.

The present article is an attempt to provide upto date information on the recent developments and applications of MBR's as well as key factors on which performance of MBR depends like membrane types, membrane fouling, backwashing etc.

Keywords: *MBR, Wastewater, Activated sludge process, Membrane fouling.*

Introduction

Today, the level of pollution is more when compared to past and even water not remained untouched by this. Every community produces wastes in all the existing forms namely solid, liquid and gaseous. Among these, liquid wastes generally we say wastewater is obtained after various forms of applications by the community. From various resources of generation, we may define wastewater as a combination of the liquid or water carried wastes removed from various sectors such as residences, institutions and commercial and industrial establishment together. This untreated wastewater when allowed to accumulate and to go septic in an open place, the organic matter present in it decomposes and even leads in the production of objectionable and odorous gases.

With this, untreated wastewater contains numerous pathogenic microorganisms that dwell in the human intestinal tract. Wastewater also contains nutrients, which can stimulate the growth of aquatic plants, and may contain toxic compounds that potentially may be mutagenic or carcinogenic. For these reasons, the immediate and nuisance free removal of wastewater from its sources of generation, followed by treatment, reuse or disposal into the environment is necessary to protect public health and environment

Wastewater treatment is based on the abatement of

1. Solids in suspension by physical process (settling and/or filtration) and
2. Soluble compounds mainly by biological processes when compounds are biodegradable, physical, or thermal specific processes (chemical oxidation, ion exchange, desorption, etc.) when they are not biodegradable.

However since the 1990's, it is no longer sufficient to eliminate only settable and soluble organic compounds but phosphorous and nitrogen must be removed to comply with the effluent quality standards.

For biodegradable compounds, biological process combinations are proposed to ensure the appropriate water quality and the environmental legislation in regards to organic carbon, nitrogen or phosphorous removal. Different extensive processes for small communities such as constructing wetlands or intensive processes such as trickling beds and activated sludge for large industrial plants or large cities are available.

The use of biological treatment can be traced back to the late nineteenth century. By the 1930s, it was a standard method of wastewater treatment (Rittmann, 1987). Since then, biological treatment methods involving, aerobic and anaerobic were commonly used to treat domestic and industrial wastewater.

During the course of these processes, organic matter, mainly in soluble form, is converted intoH_2O, CO_2, NH_4^+, CH_4, NO_2^-, NO_3^- and biological cells. Depending on the presence or absence of oxygen, the products may differ. Nevertheless, biological cells are always a product, although their quantity varies depending on whether it is an aerobic or anaerobic process. The biomass formed after the biological treatment must be removed to ensure the effluent quantity. A secondary settling tank is used for the solid/liquid separation and this clarification is often the limiting factor in effluent quality (Benefield and Randall, 1980).

In recent years, effluent standards play the major role in preserving existing water resources. Recycling and reuse of wastewater for secondary purposes is on the rise due to dwindling natural resources, increasing water consumption, and the capacity limitations of existing water and wastewater conveyance systems. In both cases, achieving a high level of treatment efficiency is imperative.

The quality of the final effluent from conventional biological treatment systems is highly dependent on the hydrodynamic conditions in the sedimentation tank and the settling characteristics of the sludge. Consequently, large volume sedimentation tanks offering several hours of residence time are required to obtain adequate solid/liquid separation (Fane *et al.*, 1978). At the same time, close control of the biological treatment unit is necessary to avoid conditions that lead to poor settle ability and/or bulking of sludge., however, economic constraints plays major role in limiting such options. Even with such controls, further treatment such as filtration, carbon adsorption, etc. are needed for most applications of wastewater reuse. Therefore, a solid/liquid separation method different from conventional methods is necessary.

Application of membrane separation (micro- or ultrafiltration) techniques for biosolid separation can overcome the disadvantages of the sedimentation tank and biological treatment steps. The membrane offers a complete barrier to suspended solids and yields higher quality effluent. Although the concept of an activated sludge process coupled with ultrafiltration was commercialized in the late 1960sby Dorr-Oliver (Smith *et al.*, 1969), the application has only recently started to attract serious attention (Figure 11.1), and there has been considerable development and application of membrane processes in combination with biological treatment over the last two decades.

This emerging technology, known as a membrane bioreactor (MBR), offers several advantages over the conventional processes currently available. These include excellent quality of treated water, which can be reused for industrial processes or for many secondary household purposes, small footprint size of the treatment plant, and reduced sludge production and better process reliability.

The Activated Sludge Process

The activated sludge process consists of three processes in series in which the mixture of wastewater and biomass is manipulated to perform the desired reactions, resulting in a clean effluent. The first step consists of pre-treatment to remove coarse material and other undesired substances. Usually this is followed by primary treatment, like sedimentation, in order to remove particles. Subsequently the influent is mixed with the biomass and treated under aerobic and/or anoxic conditions. The treated water is separated from the biomass in a clarifier, usually a sedimentation tank. The biomass is partially returned to the reactor and partially wasted; a schematic overview of a typical conventional activated sludge process is presented in Figure 11.1 given below.

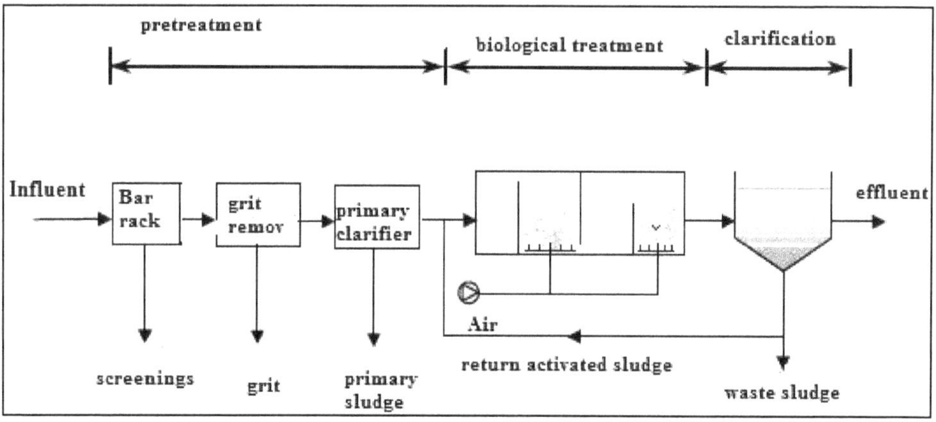

Figure 11.1: Flow Scheme of a Conventional Activated Sludge Process.

Although this conventional technique may appear robust and safe, the activated sludge process displays three major disadvantages.

- ☆ The treated water quality is dependent on the settling properties of the biological suspension. Indeed, bad flocculation of the biological suspension, resulting from dysfunction (*e.g.* variation of influent characteristics, oxygen shortage, etc.), includes bad settling with (i) a resulting presence of suspended solids in the treated water and (ii) a progressive washing out of the biomass from the bioreactor.
- ☆ According to the variation of the biomass settling rates regards to biomass concentration, the clarification is generally carried on with biomass concentration lower than 5 kgTSS/m^3 to be in favorable settling conditions.
- ☆ The insufficient germ removal or persistent organic pollutants presence limit the reuse of treated waters except (i) for specific condition of irrigation or (ii) when using tertiary treatment including extensive processes or some drastic oxidation step.

Now days, because of the shortage of water resources and continuous quality degradation, it is clearly essential that new generation of treatment processes have to be defined to achieve:

- ☆ Disinfection without any oxidation step that induces carcinogen molecule formation,
- ☆ Possibility of compactness to optimize aesthetics, environmental impact (odour and noise),
- ☆ Reliability not withstanding the influent characteristic variation,
- ☆ Standards regarding sustainability (energy, chemicals and waste production).

MBR

The Membrane Bioreactor (MBR) holds the potential to become one of the new generation treatment processes. This system is based on the combination of a suspended biomass reactor and a separation step on porous membrane filtration. Low –pressure membranes, either microfilters (MF) or ultrafilters (UF), are used for the solid/liquid separation. The membrane retains not only solids but also high – molecular weight compounds that generally escape into the treated water when settling is operated as the final separation step. Thus the retention capacity of the porous involves a very high treated degree of the treated water in terms of solids retention including germs and viruses (ref 3).

The membrane configurations are based on either plane or cylindrical geometry by using plate-and-frame membranes, tubular and hollow fibres. The membrane bundle can be configured in a side stream position or directly immersed in the bioreactor.

In side stream configuration, the operation of filtration is generally carried out under high pressure (from 1 to 4 bar) and consists in a tangential recirculation at high speed of the bacterial suspension in the membrane module (from 0.5 to 4 m/s).

In immersed configuration, the membranes are even inside the biological reactor, in total immersion. Filtration is carried out under low transmembrane pressure (from 0.05 to 0.5 bars) and of more important membrane surfaces must be so operated in order to ensure the desired productivity. Because the membrane separation is not dependant on the influent characteristics or the flocculation state of the biological suspension, the biomass concentration can be higher than in conventional activated sludge (2-5 times higher). So the reaction rates can be more successful and the quality of treated water is generally very high from sanitary point of view.

Historical Development of MBR

The first descriptions of this technique date from the late sixties. In the 1970s the technology entered the Japanese market and by 1993, 39 external MBR systems had been reported for use in sanitary and industrial applications. At this moment, MBR systems are applied widely throughout Japan for domestic wastewater treatment, reuse, and some industrial applications. In the 1980s, the Japanese government invested in the development of a low footprint, high product quality process that would be suitable for water recycling. Within this programme the Kubota plate and frame membrane was developed. Many small-scale applications were realised in Japan during the 80's and 90's.At the American continent the developments in the MBR field led to the development of the hollow fibre submerged membrane, by Zenon. In the nineties, the application of the process was extended to larger scale wastewater treatment and much of these developments took place in the USA and Europe. Due to decreasing membrane prices and improvements of membrane modules and materials, the number of MBR installations in Europe increased considerably in the past ten years. At this moment some 45 municipal MBRs are in operation throughout Europe. Many manufacturers of membranes have developed membrane configurations and membrane materials especially for application in MBR. Characteristics of some commercially available MBR configurations are given in Table 11.1.

Table 11.1: Commercially available MBR Configurations.

Supplier	*Type of Reactor*	*MBR Type*	*Membrane Configuration*	*Membrane Material and Structure*
Huber VRM	Aerobic	Submerged, UF	Plate and frame	PES composite asymmetric
Hydranautics	Aerobic	External, UF	Capillary	PES asymmetric
Kubota	Aerobic	Submerged, UF	Plate and frame	
Mitsubishi	Aerobic	Submerged, UF	Capillary	PE, symmetric
X-flow/Norit	Aerobic	External, UF	Tubular	PVDF, asymmetric
Puron	Aerobic	Submerged, UF	Capillary	PES, symmetric
Rhodia	Aerobic	External, UF	Plate and frame	PES, symmetric
Seghers	Aerobic	Submerged, UF	Plate and frame	PVDF, asymmetric
Zenon	Aerobic	Submerged, UF	Capillary	PVDF, asymmetric

Configuration of MBRs

According to the configuration, MBRs can be classified as side stream and submerged.

Membranes Submerged in the Activated Sludge Tank

This configuration was first introduced by Yamamoto *et al.* (1988). A membrane module was directly submerged in the bioreactor to avoid the recirculation pump. Only a suction pump was used on the permeate side to create transmembrane pressure (TMP). In some circumstances (*e.g.*, MF membrane and very low filtration fluxes), the permeate side is placed on a lower position, and the gravity itself is the driving force for filtration (Ueda and Hata, 1999). Air is injected at the bottom of the membrane in order to limit fouling. The configuration is shown in Figure 11.2.

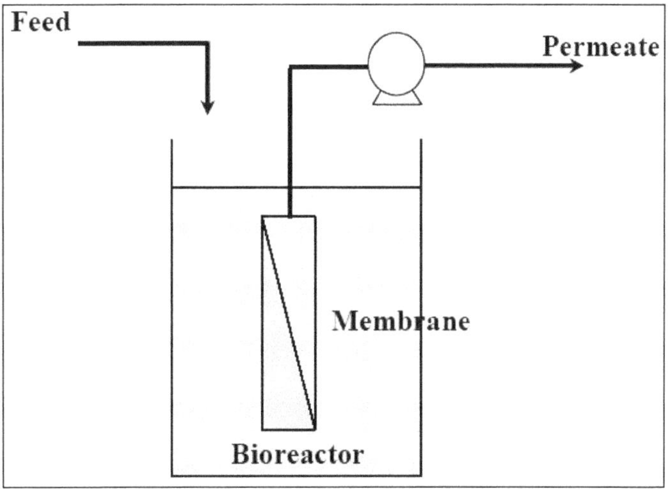

Figure 11.2: Submerged Membranes.

Recent systems are using two different aeration units. One injects fine dispersed bubbles, which is necessary for biomass growth, and another generates coarse bubbles to control fouling. The liquid flow close to the membrane is only due to permeate flow through the membrane wall and to a liquid motion induced by the bubbles. Commonly used membrane configurations are hollow-fibre and plate and frame modules.

External Membranes System (Figure 11.3)

The second possible location for the membrane systems is to be external to the activated sludge tank.

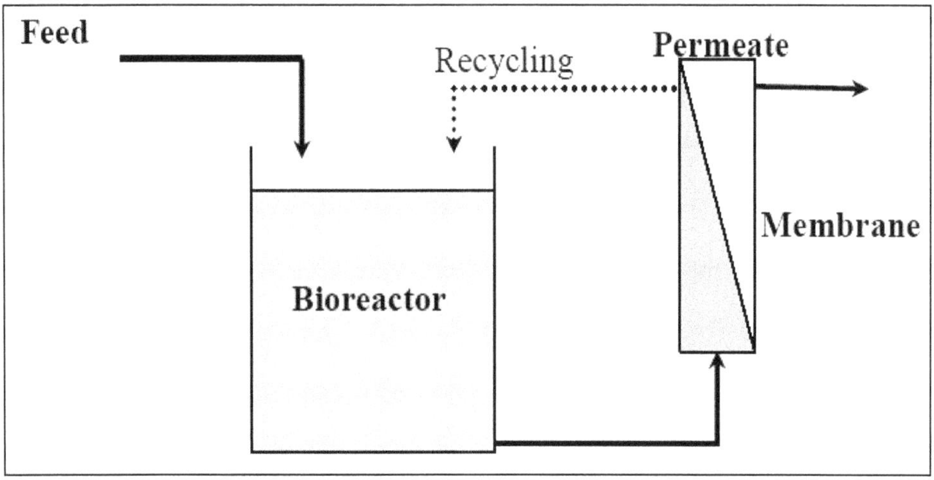

Figure 11.3: External Membrane System or "Side Stream" System.

The membrane system is thus outside the bioreactor. If the concentrate is recycled to the bioreactor a liquid velocity is generated in the membrane system, the flow patterns and the recycling rate are linked. For tubular membranes, high velocities and high recycling rates are generally imposed. In hollow fibres or flat sheet membranes, the range of liquid velocities is lower. Low liquid velocities are used to decrease the particles concentration in the membrane system. High liquid velocities are used with the aim to control the formation of the deposit. If the concentrate is not recycled, the liquid velocity in the carter is negligible, but the water recovery by the filtration system is 100 per cent, which can be called a "dead-end" system. This process is commercialised by industrial firms with hollow fibres as Polymem, Degrémont, Mitsubishi, Memcor, and Zenon.

When coming to point of comparison between submerged and external membrane systems Gander *et al.*, reviewed four side stream and four submerged MBR systems and concluded that the side stream MBRs have a higher energy cost, by up to wo orders of magnitude due to the high recycle flow velocity (1-3 m/s) and head loss with membrane module. In addition to the energy, saving the submerged MBRs suffered less fouling and can cleaned easily (Gander *et al.*, 2000).

Possible advantages and drawbacks of these two MBR configurations are presented in Table 11.2 (Till and Malia; 2001).

Table 11..2: Configuration of MBRs.

Submerged Membrane System	*External Membrane System*
Aeration cost high ~90 per cent	Aeration cost low ~20 per cent
Very low pumping costs	Very high pumping costs
Lower flux	Higher flux
Less frequent cleaning required	More frequent cleaning required
Lower operating costs	Higher operating costs
Higher capital costs	Lower capital costs

Comparison of Membrane Systems

There are very few studies showing a comparison of membrane systems. Several studies are describing MBRs operation and they are providing data on single system. But comparison of these data will be very risky. Indeed, it will be very difficult to compare these results for systems, which are operating at different conditions with different wastewaters, at different scales (from lab scale to industrial scale).

As an example, Fane (2002) proposed a qualitative comparison of some membrane modules (Table 11..3).

Table 11.3: Characteristics of Tubular Membranes, Flat Sheet Membranes, and Hollow Fibre Membranes Adapted from Fane (2002)

Characteristics	*Tubular Membranes*	*Flat Sheet Membranes*	*Hollow Fibre Membranes*
Arrangement	External-recycling	External/submerged	External/submerged
Packing density	Low	Moderate	High
Energy demand	High (turbulent flow)	Low-moderate (laminar flow)	Low
Cleaning	Efficient +physical cleaning possible	Moderate	Back washing possible
Replacement	Tubes or element	Sheet	Element

Advantages and Disadvantages of MBR (Table 11.4)

The MBR process is actually a extension or a new development of the existing conventional activated sludge (CAS) process. The invention of MBRs is to overcome few drawbacks of the activated sludge process, where the often-occurring drawback will be the biomass separation step in the secondary clarifier. In general, gravity will be the main force to settle the flocs in the secondary clarifier. As the specific gravity variation between the sludge (1.02) (Tchobanoglous *et al.*, 2003) and water is very close, the poor settling of flocs is a common phenomenon in the secondary clarifier of activated sludge process with hydraulic retention time being 2-3hrs. Generally the settling problems of sludge will be associated with size (small flocs <10μm), open structure and concentration (>5g/l). Sludge blocking will be one of the most common settling problems in activated sludge process. The causes for settling problems are very complex for *e.g.* low dissolved oxygen, nutrient deficiency, shock loading, low loading rate, nitric oxide build-up, toxic compounds, process dynamics and process configuration, etc (Casey *et al.*, 1995; Eikelboom *et al.*, 1998; Musvoto *et al.*, 1999; Jenkins *et al.*, 2004).

MBRs include the separation ability of membrane technology to eliminate the problem of biomass separation. MBR, which includes membrane filtration, has higher ability when compared to gravity, especially for separating the small flocs and colloidal particles. In the secondary clarifier the driving force being the density difference between the water and flocs, was only related to the size, structure and floc mass, which are not directly controllable. In the case of MBR, the membrane filtration depends upon Trans membrane pressure, which can be controlled with the help of a suction pump. As a result, we can use higher sludge concentrations (upto 60g/l), lower hydraulic retention times (HRT) and longer solid retention times (SRT). In addition, the other advantage being is that HRT and SRT are independent in MBRs, since the MLSS concentration can go beyond 4-5 g/L without involving any problem. The mentioned key feature will result in many other characteristics of MBRs. The advantages and disadvantages of MBR process are summarised as follows:

Table 11.4: Advantages and Disadvantages of MBRs

Advantages	*Disadvantages*
Effluent quality was excellent (reusable)	Fouling of membrane takes place
Independence between HRT and SRT	Low oxygen transfer efficiency
High loading rate	High capital cost, no economy scale
Small foot print	Complicated control system
No sludge bulking risk	
Low sludge production	
Possibility to grow specific microorganisms	
Treats wastewater under extreme conditions	
Flexible modular design	

The main driving forces of using MBRs are: 1) strict effluent discharge standard. In many cases the MBR effluent quality is so good that it can be used directly for non-potable purposes, *e.g.*, for landscape and irrigation, 2) Small footprint. As many countries suffer high population density and space limitation in constructing new or expanding existing WWTPs. in these cases MBRs beat CAS processes in space saving, 3) Continuous decreasing in membrane costs making MBR more competitive compared with CAS processes.

The investments for MBRs are low when compared with CAS systems with secondary clarification. Fouling phenomena on the membrane and within the pores reduces the long term stability of flux performance, thus increasing the demand for membrane replacements and high energy demand for hydrodynamic control of membrane fouling (judd, 2006). However fouling can be controlled by permeate back flushing and chemical cleaning, but results in losses of net filtration efficiency and membrane may get damaged due to possible chemical agents. The evolution of membrane permeability under certain operating conditions or the effect of cleaning measures cannot be predicted correctly. These uncertainties will cause difficulties in plant layout, design and operation.

Presentation of Filtration Systems in MBRs

Depending on the view of filtration, MBR systems can be classified according through key points of filtration process: membrane, filtration mode, module design and filtration process. The main definitions of the filtration system that can be useful to describe MBRs are discussed below.

Membrane and Filtration Modes (Figure 11.4)

Most MBRs use microfiltration or ultrafiltration membranes. The shape of membranes can be hollow fibres, tubular or flat sheet. In the case of hollow fibres or tubular membranes, filtration can be operated from outside to inside the membrane (which is the case for all hollow fibres) or from inside to outside the membrane (only for tubular membranes).

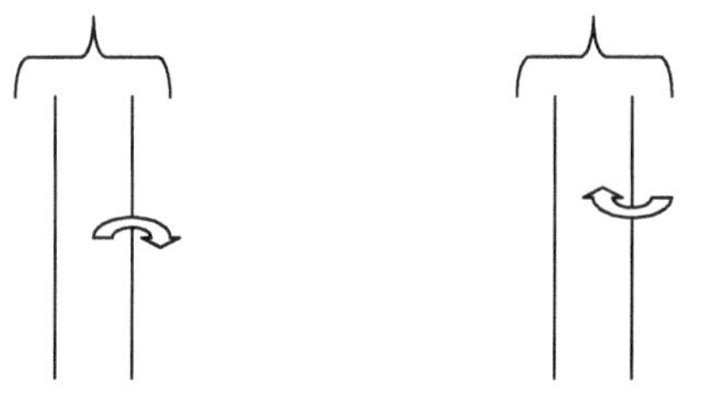

Figure 11.4: Inside/Out, Outside/in Membrane Filtration.

Module Design

The module configuration defines the membrane arrangement, packing density, the filtration mode and the hydraulic distribution throughout the membrane network (Fane 2002). It also determines the energy requirement and the ease in the process of membrane generation. In general three main types of modules are used in MBR process for wastewater treatment, which are based on flat sheet membranes, tubular membranes or hollow fibre membranes.

Flat Sheet Module

A flat sheet module is an assembly of flat sheet membranes. In this type each module is enclosed within a rectangular box, used for collecting the permeate. In general, for the application of MBRs the membrane sheets should be placed atleast 6 to 7mm apart (Sofia *et al.*, 2004). The main examples of flat sheet membranes will be the KUBOTA membranes. This kind includes the advantage of easy controlling hydrodynamics for one phase flow. But in the case of gas/liquid flows description of flows will become much more complicated (Ducom *et al.*, 2003). Mostly, this form of configuration does not allow backwashing.

Tubular Module (Figure 11.5a)

Association of more number of tubes, single membrane tube or multi-channel membrane tubes may result in the formation of Tubular module. The tubes are linear, rigid and are mainly operated in outside/in filtration. The deposition of sludge within the tubes can be minimised easily and membrane permeability can be controlled easily by maintaining high liquid velocities, shear stresses. In MBRs to prevent, tube clogging the inner tube diameter must be higher than five mm. coming to the point of tubes of MBR they can be polymeric as in STROCK or millennium pore systems or mineral as in the Degremont MBR. As an example, Mayer *et al.* (2005) used a tubular membrane module with 5.2 mm inner diameter tubes (109 tubes). Chang *et al.* (2002).

Figure 11.5a: Multi-channel Membrane Tube

Worked with a module composed of 9.5 mm inner diameter polyethersulfone tubes (20 tubes).

Hollow Fibre Module

In general, the hollow fibre modules in MBR system are composed of fibres with 1.5 to 2.5 mm inner diameter. To avoid the phenomenon of clogging these modules are operated in outside/in filtration mode. For application to MBRs, several bundle configurations have been developed and studied. For example, at lab scale the following modules can be found.

a) One-Side Potted and Freely Movable (Figure 11.5b)

In this type each fibre is glued internally at its top and the bundle is potted at the bottom. Thus fibres are free to move independently at their top and are fixed at their bottom. (Espinosa-Bouchot *et al.*, 2003)

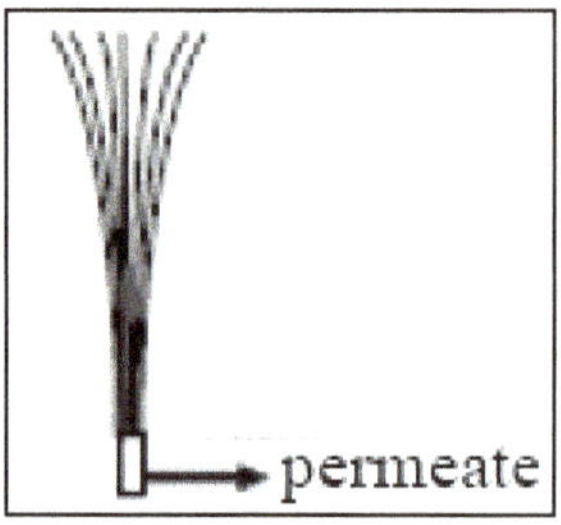

Figure 11.5b: One-side Potted and Freely Movable Membranes.

b) Right and Tended Fibres Module (Figure 11.5c)

Here in this type fibres are potted on the both sides. They are right and slightly tended and this does not allow any fibre motion (Hong *et al.*, 2002; Lim *et al.*, 2003).

Right and Released Fibres (Figure 11.5d)

These have the same configuration than the previous one but the fibres are not tended, thus they are free to move in between their fixation points. (Hong *et al.*, 2002; Delgado *et al.*, 2004).

d) Right Fibres Module in Horizontal Position (Figure 11.5e)

These are composed by right and non-tended fibres positioned horizontally. Here feed water flows vertically through the bundle. Depending on the consequence

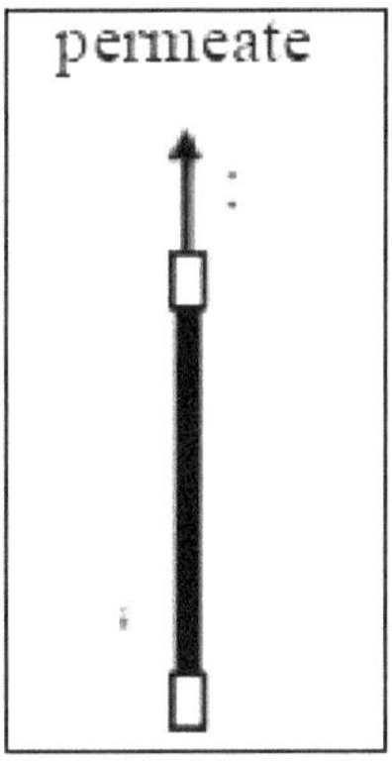

Figure 11.5c: Right and Tended Fibres.

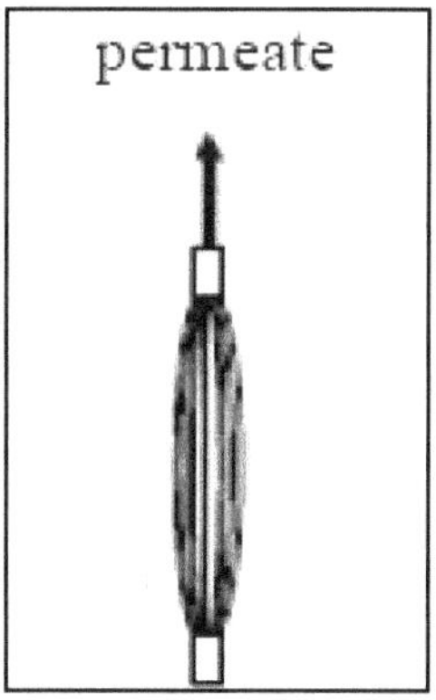

Figure 11.5d: Right and Released Fibres.

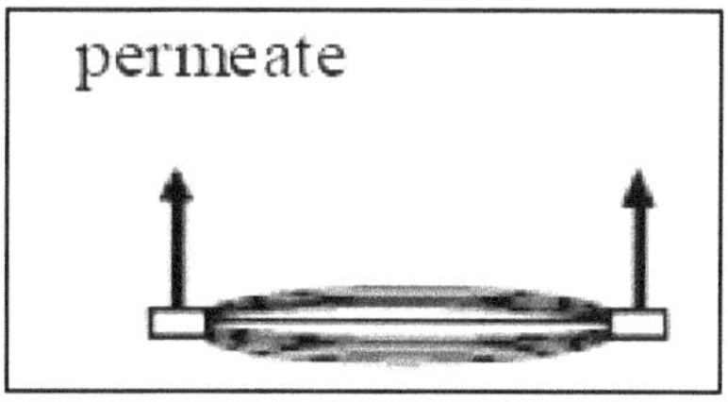

Figure 11.5e: Right Fibres Module in Horizontal Position.

hydrodynamics and fouling phenomena might be very different than in the different position. However, this module configuration might favour breaker (Ueda *et al.*, 1997).

e) U Shaped Bundle (Figure 11.5f)

It contains one or several bundles of fibres disposed in the form of 'U' and potted together at bottom of the module. This arrangement can only be developed with low inner diameter fibres and deformable fibres (Espinosa-Bouchat *et al.*, 2003; Yu *et al.*, 2003; van kaam *et al.*, 2004).

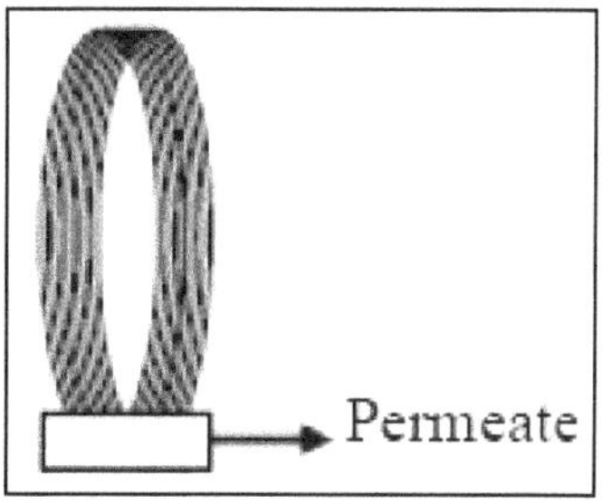

Figure 11.5f: U-Shaped Bundle.

Process Operation in MBRs)

There process operation modes in MBRs can distinguished depending on the following factors

- ✰ The use of aeration or not during filtration
- ✰ Location of membrane compared to the bioreactor
- ✰ Concentrate recycling in bioreactor area
- ✰ Liquid velocity inside the membrane system

Aeration or Not

Air can be injected at the membrane bottom in order to limit fouling of the membrane surface or clogging of the bundle. Air is introduced below the membrane assembly and is supposed to be ideally distributed to optimise the air scouring action across the membrane surface. The ideal airflow and flow pattern for this purpose are not clearly known today. Small air bubbles are required to increase oxygen transfer between the gas and the liquid phase for providing oxygen to the biomass. For preventing fouling, what sort of aeration mode to be used is not known until today. However many studies claim that big bubbles would be preferable.

Using specific air injectors, aeration for preventing fouling can be done, independently to the air injection for biomass. Operation with sequential air injection is also been used at industrial scale, for example in the ZENON systems. This operation will reduce operating costs due to energy consumption induced by air sparging.

Location of Membrane

Depending on the location of membrane, as discussed earlier we can distinguish the operation of membrane.

Factors Affecting the Performance of MBR Filtration

Different operating parameters seem to have an influence on the filtration performances of MBR.

Overview of Membrane Filtration Process (Table 11.5a)

The basic principle of all membrane operations is substances mixture separation with selective film. The transport of matter through a selective barrier is caused by a chemical potential difference between two phases *i.e.*, the feed and the permeate (Mulder, 1996). Pressure difference across the membrane is the main driving force for the pressure driven membrane filtration systems. Typically, four types of membranes are distinguished according to their separation range (molecular weight cut-off or pore size) and the applied transmembrane pressure: reverse osmosis (RO), nanofiltration (NF), ultrafiltration (UF), and microfiltration (MF).

In general MBRs, utilise a tight microfiltrarion or a lose ultrafiltration membranes. The membrane materials often include organic polymers, *e.g.*, polyethylene, polypropylene and polyvinylidene fluoride (PVDF) membranes (Judd, 2006). The materials used for blending of materials will change their surface charge or hydrophobicity (Mulder, 1996). Inorganic membranes (*e.g.*, ceramic membranes) are only used in special applications *e.g.*, solvent resistance and thermal stability are required, due to their high costs (Baker, 2004). The ultrafiltration membrane will often have a supporting layer (a microfiltration membrane), onto which a thin skin layer, *i.e.*, a true ultrafiltration membrane, is attached.

In MBRs, membrane fouling is inevitably linked to the interactions between the membrane material and the components of the activated sludge (feed components, cells, microbial metabolites).

It is generally considered that extracellular polymeric substances (EPS) are playing an important role in fouling [Chang and Lee, 1998]. EPS compounds are mainly proteins, polysaccharides and humic substances. Proteins show high affinities with hydrophobic surfaces whereas carbohydrates are relatively much more hydrophilic than proteins [Mukai *et al.*, 1999].

The most important characteristic of membranes in MBRs is thus their hydrophilic/hydrophobic comportment. The behaviour of three membranes in MBR conditions were tested by Choi *et al.* (2002): a very hydrophilic (cellulose acetate), an hydrophobic (polyethersulfone) and an intermediate (sulfonated polyethersulfone). It revealed that hydrophobic membranes were much more sensitive to fouling than

Table 11.5(a): Membrane Classification and Colloidal/Macromolecular Organic Matter in Ground Surface Water (Mallevialle *et al.*, 1996)

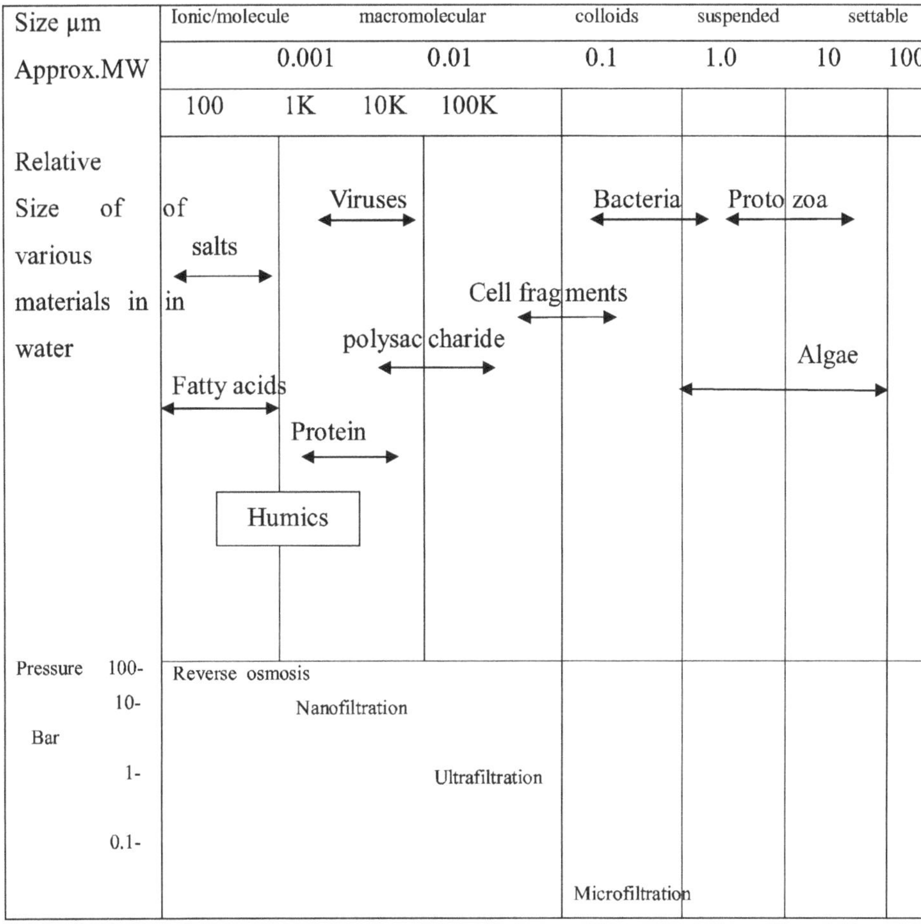

the hydrophilic membrane, where the rejection rate of soluble EPS being the same., the final fluxes were almost equal for all three membranes [Mukai *et al.*, 1999] even if the initial flux of the hydrophobic membrane was higher than the hydrophilic one. Hydrophobic interactions are stronger than hydrophilic interactions, and thus support the previous result Ma *et al.* (2000). the ratio proteins/polysaccharides is considered as more important than the total amount of EPS for the attachment of flocs and solutes on the membrane surface [Mukai *et al.*, 1999, Lee *et al.*, 2003] by some authors. Furthermore, in the case of hydrophobic membranes, at the starting of a filtration step preferential hydrophobic adsorption leads to an irreversible attachment of sludge particles and solutes on the membrane, with a high protein/carbohydrate ratio. So the membrane is getting more and more hydrophobic with time and fouling is getting more and more significant too [Lei and Zhou, 2005]. Furthermore, Yamato *et al.* (2006) compared two hydrophobic polymeric membranes, made of polyethylene (PE) and polyvinylidene fluoride (PVDF) during filtration of the same wastewater. This revealed that PVDF is superior to PE in term of prevention of irreversible fouling in MBRs: most of the fouling observed for the PE membrane was irreversible while reversible fouling was dominant for the PVDF membrane. Moreover, the origin of fouling matters was shown to be different for the two membranes: t dissolved matter causes irreversible fouling of PE membrane and carbohydrates induce reversible fouling of the PVDF membrane [Yamato *et al.*, 2006].

Membrane electric charge plays an important role in influencing membrane performances. Ma *et al.* (2000) operated filtration of an *E coli* suspension with neutral and charged membranes. It revealed that the neutral membrane surfaces were most effective in reducing *E. coli* fouling. As *E. coli* has both negatively and positively charged surface groups the result was not that much successful. So that, charged membrane surfaces develop stronger interactions with charged feed than neutral membrane surfaces.

Nevertheless, it seems that influences of membrane surfaces properties and thus of membrane material are important mainly in the case of short filtration duration [Ma *et al.*, 2000]. Actually, Ma *et al.* (2000) reported that after 1 hour of filtration, membranes are sufficiently fouled so that deposit controls the filtration behaviour and in this way, membrane properties have less influence over longer durations.

Finally, we can conclude that microorganisms have the ability to biodegrade cellulose acetate (CA) membranes for that reason; due to this CA, membranes are not used further in MBRs.

As a conclusion, membrane material (hydrophily/hydrophoby, surface charge) has a strong influence on fouling in MBRs and on its reversibility. It mainly influences the first filtration instants.

Effect of Permeation Flux on Membrane Performance

The permeation flux remains as one of the important parameter for the MBR process optimisation [Gui *et al.*, 2003]. During the time of filtration, particles, colloidal compounds and a part of soluble compounds are retained by both membrane and filtration cake. The material deposition onto the membrane takes place during filtration and causes a fouling, which increases with increasing the permeation flux [Yang *et al.* (2006)]. Thus, more permeation flux involves in more fouling.

According to the Darcy's law, increasing the permeation flux not only leads to fouling but also increases the TMP and hence due to the cake compressibility, to reduce the material deposit porosity and permeability reducing the filtration performance.

The concept of critical flux was first introduced by Field *et al.* (1995), it is defined for microfiltration like flux below which a decline of permeability with times does not occur and above which fouling is observed. Finally, it was recommended to use a permeation flux below the critical flux in the view to reduce the membrane fouling.

However, even with sub-critical flux filtration conditions, membrane fouling occurs due to irreversible adsorption of mixed liquor soluble compounds and a decrease of TMP is observed after some critical filtration period [Pollice *et al.*, 2005]. Actually, the value of the critical flux is strongly dependant of operating conditions such as suspension properties, membrane properties and hydrodynamic conditions [Pollice *et al.*, 2005, Bacchin (2004)]. So, a value of critical flux is related to process configuration, plant size and nature of influent. These parameters affect upon the shear stress and the foulant potential of mixed liquor, and thus, influence the value of critical flux.

Membrane Fouling (Figures 11.6a and b)

Characteristics of Membrane Fouling and its Importance in MBRs

Membrane fouling is a major obstacle that hinders faster commercialisation of MBRs. Membrane fouling refers to the deposition or adsorption of material on the surface of the membrane or within the pores. It is a common and most costly problem in membrane filtration applications. Fouling may cause a decline in permeate flux, increase in TMP, loss of permeate quality and membrane deterioration etc. In conjunction with above formation of filter cake, shift in the effective pore size or molecular weight cut off is common, which can result in a MF process displaying the characteristics of UF membranes (Lee *et al.*, 2001, LaPara *et al.*, 2006).

With respect to MBRs, membrane fouling occurs due to the following mechanisms: (1) solutes or colloids adsorption within/on membranes; (2) sludge flocs deposition onto the membrane surface; (3) cake layer formation on the membrane surface; (4) foulants detachment attributed mainly to shear forces; (5) the spatial and temporal changes of the foulant composition during the long-term operation (*e.g.*, the change of bacteria community and biopolymer components in the cake layer). In other words, fouling can be defined as the undesirable deposition and accumulation of microorganisms, colloids, solutes, and cell debris within/on membranes.

Generally, as shown in Figure 11.6a, a three stage fouling history might be proposed (Cho and Fane, 2002; Zhang *et al.*, 2006a):

- ☆ Stage 1: an initial short-term rapid rise in TMP;
- ☆ Stage 2: a long-term weak rise in TMP;
- ☆ Stage 3: Sharp increases in dTMP/dt, also known as TMP jump (Cho and Fane, 2002).

The schematic illustration shows the occurrence of TMP jump. The TMP jump is believed to be the consequence of severe membrane fouling (Cho and Fane, 2002)

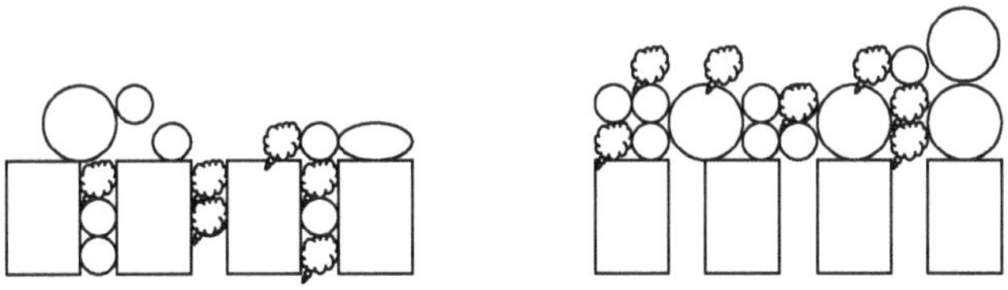

(i) (ii)

Figure 11.6a: Membrane Fouling Processes in MBR (i) Pore blocking; (ii) Cake layer.

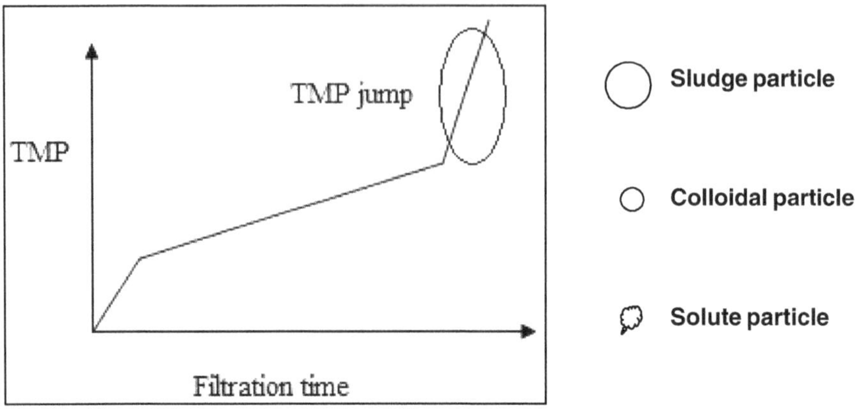

Figure 11.6b: Schematic Representation of the Occurrence of TMP Jump.

attributed the TMP jump to the changes in the local flux due to fouling eventually causing local fluxes to be higher than the critical flux. Latterly, Zhang *et al.* (2006a) reported that the sudden jump was possibly not only due to the local flux effect, but also caused by sudden changes of the biofilm or cake layer structure. Due to oxygen transfer limitation, the bacteria in the inner biofilms tend to die and release more extracellular polymeric substances (EPS). A more recent investigation also confirmed that the sudden jump of TMP was closely related to the sudden increase in the concentration of EPS at the bottom of cake layer, which might be attributed to the death of bacteria in the inner of cake layer (Hwang *et al.*, 2008).

The occurrence of the TMP jump also depends on operating conditions. Different studies have shown that the jump of TMP varies with time, fluxes (Zhang *et al.* (2006 a). Pollice *et al.* (2005) reported that the TMP jump was more frequently observed in small-scale experiments. It should be borne in mind that fouling rates measured in lab-scale are in appropriate to describe long-term full-scale operation due to distinct and inherent differences between them (Kraumeet al., 2009).

From above studies, it can be concluded that the relation between TMP jump and these operating parameters are very complex, and TMP jump occurs inevitably during long-term operation of MBRs. Thus, the overall goal of fouling control is to retard the occurrence of the TMP jump via modifying sludge characteristics or decreasing membrane flux *(e.g.,* operation below critical flux).

Classification of Membrane Fouling

Removable and No removable Fouling

Definition of Removable and Irremovable Fouling

Fouling of a membrane is very complicated phenomenon and results from multiple causes. Particle sizes of sludge flocs, colloids and solutes in mixed liquor may strongly affect fouling mechanisms in a membrane filtration system. Adsorption on pore wall and pore blocking may occur, if foulants are comparable with the membrane pores (*i.e.*, colloids), or smaller than the membrane pores (*i.e.*, solutes), However, a cake layer on the membrane surface will be formed if foulants (*i.e.*, sludge flocs and colloids) are much larger than the membrane pores.

Generally, the irreversible fouling is that kind of fouling which can't be removed by any methods including chemical cleaning. However, some previous revealed that irreversible fouling can be removed by chemical cleaning but not by physical cleaning. Here, we define three types of fouling removable fouling, irremovable fouling and irreversible fouling. The removable fouling can be easily eliminated by implementation of physical cleaning (*e.g.*, backwashing) while the irremovable fouling needs chemical cleaning to be eliminated.

The removable fouling and reversible fouling are the same. The removable fouling is caused by loosely attached foulants; however, irremovable fouling is caused by pore blocking and strongly attached foulants during filtration. The irreversible fouling is a permanent fouling which cannot be removed by any approaches. In general, removable fouling is attributed to the formation of cake layer, and the irremovable fouling is attributed to pore blocking.

Formation of the Cake Layer

In many cases, cake layer formation linked with removable fouling was considered as the major contributor to membrane fouling in MBRs. Lee *et al.* (2001) reported that the filtration resistances included membrane resistance (12 per cent), cake resistance (80 per cent), blocking and irremovable fouling resistance (8 per cent), indicating that the formation of cake layer is the main cause leading to membrane fouling. The cake layer formation and its effects on fouling has been studied by many researchers and they are as follows.

Chu and Li (2005) reported that the cake layer was not uniformly distributed on the entire surface of all of the membrane fibers. The membranes were covered partially by a static sludge cake that could not be removed by the shear force due to aeration, and partially by a thin sludge film that was frequently washed away by aeration turbulence. Di Bella *et al.* (2007) found that the cake in anaerobic MBR had a mainly removal nature. These investigations suggest that the cake layer formed with aerobic sludge and anaerobic sludge might have different removability. Jeison and van Lier (2007) performed a study on a lab-scale anaerobic submerged membrane bioreactor (AnMBR) for over 200 days, and observed that cake formation was removable on a short-term basis, however, cake consolidation was observed when a long-term operation was performed at a flux close to the critical flux.

A cake layer can be described as a porous media with a complex system of inter connected inter-particle voids. Yanget al. (2007) simulated the intra-layer flow field by using the three-dimensional volumetric grid model and confocal laser scanning microscopy (CLSM) analysis, and observed that there was a very complex flow pattern in the fouling layer. Because of the inter-connectivity of the neighboring pores, the flow direction may even be the reverse of that of the pressure gradient (dP/dx).

Irremovable Fouling in MBRs

For long-term and sustainable operation of MBRs control of irremovable membrane, fouling is of great importance.colloids, solutes and microbial cells pass through and precipitate inside the membrane pores during initial filtration. However, during the long-term operation of MBRs, the deposited cells multiply and yield EPS, which clog the pores and form a strongly attached fouling layer.

Operation below the critical flux is an effective approach to avoid severe fouling including removal and irremovable fouling within a given filtration system. Field *et al.* (1995)introduced critical flux concept, operation below the critical flux concept is called sub-critical flux or non-fouling operation and is expected to lead to little irremovable fouling As a matter of fact, for a long-term operation of MBRs, irremovable fouling can occur even if they are operated below the critical flux. Ognier *et al.* (2004) reported that despite the initial choice of sub-critical flux filtration conditions, gradual fouling was seen to develop which, after long periods of operation without intermediary membrane regeneration, proved to be hydraulically irremovable Further discussion of critical flux can be found in recent review articles (Pollice *et al.*, 2005; Bacchin *et al.*, 2006). The concept of critical flux has been popularly used in the study of MBR fouling (Guglielmi *et al.*, 2007b; Lebegue *et al.*, 2008; Wanget al., 2008b).

However, most of the investigations on the determination of critical flux are based on ex-situ measuring, which cannot offer the real fouling propensity. Recently, an insitu method was developed by de la Torre *et al.* (2008), which can provide more reliable information about critical flux than ex-situ methods. Huyskens *et al.* (2008) developed an on-line measuring method, which was used to evaluate the removable and irremovable fouling propensity of MBR mixed liquor in a reproducible way. These studies imply that it is possible to develop on-line or in situ method to determine critical flux or removable/irremovable fouling. It is also of high interest to develop a unified measuring method or apparatus.

Types of Fouling and their Characteristics

The compounds that cause fouling can be categorised into four types: particulate fouling caused by colloids, biofouling caused by de3position or growth of microorganism, organic fouling caused by adsorption of organic matter, and scaling caused by salt precipitation as shown in Table 11.5b.

Particulate Fouling

Formation of a filter cake by accumulation of small particles on the membrane surface, is referred as particulate fouling. Suspended solids, colloids and even microorganisms are the major particulates. In most UF and MF membranes particulate fouling is the dominant type of fouling. However, MBRs using MF and UF membranes suffer more colloidal and organic fouling. than biopolymers. More recently, Wang *et al.* (2008b) observed that the cake layer was formed by organic substances and inorganic elements such as Mg, Al, Fe,Ca, Si, etc. The organic foulants coupled with the inorganic precipitation enhance the formation of a cake layer. Lyko *et al.* (2007) also found that metal substance was a more significant contributor to membrane fouling than biopolymers. Sometimes, the fouling caused by inorganic scaling is not easy to be eliminated even by chemical cleaning (You *et al.*, 2006). These findings indicate that particulate fouling has become more and more important in MBRs. However, the understanding of particulate fouling is still not clear. The investigation on the limiting concentration of metal ions in the feed wastewater that can lead to particulate fouling will be of great interest, since the chemical composition of the wastewater is in close relation with the formation of precipitation.

Table 11.5b: Fouling of Membrane.

	Particulate Fouling	*Organic Fouling*	*Biofouling*	*Scaling*
Foulants	Colloids Suspended solids	Organic matter	Microorganisms	Salt Metal cations
Major factors affect fouling	Concentration Particle size distribution Compressibility of particles	Concentration Charge Hydrophobicity pH Ionic strength Calcium	Temperature Nutrients	Temperature Concentration pH
Indicator of fouling prediction	Silt density index (SDI) Modified fouling index (MFI) Specific resistance to fouling (SRF)	$DOCUV_{254}$ SUVA	Assimilable organic carbon (AOC) Biofilm formation rate (BFR)	Solubility
Feed water treatment	Coagulation MF and UF	Adjustment of pH Coagulation	Sand filtration Bio-filter Coagulation Flocculation UF and MF	Acid Anti-scalent

Organic Fouling

Due to intermolecular interactions between the membrane and organic matter, dissolved organic substances get adsorped on the membrane surface, leading to Organic fouling (Zeman and Zydney, 1996). The well-known problem of this kind will be Natural organic matter (NOM) in drinking water filtration processes (Combe *et al.*, 1999; Lee *et al.*, 2004, Jones *et al.*, 2000). Major fraction of NOM will be the humic substance. However, the filtration of wastewater and activated sludge has been applied more recently and soluble microbial products (SMP) fouling has been the main concern. Recently, in order to provide a unique insight into the composition (protein and carbohydrate),Metzger *et al.* (2007) have performed a more detailed study to characterise deposited biopolymers in MBRs.

Biofouling

Loss of membrane performance due to adhesion and growth of microorganisms on the membrane surface (formation of a biofilm) results in Biofouling. As bacteria have developed elaborate adhesion mechanisms, biofilm can occur on all kinds of surfaces, natural and synthetic. Due to their low flux and limited membrane cleaning options RO and NF processes suffer more of biofouling (Flemming *et al.*, 1996; Flemming, 1997; Baker and Dudely, 1998).

Scaling

Formation of scaling may occur if dissolve salts exceed their solubility product. Typically, over saturation is of concern in reverse osmosis and nanofiltration operations with regard to $CaCO_3$, $CaSO_4$, $BaSO_4$, $SrSO_4$, $MgCO_3$, and SiO_2 Baker, 2004. However, RO plants can operate at super-saturation condition (*e.g.*, $BaSO_4$) without scaling (Bonne *et al.*, 2000). Scaling is not dominating in MBR fouling. However, iron or calcium precipitation may occur in some cases. Acid cleaning should be considered if oxidant cleaning is not sufficient to restore the membrane permeability (te poele and van der Graaf, 2005).

However, there may be overlaps in the above four types of foulants, *e.g.*, organic fouling due to deposition of suspended solids can be particulate fouling, so is the biofouling due to the seeding and growing of a biofilm. In addition, the different types of fouling can occur simultaneously and form hybrid fouling, which can be more difficult to clean (te poele and van der Graaf, 2005).

Fouling Factors and Control Strategies

The factors affecting membrane fouling can be classified into four groups (Le-Clech *et al.*, 2006): membrane materials, biomass characteristics, feed water characteristics, and operating conditions (Table 11.5c). The complex interactions between these aspects complicate the understanding of membrane fouling. For a given MBR process, the fouling behaviour is directly determined by sludge characteristics and hydrodynamic conditions. But, operating conditions (*i.e.*, SRT, HRT and F/M)and feed water have indirect actions on membrane fouling by modifying sludge characteristics. Table 11.5c gives the relationship between various fouling factors and membrane fouling on the basis of recent literatures.

Table 11.5c: Relationship between Various Fouling Factors and Membrane Fouling.

Sludge Condition	*Effect on Membrane Fouling*	*Reference*
MLSS	– MLSS↑→ normalized permeability ↓	(Trussell *et al.*, 2007)
	– MLSS↑→ fouling potential↑	(Psoch and Schiewer, 2006a)
	– MLSS↑→cake resistance↑ specific cake resistance↓	(Chang and Kim, 2005)
Viscosity	– Viscosity↑→membrane permeability↓	(Li *et al.*, 2007a)
	– MLSS/Viscosity↑→membrane permeability↓	(Trussell *et al.*, 2007)
	– Viscosity↑→membrane resistance↑	(Chae *et al.*, 2006
F/M	– F/M↑→fouling rates↑	(Trussell *et al.*, 2006
	– MLSS (2–3 g/L): F/M↑→irremovable fouling↑	(Watanabe *et al.*, 2006
	– MLSS (8–12 g/L): F/M↑→removable fouling↑	(Kimura *et al.*, 2005
	– F/M↑→Protein in foulants↑	(Drews *et al.*, 2006)
EPS	– Polysaccharide↑→fouling rate↑	(Cho *et al.*, 2005c)
	– Bound EPS influences on specific cake resistance	(Lesjean *et al.*, 2005)

Contd...

Table 11.5c–*Contd...*

Sludge Condition	*Effect on Membrane Fouling*	*Reference*
	– Polysaccharide↑→fouling rate↑	(Lesjean *et al.*, 2005)
	– Bound EPS↑→membrane resistance↑	(Chae *et al.*, 2006)
	– The loosely bound EPS contributes to most of the filtration resistance of the whole sludge	(Ramesh *et al.*, 2007)
SMP	– SMP is more important than MLSS	(Zhang *et al.*, 2006b)
	– Colloidal TOC relates with permeate flux	(Fan *et al.*, 2006)
	– Filtration resistance is determined by SMP)	(Jeong *et al.*, 2007)
	– SMP is probably responsible for fouling)	(Sperandio *et al.*, 2005
	– Polysaccharide is a possible indicator of fouling	Le-Clech *et al.*, 2005)
	– SMP↓→fouling index↓	(Jang *et al.*, 2006)
	– Fouling rates correlate with SMP	(Trussell *et al.*, 2006
Filamentous	– Filamentous bacteria↑→sludge viscosity↑	(Meng *et al.*, 2007a)
bacteria	– Bulking sludge could cause a severe fouling)	(Sun *et al.*, 2007
	– Filamentous-bacteria↓→cake resistance↓	(Kim and Jang, 2006)
Operating condition	– SRT decrease from 100 to20d→TMP↑	(Ahmed *et al.*, 2007)
SRT	– SRT decrease from 30 t10d→fouling↑	Zhang *et al.*, 2006b
	– SRTs↑→fouling potentials of SMP↑	(Liang *et al.*, 2007)
	– SRTdecrease from 5 to 3 d→fouling↑	(Ng *et al.*, 2006c)
HRT	– HRT↓→membrane fouling↑	(Meng *et al.*, 2007a)
	– HRT↓→membrane fouling↑	(Chae *et al.*, 2006
	– HRT↓→membrane fouling↑	(Cho *et al.*, 2005a)
Aeration	– Aeration intensity↑→permeability↑	(Trussell *et al.*, 2007)
	– Air-sparging improves membrane flux larger bubbles for fouling control are preferable)	(Psoch and Schiewer, 2006a)
	– Air backwashing for fouling control is preferable (- bubble-induced shear reduces fouling significantly)	(Phattaranawik *et al.*, 2007) (Chae *et al.*, 2006)
	– Air scouring can prolong membrane operation	(Wicaksana *et al.*, 2006 (Sofia *et al.*, 2004)
Permeate flux	– Sub-critical flux mitigates irremovable fouling)	(Lebegue *et al.*, 2008
	– sub-critical flux mitigates fouling	(Guo *et al.*, 2007)

Removal of Fouling

Physical Cleaning

Physical cleaning techniques for MBRs include mainly membrane relaxation (where filtration is paused) and membrane backwashing (where permeate is pumped in the reverse direction through the membrane). These procedures have been incorporated in most MBR designs as standard operating strategies to limit fouling; although vigorous backwashing is not an option for flat plate submerged membranes.

Backwashing (backflushing) has been found to effectively remove most of the reversible fouling due to pore blocking, transport it back into the bioreactor, and partially extricate loosely attached sludge cake from the membrane surface. In some cases, clogging near the membrane surface may also be partially loosened or removed by backwashing. The efficiency of backwashing has been studied in detail (E.H. Bouhabila *et al.*, Psoch *et al*). Key parameters in the design of backwashing are its frequency, duration, the ratio between those two parameters and its intensity. For example, less frequent, but longer backwashing (600 s filtration/45 s back washing) was found to be more efficient than more frequent backwashing (200 s filtration/15 s backwashing) (T. Jiang *et al*). In another study based on factorial design, suction time (between8 and 16 min) was found to have more effect on fouling removal than both the aeration intensity (0.3–0.9m3/m2 h) and the backwash time (25–45s) (P. Schoeberl *et al*). Although more fouling is expected to be removed when backwashing duration and frequency are increased, optimization of backwashing is required in regard to energy and permeate consumptions. This was achieved by the design of a generic control system which automatically optimized the duration of the backwash according to the monitored value of TMP (P.J. Smith *et al.*).

This anti-fouling operation obviously affects operating costs as energy is required to achieve a pressure suitable for flow reversion. Moreover, between 5 and 30 per cent of the produced permeate is used in the process. Air can also be used as the backflushing medium (Y. Sun *et al*). Although improving the flux by nearly 400 per cent (compared to continuous operation), 15 min of air backwash was required every15 min of filtration to obtain this result [C. Visvanathan *et al*].However, air backwashing is an efficient method for flux recovery, it may also present potential issues of membrane breakage and rewetting. Membrane relaxation (or non-continuous operation of the membrane) significantly improves membrane productivity. Under relaxation, back transport of foulants is naturally enhanced as non-irreversibly attached foulants can diffuse away from the membrane surface through the concentration gradient. The fouling removal efficiency of this method can be further increased when air scouring is applied during relaxation (S.P. Hong *et al.*, H.C. Chua *et al*). Detailed studies of the TMP behaviour during this type of operation revealed that although the fouling rate is generally higher than for continuous filtration, membrane relaxation allows filtration to be maintained for longer period of times before the need for cleaning (C.A. Ng, D. Sun *et al*). Although some have reported that this type of operation may not be economically feasible for large-scale MBRs, further cost and productivity analysis are probably required to compare this method against backwashing. Recent studies assessing alternative strategies for fouling mitigation tend to combine intermittent operation with frequent backwashing for optimum results (S.T. Zhang *et al.*, M.V.G. Vallero *et al.*).

Chemical Cleaning

It is expected that membrane relaxation and backwashing effectiveness tend to decrease with operation time as more irreversible fouling accumulates on the membrane surface. Therefore, in addition to the physical cleaning strategies, different types/intensities of chemical cleaning may also be recommended. They include:

- Chemically enhanced backwash (on a daily basis),
- Maintenance cleaning with higher chemical concentration (weekly), and
- Intensive (or recovery) chemical cleaning (once or twice a year).

Maintenance cleaning is used to maintain design permeability and helps to reduce the frequency of intense cleaning. Intensive cleaning is generally carried out when further filtration is no longer sustainable because of an elevated TMP. Each of the four main MBR suppliers (Kubota, Memcor, Mitsubishi and Zenon) proposes their own chemical cleaning recipes, which differ mainly in terms of concentration and methods. Under normal conditions, the prevalent cleaning agents remain sodium hypochlorite (for organic foulants) and citric acid (for inorganics).Sodium hypochloride hydrolyzes the organic molecules, and therefore loosen the particles and biofilm attached to the membrane. The effects of cleaning chemical agents like NaOCl on microbial community have been also recently studied for modelled MBR processes (B.-R. Lim *et al*). It is also common for MBR suppliers to adapt specific protocols for chemical cleanings *(i.e.* chemical concentrations and cleaning frequencies) for individual facilities (T. Sato *et al.*, L.S.D.M. Kox *et al.*, P. Le-Clech *et al*) It also has been mentioned that the level of pollutants (measured as TOC) in the permeate rises just after the chemical cleaning episodes (T. Sato *et al*). This is important for MBRs used in reclamation process trains *(i.e.* upstream of RO for example). So far, no systematic studies on cleaning agents or procedures have been published (B.Q. Liao *et al*). This is probably due to the site-specific nature of the MBR fouling.

Maintenance cleaning, taking up to 30 min for a complete cycle, is normally carried out every 3–7 days at a moderate reagent concentration of 0.01 wt. per cent NaOCl. Recovery cleaning employs rather higher reagent concentrations of 0.2–0.5 wt. per cent

NaOCl coupled with 0.2–0.3 wt. per cent citric acid or 0.5–1 wt. per cent oxalic acid. Research on the efficiency of sonification for removing cake layers in MBRs has also been carried out (H.H.P. Fang, *et al*). The sonification cleaning process is based on the breakdown of the fouling cake into smaller fragments. Although sonification can successfully remove the cake from the membrane surface, this cleaning method was not effective on all types of fouling due to pore blocking and may even worsen this type of fouling. A combination of sonification with backwashing and chemical cleaning appeared to achieve almost complete flux recovery (A.L. Lim *et al*). More details of the cleaning mechanisms are available in (A.L. Lim *et al*). However, sonification would be difficult to apply at a large-scale due to the focused nature of the sonic energy.

Limitation of Fouling (Table 11.5d)

Table 11.5d: Summarization of Fouling Control Strategies Based on Various Fouling Factors.

Control Strategy	*Control Item and its Effect*
Hydraulic Control	– HRT↓→sludge viscosity↑ (Meng et al., 2007a), EPS↑(Chae *et al.*, 2006)
	– Aeration↑→permeability↑ (Psoch and Schiewer, 2006a; Trussell *et al.*, 2007), fiber movement ↑(Wicaksana et al., 2006), cake-removing efficiency↑ (Chang and Judd, 2003), and cake resistance↓ (Psoch and Schiewer, 2006b; Fan and Zhou, 2007)
	– Periodical backwashing/flux↑ (Psoch and Schiewer, 2006a) and operation period↑(Chae *et al.*, 2006), total resistance↓ (Psoch and Schiewer, 2006b)
	– Subcritical/lowfluxoperation/sustainable operation→ (Bacchin *et al.*, 06; Guglielmi *et al.*, 2007a)
Chemical Control	– Powdered activated carbon→EPS↓ (Ying and Ping, 2006), irremovable fouling↓ (Ng *et al.*, 2006a; Ying and Ping, 2006)
	– Membrane fouling reducer/cake porosity↑, soluble EPS↓ (Hwang *et al.*, 2007; Lee *et al.*, 2007)
	– Flocculation→coagulation/organic matter↓ (Zhang *et al.*, 2008c)
	– Chemically enhanced backwashing→remove fouling (Kim *et al.*, 2007
Biological Control	– SRT↑→bound EPSY (Ahmed et al., 2007), SMP↓ (Liang *et al.*, 2007)
	– MLSS→viscosity↓→permeate flux↑(Li *et al.*, 2007a; Trussell *et al.*, 2007), cake fouling↓ (Chang and Kim, 2005; Chae *et al.*, 2006)
	– F/M ratio↓/fouling resistance↓ (Trussell *et al.*, 2006)
	– Filamentous bacteria↓→bound EPS↓ (Meng *et al.*, 2006a)

It may also be possible to prevent fouling before its occurrence by (1) improving the anti-fouling properties of the membrane, (2) operating the MBR under specific non-or-little fouling conditions and/or (3) pre-treating the biomass suspension to limit its fouling propensity.

Applications in Various Sectors

By the mid 1990s, the development of less expensive submerged membranes made MBRs a real alternative for high flow, large-scale municipal wastewater applications. Over 1,000 MBRs are currently in operation around the world with approximately 66 per cent in Japan, and the remainder largely in Europe and North America. Out of these installations, about 55 per cent use submerged membranes while the rest have external membrane modules (Van de Roest *et al.*, 2002).

Applications in Industrial Wastewater Treatment

High organic loadings and very specific and difficult to treat compounds are two major characteristics of industrial waste streams that render alternative treatment techniques such as the MBR desirable. Since, traditionally wastewater with high COD content was treated under anaerobic conditions, initial attempts of MBR applications for industrial wastewater were in the field of anaerobic treatment.

Applications in Fields of Landfill Leachate, Sludge Digestion, and Human Excrement

In addition to municipal and industrial wastewater treatment, MBRs have been utilized in a number of other areas. One such area is the treatment of landfill leachates. Landfill leachate usually contain high concentrations of organic and inorganic compounds. Conventionally, the treatment of leachates involves a physical, biological, or membrane filtration process (or a combination of them). MBR systems have been successfully utilized with an additional treatment step for inorganics and heavy metal removal, such as reverse osmosis (RO). Several industrial scale plants, combining a MBR and a reverse osmosis system, are presently operated (Manem 1996; Wehrle 1997, 1998).

The MBR system was also used in the treatment of human excreta in domestic wastewater. These applications, also known as night soil treatment systems, were typified by the high strength of the waste and the need for on-site treatment. The MBR system replaced a rather complex set of treatment systems which incorporated denitrification, coagulation, filtration, and activated carbon treatment (Magara and Itoh 1991; Manem1996). Another application of the MBR is in the area of sludge treatment. Conventionally, a single pass, anaerobic digester, achieves sludge stabilization in wastewater treatment plants. Since the HRT and the SRT are identical in these systems, the capacity is limited and long solid retention times are required for effective solids destruction. Pillay *et al.* (1994) showed that a microfiltration unit enhances the performance of the digester by decoupling the HRT and the SRT and, thereby, allowing higher volumetric throughput.

Applications in Agricultural Wastewater Treatment

A Self-Sustaining Waste Treatment System for Intensive Livestock Operations

World's livestock industry is experiencing rapid growth with an increasing number of large-scale confinement livestock operations. The growing concern is the environmental impact of waste generated in these facilities in the form of manure, wastewater, unpleasant odours, ammonia, and methane. New large scale farming facilities require large crop areas for nutrient application and in some regions nutrients in livestock waste exceed available cropland capacity to receive them in agronomic rates (Miller, 2000). The public is becoming increasingly concerned with the livestock industries potential impact on water and air quality, which will ultimately increase pressure for stricter government regulations.

Intensive livestock operations commonly combine solid and liquid waste in a manure slurry form and have extensive ventilation systems for heat, moisture, and air quality control. If a completely self-sustaining system is desired, the technology selected would have to effectively ameliorate both waste streams, produce reusable water, eliminate unpleasant odours, and be easily upgradeable. A submerged aerobic membrane bioreactor may be used as the centrepiece of such a treatment process complimented by an anaerobic digester or anaerobic lagoon as a pre-treatment step. Since one third of the organics and the majority of nutrients and metals remain in the effluent of anaerobic digesters, a submerged MBR system that facilitates nutrient and organics removal may be utilized. Nitrogen is usually the key nutrient in livestock waste management and a treatment process incorporating nitrification and denitrification is essential. Either intermittent aeration or an anoxic department within the bioreactor can be employed for total nitrogen removal (Cheng and Liu 2001). Metal salts can also be added to reduce phosphorus content in the final effluent. MBRs are capable of producing effluent free of suspended solids, bacteria, and pathogens allowing direct reuse of the product water in the livestock facility.

Aerobic activated sludge reactors have been used on a limited scale as bio-scrubbers for the treatment of odorous air (Bowker, 2000). Despite numerous positive reports from full scale applications in North America, little data are available on the actual performance of these systems with wide ranging concerns on reduction of settling efficiency due to changes in filamentous organisms and bacterial flocs (Burgess *et al.,* 2001).These concerns are alleviated in MBRs where gravitational settling of the microbial solution is replaced by physical filtration. Also, the diffusion and bioconversion of odorous gases are a function of contact time, bubble size, and reactor configuration (Burgess *et al.,* 2001). Submerged MBRs incorporate the membrane unit within the bioreactor and rely on gas and liquid scouring to clean the membrane surface. Since modern livestock operations are equipped with blowers and ventilation systems, booster fans could be added to increase outflow pressure. This concept was explored in past research efforts when bio-filter beds (compost and wood chips) were tested for odour removal (Mann *et al.,* 2002). The outlet gas stream could be introduced into an aerobic submerged MBR which would facilitate aeration, agitation, and membrane scouring while significantly reducing the release of odorous gases. The ultimate goal would be to design a process that would reduce the dependency of intensive livestock producers on cropland, remove unpleasant odours from intensive livestock operations, reuse water on-site, and thereby substantially reduce water use and reduce potential environmental risks associated to the incorrect storage, handling, and application of manure.

Food Processing Wastewater

From an environmental perspective, the majority of food processing facilities are characterized by very high water consumption and high organic strength wastewater generation (Parsons, 2001). Major waterborne pollutant loadings are biological/chemical oxygen demand, total suspended solids, fats-oils-greases, and nutrients. Most facilities employ on-site primary treatment prior to sending their wastewater to municipal wastewater treatment plants. Large volumes of high strength wastewater both increase the cost of disposal for food processing facilities and present difficult challenges for the municipal wastewater treatment plant operators. Since MBRs are capable of treating high strength wastewater, attempts were made to evaluate their effectiveness with food processing effluents. At the times of these applications, external membranes were thought to be more suitable for high temperature, high organic strength, and difficult to filter waste streams. High pressure requirements and capital investment costs resulted in the lack of large scale implementation of many such systems.

The emergence of submerged MBRs that utilize fairly economical polymer-based membranes and require less energy than external MBRs has revolutionized municipal wastewater treatment and has tremendous potential in larger scale, high volume throughput facilities across the globe. The potential of reusing the MBR product water on-site for washing or transport purposes offers many cost benefits such as reduced freshwater requirements, lower sewer costs, and possibility for direct discharge to surface waters. Depending on the wastewater characteristics and effluent requirements, both aerobic and anaerobic submerged MBRs could be employed. Industries such as slaughter houses, fermentation plants, meat, dairy, egg, and potato processing facilities, and liquor production plants could utilize this technology. Pilot-scale testing and optimization of the process would be required on a case by case basis. The fouling and flux behaviour of submerged membranes when exposed to specific waste streams would require detailed evaluation. However, intrinsic characteristics of the MBR technology such as the ability to treat high strength, greatly fluctuating wastewater, resilience in the face of shock loads and toxic chemicals, and production of superior quality effluent would justify consideration of the process in food processing facilities.

Nitrate Removal in Drinking Water

Denitrification and removal of natural organic matter are two main treatment requirements for drinking water. Nitrate is the most common groundwater contaminant worldwide (Kapoor and Viraraghavan 1997). Nitrate is a stable and highly soluble nitrogen species, easily transported and accumulated in groundwater systems. These properties, coupled with increased anthropogenic discharges of nitrogen containing compounds from point and non-point sources, have resulted in elevated nitrate concentrations in ground and surface waters. Non-point sources may have a larger impact on groundwater and are associated with agricultural and livestock practices and residential septic tank effluents (Bouchard *et al.*, 1992; Kapoor and Viraraghavan 1997).

Nitrates can be removed either biologically or by physicochemical treatment techniques such as reverse osmosis, ion exchange, and electrodialysis. Natural organic matter can be treated biologically or through activated carbon adsorption. Biological removal of nitrates and organic matter is receiving more attention due to the complete conversion of nitrate into nitrogen gas and relative ease of operation (Falk 2002).Conventional physico-chemical treatment methods only concentrate nitrate into solutions which still require disposal. In typical biological dentitrification processes, however, post treatment processes such as sand filtration, activated carbon adsorption, and disinfection are required to remove biological entities and excess organic matter and colour. The number of post-treatment processes can be significantly reduced by using a MBR for biological denitrification. All biological entities as well as some dissolved organic matter will be retained in the bioreactor while long denitrifying culture retention times and short hydraulic retention times can be maintained (Nuhoglu *et al.*, 2002).

MBRs have been investigated on an experimental scale for heterotrophic denitrification of groundwater and drinking water using two significantly different configurations. One configuration employed the membrane as a cell recycle tool in an external MBR set-up (Barreiros *et al.*, 1998), whereas the other configuration used the membrane as a semi-permeable ion exchange barrier for nitrate transfer (Fonseca *et al.*, 2000; Velizarov *et al.*, 2000). Up to 99 per cent nitrate removal, despite unusually high nitrate loadings and low hydraulic retention times were reported in these studies. Further investigation and optimization on larger scale systems are required to determine the economical feasibility of such processes.

MBR Markets and Applications (Figures 11.7 and 11.8)

Historically, the market for MBRs has been dominated by activity in Asia. This started in Japan in the 1980's and was followed by an enthusiastic uptake in south Korea in the 1990's and more recently in china. Currently, the share by number of plants between major worldwide markets had been increased enormously and mentioned in the below figure.

Wastewater treatment applications are founding a broad range of industries. In some cases the driver for treatment is to meet discharge consents; in other cases, it is to provide are source for reuse. For wastewater feeds that are easy to treat, such as from the municipal sector, conventional treatment is likely to be sufficient to meet discharge standards, except in particularly environmentally sensitive areas. Most membranes applications in the municipal sector are reuse applications.

MBRs tend to be applied either in more difficult applications within the industrial wastewater treatment sector, or to those applications where reuse is the target. The

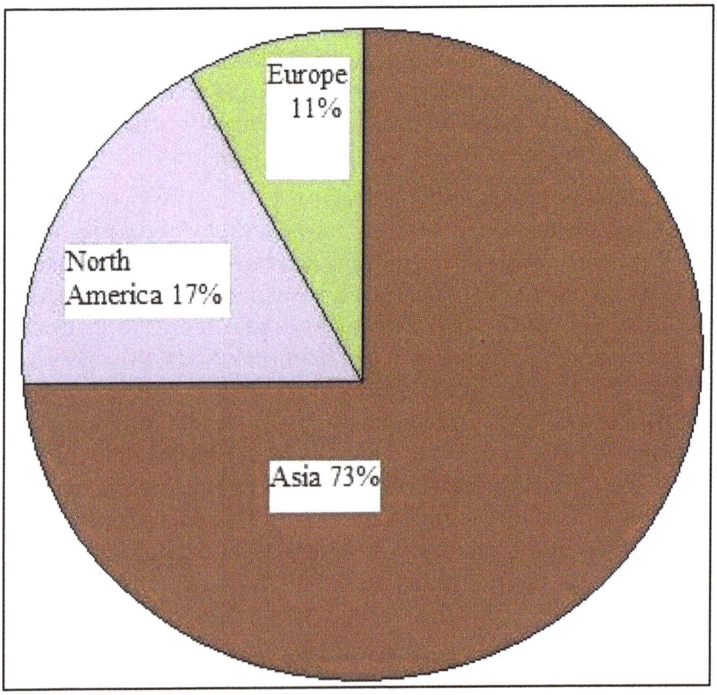

Figure 11.7: Regional Share of MBR Market.

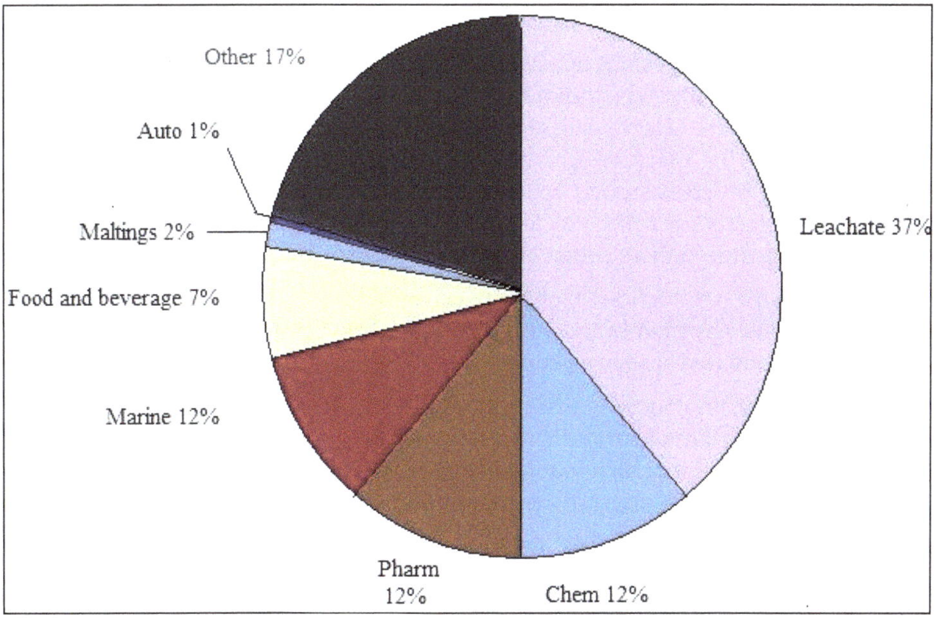

Figure 11.8: Industrial Wastewater Applications.

leachate and marine sectors are examples of areas where MBRs are relatively well used.

For industrial wastewater feeds that are difficult to treat, such as landfill leachate, it is quite common to use membranes simply to meet discharge standards. Most industrial wastewaters other than landfill leachate fall between these two extremes. With the enduse either for discharge or for reuse, though more commonly the latter. The membrane technology used in any of these applications described above could be UF/MF after conventional treatment, or an MBR.

MBRs tend to be applied either in more difficult applications within the industrial wastewater treatment sector, or to those applications where reuse is the target. The leachate and marine sectors are examples of areas where MBRs are relatively well used.

The applications for which membranes have been preferred ton conventional treatment can be summarised as follows:

- Surface water discharge (special circumstances) (MBRs)
- Urban reuse (opportunity for MBRs, MF,UF)
- Groundwater recharge (UF or MF/RO)
- Industrial reuse (MBR/RO from industrial source, UF or MF/RO from municipal).

Conclusion

Over the past decades, the appliance of Membrane Bioreactors for the treatment of municipal wastewater has drawn a lot of attention. Most important reason for this interest is the expected increase in effluent quality, achieved by the membrane. The effluent of an MBR plant would be very suitable for further upgrading with nanofilteration/reverse osmosis up to drinking water quality. Moreover, because of its modular property the MBR may also be favourable developing a new wastewater network configuration in an urban area.

Furthermore, a reduction of footprint can be achieved by the replacement of the secondary clarifier by membranes and the higher MLSS concentration compared to conventional activated sludge systems.

One of the main concerns when operating an MBR is the performance of the membranes. Due to the nature and composition of the mixed liquor, membrane fouling may occur. The extent to which fouling occurs varies with time, depending on inflow conditions, membrane configuration, membrane operation, etc.

Current MBR energy needs are higher than conventional processes even if followed by membrane filtration, though gradual improvements are being made. MBRs are now being implemented for municipal applications, with significant growth rates and other water stressed areas, and growth potential predicted in world wide markets.

In the past, both capital and operational cost of membrane modules seemed to be the major hindrance to industrial-scale application of this technology. More recently, due to increased membrane production and improvement of automated systems, there has been a significant reduction in the cost of membrane systems during the last few years. Meanwhile, increasing interest in reuse of treated water and the need for more compact systems coupled with progress in research is expected to lead to more widespread application of the MBR process. Among those processes, the submerged configuration seems likely to see the widest application. Given these developments and considering that, in general, membrane costs are decreasing, it can be expected that the MBR technology will become more economically competitive with conventional processes and find increasingly wider applications in wastewater treatment (ref 13).

References

1. Ahmed, Z., Cho, J., Lim, B.-R., Song, K.-G., Ahn, K.-H., 2007. Effects of sludge retention time on membrane fouling and microbial community structure in a membrane bioreactor. *Journal of Membrane Science* 287 (2), 211–218.

2. Barreiros, A.M., C.M. Rodrigues, J.P.S.G. Crespo and M.A.M.Reis. 1998. Membrane bioreactor for drinking water denitrification. *Bioprocess Engineering* 18 (4): 297-302.

3. Benefield, L. D. and Randall, C. W. (1980). *Biological Process Design for Wastewater Treatment*, Prentice-Hall, New Jersey.

4. Bouchard, D.C., M.K. Williams and R.Y. Surampalli. 1992. Nitrate contamination of groundwater: Sources and potential health effects. *Journal American Water Works Association* 84 (9): 85-90.

5. Burgess, J.E., S.A. Parsons and R.M. Stuetz. 2001. Developments in odour control and waste gas treatment biotechnology: A review. *Biotechnology Advances* 19 (1): 35-63.

6. Baker, J.S. and Dudely, L.Y. (1998). Biofouling in membrane systems – A review. *Desalination* 118 (1-3), 81-89.

7. Bouhabila E.H. Ben Aim R, H. Buisson, (2001). Fouling characterisation in membrane bioreactors, *Sep. Purif. Technol.* 22–23 123–132.

8. Baker, R.W. (2004) Membrane technology and applications, J. Wiley, Chichester; New York.

9. Bonne, P.A.C., Hofman, J.A.M.H. and van der Hoek, J.P. (2000) Scaling control of RO membranes and direct treatment of Surface Water. *Desalination* 132 (1-3), 109-119.

10. Bacchin P. 2004 Jan 15.A possible link between critical and limiting flux for colloidal systems: consideration of critical deposit formation along a membrane. *Journal of Membrane Science.* 228 (2): 237-241.

11. Casey, E., Glennon, B. and Hamer, G. (1999) Oxygen mass transfer characteristics in a membrane- aerated biofilm reactor. *Biotechnology and Bioengineering* 62 (2), 183-192.

12. Chang, I.S., Kim, S.N., 2005. Wastewater treatment using membrane filtration – effect of biosolids concentration on cake resistance. *Process Biochemistry* 40 (3–4), 1307–1314.

13. Chang S., Fane A., and Vigneswaran S. 2002. Experimental assessment of filtration of biomass with transverse and axial fibres. *Chemical Engineering Journal.* May 28; 87 (1): 121-127.

14. Christelle Wisniewski, 2007. Membrane Bioreactor for water reuse, *Desalination,* 15-19.

15. Cho, B.D. and Fane, A.G. (2002) Fouling transients in nominally sub-critical flux operation of a membrane bioreactor. *Journal of Membrane Science* 209 (2), 391-403.

16. Chu, H.P., Li, X.Y., 2005. Membrane fouling in a membrane bioreactor (MBR): sludge cake formation and fouling characteristics. *Biotechnology and Bioengineering* 90 (3), 323–331.

17. Chae, S.-R., Ahn, Y.-T., Kang, S.-T., Shin, H.-S., 2006. Mitigated Membrane fouling in a vertical submerged membrane bioreactor (VSMBR). *Journal of Membrane Science* 280 (1–2), 572–581.

18. Chua H.C., Arnot T.C., Howell J.A, (2002). Controlling fouling in membrane bioreactors operated with a variable throughput, *Desalination* 149 225–229.

19. Combe, C., Molis, E., Lucas, P., Riley, R. and Clark, M.M. (1999). The effect of CA membrane properties on adsorptive fouling by humic acid. *Journal of Membrane Science* 154 (1), 73-87.

20. Cheng, J. and B. Liu. 2001. Nitrification/denitrification in intermittent aeration process for swine wastewater treatment. *Journal of Environmental Engineering-ASCE* 127 (8): 705-711.

21. Di Bella, G., Durante, F., Torregrossa, M., Viviani, G., Mercurio, P., Cicala, A., 2007. The role of fouling mechanisms in a membrane bioreactor. *Water Science and Technology* 55,455–464.

22. Ducom G., Puech F. P., and Cabassud C. 2003. Gas/liquid two-phase in a flat sheet filtration module: measurement of local wall shear stresses. *Canadian Journal of Chemical Engineering.* 811-5.

23. Delgado S., Diaz F., Vera L., Diaz R., and Elmaleh S. 2004 Jan 1. Modelling hollow-fibre ultrafiltration of biologically treated wastewater with and without gas sparging. *Journal of Membrane Science.* 228 (1): 55-63.

24. de la Torre, T., Lesjean, B., Drews, A., Kraume, M., 2008. Monitoring of transparent exopolymer particles (TEP) in a membrane bioreactor (MBR) and correlation with other fouling indicators. *Water Science and Technology* 58 (10), 1903–1909.

25. Drews, A., Vocks, M., Iversen, V., Lesjean, B., Kraume, M., 2006. Influence of unsteady membrane bioreactor operation on EPS formation and filtration resistance. *Desalination* 192 (1–3), 1–9.

26. Eikelboom, D.H., Andreadakis, A. and Andreasen, K. (1998) Survey of filmanteous populations in nutrient removal plants in four European countries. *Water Science and Technology* 37 (4-5), 281-289.

27. Espinosa-Bouchot M. and Cabassud C. 2003. Gas sparging in a membrane bioreactor for wastewater treatment. IMSTEC'03, Sydney-Australia.

28. Fane, A.G., Fell, C. J. D., and nor, M. T. (1978). Wastewater treatment using a combined activated sludge and ultrafiltration system, *Proceedings of Asian Pacific Confederation of Chemical Engineering (APCCHE)*, Jakarta, November 1978. 117–129.

29. Fane A., 2002; Membrane bioreactors: design and operational options. *Filtration and Separation.* 39 (5): 26-29.

30. Fang H.H.P, X. Shi, (2005). Pore fouling of microfiltration membranes by activated sludge, *J. Membr. Sci.* 264 161–166.

31. Field, R.W., Wu, D., Howell, J.A., 1995. Critical flux concept for microfiltration fouling. *Journal of Membrane Science*, 100259–100272.

32. Flemming, H.C., Griebe, T. and Schaule, G. (1996). Antifouling strategies in technical systems- A short review. *Water Science and Technology* 34 (5-6), 517-524.

33. Flemming, H.C. (1997). Reverse osmosis membrane biofouling. *Experimental Thermal and Fluid Science* 14 (4), 382-391.

34. Fan, F., Zhou, H., Husain, H., 2006. Identification of wastewater sludge characteristics to predict critical flux for membrane bioreactor processes. *Water Research 40* (2), 205–212.

35. Fonseca, A.D., J.G. Crespo, J.S. Almeida and M.A. Reis. 2000. Drinking water denitrification using a novel ion-exchange membrane bioreactor. *Environmental Science and Technology* 34 (8): 1557-15.

36. Gander, M., Jefferson, B. and judd, S. (2000) Aerobic MBRs for domestic wastewater treatment: a review with cost considerations. *Separation and purification Technology* 18 (2), 119-130.

37. Gui P., Huang X., Chen Y., and Qian Y., 2003 Jan 10. Effect of operational parameters on sludge accumulation on membrane surfaces in a submerged membrane bioreactor. *Desalination*. 151 (2): 185-194.

38. Guglielmi, G., Saroj, D.P., Chiarani, D., Andreottola, G., 2007b. Subcritical fouling in a membrane bioreactor for municipal wastewater treatment: experimental investigation and mathematical modelling. *Water Research* 41 (17), 3903–3914.

39. Guo, W.S., Vigneswaran, S., Ngo, H.H., Xing, W., 2007.Experimental investigation on acclimatized wastewater for membrane bioreactors. *Desalination* 207, 383–391.

40. Hong S. P., Bae T. H., Tak, T. M., Hong S. and Randall, A. Fouling control in activated sludge submerged hollow fiber membrane bioreactors. *Desalination*. 2002 Jun 10; 143 (3): 219-228.

41. Hwang, B.K., Lee, W.N., Yeon, K.M., Park, P.K., Lee, C.H., Chang, I.S.,Drews, A., Kraume, M., 2008. Correlating TMP increases with microbial characteristics in the bio-cake on the membrane surface in a membrane bioreactor. *Environmental Science and Technology* 42 (11), 3963–3968.

42. Huyskens, C., Brauns, E., Van Hoof, E., De Wever, H., 2008. A new method for the evaluation of the reversible and irreversible fouling propensity of MBR mixed liquor. *Journal of Membrane Science* 323 (1), 185–192.

43. Herman evenblij, 2006. *Filtration Characteristics in Membrane Bioreactors*. IWA publishing, Rotterdam.

44. Jang, N., Ren, X., Cho, J., Kim, I.S., 2006. Steady state modelling of bio-fouling potentials with respect to the biological kinetics in the submerged membrane bioreactor (SMBR). *Journal of Membrane Science* 284 (1–2), 352–360.

45. Jiang T.,.Kennedy M.D, Guinzbourg B.F., Vanrolleghem P.A., Schippers J.C. (2005). Optimising the operation of a MBR pilot plant by quantitative analysis of the membrane fouling mechanism, *Water Sci. Technol*. 51 19–25.

46. Jenkins, D., Richard, M.G., Daigger, G.T. and Jenkins, D. (2004) Manual on the causes and control of activated sludge bulking, foaming, and other solids separation problems, Lewis Publishers, Boca Raton, Fla.

47. Jeison, D., van Lier, J.B., 2007. Cake formation and consolidation: main factors governing the applicable flux in anaerobic submerged membrane bioreactors

(AnSMBR) treating acidified wastewaters. *Separation and Purification Technology* 56 (1), 71–78.

48. Jeong, T.-Y., Cha, G.-C., Yoo, I.-K., Kim, D.-J., 2007. Characteristics of bio-fouling in a submerged MBR. *Desalination* 207 (1–3), 107–113.
49. Jones, M.L, Liehr, S.K., Classen, J.J. and Robarge, W. 2000. Mechanisms of dinitrogen gas formation in anaerobic lagoons. *Advances in Environmental Research* 4 (2), 133-139.
50. Judd, S. (2006) The MBR book: Principles and applications of membrane bioreactors, Elsevier, Boston, MA.
51. Kapoor, A. and T.Viraraghavan. 1997. Nitrate removal from drinking water–Review. *Journal of Environmental Engineering-ASCE* 123 (4): 371-380.
52. Kimura, K., Yamato, N., Yamamura, H., Watanabe, Y., 2005.Membrane fouling in pilot-scale membrane bioreactors (MBRs) treating municipal wastewater. *Environmental Science and Technology* 39 (16), 6293–6299.
53. Kox, L.S.D.M 2004. Membrane bioreactor in Varsseveld, the Dutch approach, in: Proceedings of the Water Environment-Membrane Technology Conference, Seoul, Korea.
54. Kraume, M., Wedi, D., Schaller, J., Iversen, V., Drews, A., 2009.Fouling in MBR – what use are lab investigations for full-scale operation? *Desalination*, 236, 94–103.
55. Lee W., Kang S., and Shin H., 2003 May 1 Sludge characteristics and their contribution to microfiltration in submerged membrane bioreactors. *Journal of Membrane Science*. 216 (1-2): 217-227.
56. Lim A. L. and Bai R., 2003 May 1. Membrane fouling and cleaning in microfiltration of activated sludge wastewater. *Journal of Membrane Science*. 216 (1-2): 279-290.
57. Lee. J., Ahn, W.Y. and Lee, C.H. (2001b) comparison of filtration characteristics between attached and suspended growth microorganisms in submerged membrane bioreactor. *Water Research* 35 (10), 2435-2445.
58. LaPara, T.M., Klatt, C.G. and Chen, R. (2006). Adoptions' in bacterial catabolic enzyme activity and community structure in membrane-coupled bioreactors fed simple synthetic wastewater. *Journal of Biotechnology* 121 (3), 368-380.
59. Lebegue, J., Heran, M., Grasmick, A., 2008. Membrane bioreactor: distribution of critical flux throughout an immersed HF bundle. *Desalination* 231 (1–3), 245–252.
60. Lyko, S., Al-Halbouni, D., Wintgens, T., Janot, A., Hollender, J., Dott, W., Melin, T., 2007. Polymeric compounds in activated sludge supernatant – characterisation and retention mechanisms at a full-scale municipal membrane bioreactor. *Water Research* 41 (17), 3894–3902.
61. Le-Clech, P., Jefferson, B., Judd, S.J., 2005. A comparison of submerged and side stream tubular membrane bioreactor configurations. *Desalination* 173 (2), 113–122.

62. Li, M.S., Zhao, Y.J., Zhou, S.Y., Xing, W.H., Wong, F.S., 2007a.Resistance analysis for ceramic membrane microfiltration of raw soy sauce. *Journal of Membrane Science* 299, 122–129.

63. Lesjean, B., Rosenberger, S., Laabs, C., Jekel, M., Gnirss, R., Amy, G. 2005. Correlation between membrane fouling and soluble/colloidal organic substances in membrane bioreactors for municipal wastewater treatment. *Water Science and Technology* 51 (6–7), 1–8.

64. Liang, S., Liu, C., Song, L., 2007. Soluble microbial products in membrane bioreactor operation: behaviours, characteristics, and fouling potential. *Water Research* 41 (1), 95–101.

65. Lebegue, J., Heran, M., Grasmick, A., 2008. Membrane bioreactor: distribution of critical flux throughout an immersed HF bundle. Desalination 231 (1–3), 245–252.

66. Lim, K.-H. Ahn, K.-G. Song, C. Jin-Woo, 2004. Microbial community in biofilm on membrane surface of submerged MBR: effect of in-line cleaning chemical agent, in: Proceedings of the Water Environment-Membrane Technology Conference, Seoul, Korea.

67. Le-Clech P., Fane A., G. Leslie, A. Childress, (2005). The operator's perspective, Filt. Sep. 42 20–23.

68. Lim A.L., R. Bai, (2003). Membrane fouling and cleaning in microfiltration of activated sludge wastewater, *J. Membr. Sci.* 216 279–290.

69. Liang, S., Liu, C., Song, L., 2007. Soluble microbial products in membrane bioreactor operation: behaviours, characteristics, and fouling potential. *Water Research* 41 (1), 95–101.

70. Manem, J.A.S. 1996. Membrane bioreactors. In *Water Treatment Membrane Processes,* ed. J. Mallevialle, P.E.Odendaal and M.R. Wiesner, 17.1-17.31. New York.

71. Metzger, U., Le-Clech, P., Stuetz, R.M., Frimmel, F.H., Chen, V., 2007. Characterisation of polymeric fouling in membrane bioreactors and the effect of different filtration modes. *Journal of Membrane Science* 301 (1–2), 180–189.

72. Metcalf and Eddy wastewater treatment and engineering technologies, fourth edition, 2003.

73. Musvoto, E.V., Lakay, M.T., Casey, T.G., Wentzel, M.C. and Ekama, G.A. (1999). Filamentous organism bulking in nutrient removal activated sludge systems – paper 8: The effect of nitrate and nitrite. *Water Sa* 25 (4), 397-407.

74. Mayer, M.; Braun, R., and Fuchs, W. 2005. Comparison of various aeration devices for air sparging in crossflow membrane filtration. *Journal of Membrane Science.* In Press, Corrected Proof.

75. Mallevialle, J., Odendaal, P. and Wiesner, M. (1996) Water treatment membrane processes, McGraw-Hill, New York.

76. Meng, F., Shi, B., Yang, F., Zhang, H., 2007a. Effect of hydraulic retention time on membrane fouling and biomass characteristics in submerged membrane bioreactors. *Bioprocess and Biosystems Engineering* 30, 359–367.

77. Mulder, M. (1996) Basic principles of membrane technology, Kluwer Academic, Dordrecht; Boston.

78. Ma H., Bowman C. and Davis R., 2000 Jul 15 Membrane fouling reduction by back pulsing and surface modification. *Journal of Membrane Science*. 173 (2): 191-200.

79. Mukai T., Takimoto K., Kohno, T., and Okada M. 2000 Feb 15 Ultrafiltration behaviour of extracellular and metabolic products in activated sludge system with UF separation process. *Water Research*. 34 (3): 902-908.

80. Miller, G. 2000. 1999-2000 Annual Report of the Environmental Commissioner of Ontario. Toronto, ON: Environmental Commissioner of Ontario. Membrane biological reactor network (an association in Germany for developing MBRs).

81. Nuhoglu, A., T. Pekdemir, E. Yildiz, B. Keskinler and G. Akay.2002. Drinking water denitrification by a membrane bioreactor. *Water Research* 36 (5): 1155-1166.

82. Ng C.A, Sun D., Zhang J.,. Chua H.C, Bing, W. S. Tay, A, 2005. Fane, Strategies to improve the sustainable operation of membrane bioreactors, in: Proceedings of the International Desalination Association Conference, Singapore.

83. Ognier, S.,Wisniewski, C., Grasmick, A., 2004.Membrane bioreactor fouling in sub-critical filtration conditions: a local critical flux concept. *Journal of Membrane Science* 229 (1–2), 171–177.

84. Pollice A, Brookes A., Jefferson B. and Judd S. 2005 Apr 20. Sub-critical flux fouling in membrane bioreactors — a review of recent literature. *Desalination*. 174 (3): 221-230.

85. Psoch, C., Schiewer, S., 2006a. Anti-fouling application of air sparging and back flushing for MBR. *Journal of Membrane Science* 283 (1–2), 273–280.

86. Phattaranawik, J., Fane, A.G., Pasquier, A.C.S., Bing, W., 2007.Membrane bioreactor with bubble-size transformer: design and fouling control. *AICHE Journal* 53, 243–248.

87. Psoch C. Schiewer S. (2006). Resistance analysis for enhanced wastewater membrane filtration, *J. Membr. Sci*. 280 284–297.

88. Pillay, V.L., B. Townsend and C.A. Buckley. 1994. Improving the performance of anaerobic digesters at wastewater treatment works: The coupled cross-flow microfiltration/digester process. *Water Science and Technology* 30 (12): 329-337.

89. Parsons, R.V. 2001. Making dollars and sense of waste and wastewater in the food processing industry. Managing Water and Wastewater in the Food Processing Industry Seminar. Winnipeg, MB: Manitoba Hydro.

90. Ramesh, A., Lee, D.J., Lai, J.Y., 2007. Membrane biofouling by extracellular polymeric substances or soluble microbial products from membrane bioreactor sludge. *Applied Microbiology and Biotechnology* 74, 699–707.

91. Rittmann, B. E. (1987). Aerobic biological treatment. *Environ. Sci. Technol*. 21 (2), 128–136.

92. Smith, C. W., Di Gregorio, D., and Talcott, R. M. (1969). The use of ultrafiltration membrane for activated sludge separation. *Proceedings of the 24th Annual Purdue Industrial Waste Conference,* Purdue University, West Lafayette, Indiana, USA. 1300–1310.

93. Sato T., Ishii Y. (1991). Effects of activated-sludge properties on water flux of ultrafiltration membrane used for human excrement treatment, *Water Sci.Technol.* 23 1601–1608.

94. Sofia A., Ng W. J., and Ong S. L. 2004. Jan 5Engineering design approaches for minimum fouling in submerged MBR. *Desalination.* 160 (1): 67-74.

95. Sperandio, M., Masse, A., Espinosa-Bouchot, M.C., Cabassud, C.2005. Characterization of sludge structure and activity in submerged membrane bioreactor. *Water Science and Technology* 52 (10–11), 401–408.

96. Sun, Y., Wang, Y., Huang, X., 2007. Relationship between sludge settleability and membrane fouling in a membrane bioreactor. *Frontiers of Environmental Science and Engineering in China1* (2), 221–225.

97. Stephenson T., S. Judd, B Jefferson and K Brindle, Membrane Bioreactors for wastewater treatment. IWA Publishing, London, 2006

98. Schoeberl P., Brik M., Bertoni, M. Braun R, Fuchs W. (2005).Optimization of operational parameters for a submerged membrane bioreactor treating dye house wastewater, *Sep. Purif. Technol.* 44 61–68. [240] J.S. Pa.

99. Smith P.J., Vigneswaran S., Ngo H.H., Ben-Aim R., Nguyen H. (2005). Design of a generic control system for optimising back flush durations in a submerged membrane hybrid reactor, *J. Membr. Sci.* 255 99–106.

100. Sun Y., Huang, F. Chen, X. Wen, 2004. A dual functional filtration/aeration membrane bioreactor for domestic wastewater treatment, in: Proceedings of the Water Environment-Membrane Technology Conference, Seoul, Korea.

101. Till S. and Mallia H. 2001. Membrane bioreactors: wastewater treatment applications to achieve high quality effluent. 64th Annual Water Industry Engineers and Operators' Conference.

102. Tchobanoglous, G., Burton, F.L. and Stensel, H, D. (2003) *Wastewater engineering*: treatment and reuse, McGraw-Hill, Boston.

103. Te Poele, S.T., Roorda, J.H. and van der Graaf, J.H.J.M (2004) Influence of the size of membrane foulants on the filterability of WWTP-effluent. *Water Science and Technology* 50 (12), 111-118.

104. Trussell, R.S., Merlo, R.P., Hermanowicz, S.W., Jenkins, D., 2007.Influence of mixed liquor properties and aeration intensity on membrane fouling in a submerged membrane bioreactor at high mixed liquor suspended solids concentrations. *Water Research* 41 (5), 947–958.

105. Ueda T., Hata K., Kikuoka Y. and Seino O. 1997 Mar. Effects of aeration on suction pressure in a submerged membrane bioreactor. *Water Research*; 31 (3): 489-494.

106. Van Kaam R., Hernadez Rojas M. E., Schetrites S., Lorain O., and Albasi C. 2004. Influence of process parameters in a sequenced membrane bioreactor to prevent membrane fouling in wastewater treatment–interest of external test. Ninth World Filtration Congress, New Orleans-USA.

107. Visvanathan C.,. Yang B.S, Muttamara S., Maythanukhraw R. (1997). Application of air back flushing technique in membrane bioreactor, *Water Sci.Technol.* 36 259–266.

108. Vallero M.V.G., Lettinga G, Lens P.N.L. (2005). High rate sulphate reduction in a submerged anaerobic membrane bioreactor (SAMBaR) at high salinity, J. Membr. Sci. 253 217–232.

109. Velizarov, S., C.M. Rodrigues, M.A. Reis and J.G. Crespo.2000. Mechanism of charged pollutants removal in an ion exchange membrane bioreactor: Drinking water denitrification. *Biotechnology and Bioengineering* 71 (4): 245-254.

110. Van de Roest, H.F., D.P. Lawrence and A.G.N. Van Bentem.2002. *Membrane Bioreactors for Municipal Wastewater Treatment*. STOWA Report. London, UK: IWA Publishing.

111. Wehrle. 1997. MBR Case Study: EAL-Sheinberg Treatment Plant. http: // www.wehrle-env.co.uk/pdf.

112. Wehrle. 1998. MBR Case Study: Trienekens GmbH.http: //www.wehrle-env.co.uk/pdf/.

113. Wang, Z., Wu, Z., Yin, X., Tian, L., 2008b. Membrane fouling in a submerged membrane bioreactor (MBR) under sub-critical flux operation: membrane foulant and gel layer characterization. *Journal of Membrane Science*.

114. Watanabe, Y., Kimura, K., Itonaga, T., 2006. Influence of dissolved organic carbon and suspension viscosity on membrane fouling in submerged membrane bioreactor. *Separation Science and Technology* 41 (7), 1371–1382.

115. Yang, Z., Peng, X.F., Chen, M.-Y., Lee, D.-J., Lai, J.Y., 2007. Intra layer flow in fouling layer on membranes. *Journal of Membrane Science* 287 (2), 280–286.

116. Yamamoto, K., Mahmood T., and Vigneswaran S. 1988. Applications of direct membrane separation to activated sludge process. Partec'88, Kuala Lumpur-Malaysia.

117. Yu K., Wen X., Bu Q., and Xia H. 2003 Oct 15. Critical flux enhancements with air sparging in axial hollow fibers cross-flow microfiltration of biologically treated wastewater. *Journal of Membrane Science.* 224 (1-2): 69-79.

118. Yamato N., Kimura K., Miyoshi T., and Watanabe Y. 2006. Difference in membrane fouling in membrane bioreactors (MBRs) caused by membrane polymer materials. *Journal of Membrane Science.* In Press, Corrected Proof.

119. Yang W., Cicek N., and Ilg J. 2006 Feb 15 State-of-the-art of membrane bioreactors: Worldwide research and commercial applications in North America. *Journal of Membrane Science.* 270 (1-2): 201-211.

120. You, H.S., Huang, C.P., Pan, J.R., Chang, S.C., 2006. Behaviour of membrane scaling during crossflow filtration in the anaerobic MBR system. Separation Science and Technology 41 (7), 1265–1278.

121. Zeman, L.J. and Zydney, A.L. (1996). Microfiltration and ultrafiltration: Principles and applications, M.Dekker, New York.

122. Zhang, J., Chua, H.C., Zhou, J., Fane, A.G., 2006a. Factors affecting the membrane performance in submerged membrane bioreactors. *Journal of Membrane Science* 284 (1–2), 54–66.

123. Zhang, J.S., Chuan, C.H., Zhou, J.T., Fane, A.G., 2006b. Effect of sludge retention time on membrane bio-fouling intensity in a submerged membrane bioreactor. *Separation Science and Technology* 41 (7), 1313–1329.

124. Zhang S.T.,. Qu Y.B, Y.H. Liu, F.L. Yang, X.W. Zhang, K. Furukawa, Y.Yamada, (2005). Experimental study of domestic sewage treatment with a metal membrane bioreactor, *Desalination* 177 83–93.

Chapter 12

Design of Wastewater Treatment Plant: 'WWTP' Software

J.K. Bassin

CSIR-NEERI Delhi Zonal Laboratory,
A-93/94, Naraina Industrial Area, Phase-I,
New Delhi

Introduction

Water, a prime natural resource so vital to survival of human being, is becoming increasingly scarce requiring careful engineering and environmental management. Rapid population growth and urbanization in developing countries like India worsen the situation. Water is the major constituent of the human body, the animal and the vegetable kingdom of our Mother Earth, and of the atmosphere around the earth. Water has been recognized as the cardinal element of nature and has been worshipped as the life-giving and life-sustaining medium in ancient and modern religions of the world. Owing to its importance, the rural and urban developments in ancient and modern history have centered around and near sources of water like rivers and lakes, such as the Ganges, the Indus, the Krishna, the Nile, the Mekong, the Volga, the Danube, and the Tiber, etc.

Water is fundamentally important resource because it is essential for survival; because its use underlies all agriculture and industrial processes; and because it can not be substituted for technologically. Industrialization and the accompanying urbanization have not only polluted the watercourses, but also created a growing demand for large quantities of protected water for public and for the industries. This, coupled with growing population and non-availability of adequate facilities for proper treatment of water, has resulted in the use of available waters, often not pure, for human consumption, resulting in health hazards. Analysis of the available information reveals that two-thirds of all illnesses in the developing countries, particularly in India, are related to water-borne diseases such as cholera, typhoid, diarrhea, infectious hepatitis (jaundice), dysentery, etc.

In most developed countries city residents used to put night soil in buckets along the streets where workers emptied the waste into 'honey wagon' tanks. The waste was transported for disposal over agricultural lands. For rural communities in many under developed countries, this method is still in use [Sincero, 1996]. The 'honey wagon' method of disposal was adequate then, since the volume of waste generated was relatively small. However, the development of flush toilet, which produced large volumes of wastewater in nineteenth century overtaxed agricultural lands as a final receptor of the waste. This induced people to use natural channels to convey into the nearest watercourse the large volumes of liquids generated by flush toilets. Natural consequence of use of open channels was the construction of large sewers. These sewers directly discharged into the streams, rivers, lakes and estuaries. This practice resulted in gross pollution of these receiving water bodies. Over a period of time, most pollution has been the result of indiscriminate disposal of wastes, including the unavoidable domestic wastes like sewage, into our water bodies. The rise in industrial and agricultural activities resulted in further pollution of watercourses and reservoirs. The situation so worsened that the government had to launch large-scale projects of the order of Ganga Action Plan, for improvement of water quality of major rivers at a cost of several crores of Rupees.

Thus, the need for development and proper management of water resources can not be emphasized more. Drinking water has clearly to meet certain fairly stringent quality standards but water quality is also important for agricultural and industrial use. Each of the water uses in turn tends to degrade the quality of water or in other words pollutes the water and thus generating the wastewater.

Wastewater and its Impact

Water, after it has been used for the designated purpose, becomes wastewater, *e.g.* more than 80 per cent of water used in domestic sector is returned to be discharged into the water bodies as sewage with highly degraded quality. Similarly, almost similar proportion of industrial process water is returned as wastewater. The degree of deterioration is indicated by the characteristics of particular industrial wastewater. More often than not, these industrial wastewaters are also discharged into public sewers and are received along with the domestic sewage. The main sources of wastewater are domestic sewage, industrial wastewaters, agricultural runoff, and storm water and urban runoff. According to a survey by the Central Pollution Control Board (CPCB), the estimated wastewater generation from 212 Class-I cities of India is 12,145 MLD of which only 2485 MLD (about 20 per cent) receives treatment, in some cases only partially. An estimated 1298 MLD of wastewater is generated from 241 out of 270 Class-II towns in India as per 1981 census. Further, it is reported that less than 5 per cent of the total wastewater generated in Class-II towns is collected and only 2 per cent is receiving some kind of treatment. The situation has only deteriorated further in time. For example, the river Ganga has, for years, been subjected to tremendous pressures. Most of its water in the upper reaches is diverted to canals and untreated domestic and industrial wastewaters are dumped into the river at several places. It is estimated that 900 million litres of sewage is discharged into the river everyday and this accounts only for 75 per cent of total pollution reaching the river [NEERI, 1992]. Residues of pesticides and insecticides used in agriculture also contribute non-point pollution in the river.

Fate of pollutants discharged by man as a result of his industrial and urban activities is shown in Figure 12.1. There are no boundaries between the soil, water and gaseous phases of the environment. Pollution occurring in one phase can get transferred to another in future as shown in the figure. Pollutants entering the environment may degrade with time or remain inert and thus undiminished with time. This condition will result in adverse economic and health effects. Many of these effects become apparent only in the long term and hence tend to be neglected until it is too late [Arceivala, 1998].

Wastewater Treatment Objectives and Methods

Wastewater treatment methods were first developed in response to the concern for public health and the adverse conditions caused by the discharge of wastewater to the environment. Also important, as the cities became larger, was the limited availability of land required for wastewater treatment and disposal, principally by irrigation and intermittent filtration, methods commonly used in the early 1900s. The purpose of developing other methods of treatment was to accelerate the forces of nature under controlled conditions in treatment facilities of comparatively smaller size. [Metcalf and Eddy, 1997]. The treatment objectives were concerned with (1) the removal of suspended and floatable material, (2) the treatment of biodegradable organics, and (3) the elimination of pathogenic organisms. Since 1980, because of increased scientific knowledge and an expanded information base, wastewater treatment has begun to focus on the health concerns related to toxic and potentially toxic chemicals released to the environment [Metcalf and Eddy, 1997]. The contaminants in wastewater (unwanted substances termed as pollutants) are removed by physical, chemical, biological or other means and the treated wastewater can be used for consumptive purposes or can be discharged into the water bodies without causing quality degradation beyond those permitted.

Wastewater treatment plants utilize a number of treatment processes to achieve the desired degree of treatment. In addition, the environmental design engineer must evaluate numerous other important factors in selection of the treatment processes. These include constituents treated, effluent limitations, proximity to build-up areas, hydraulic requirements, sludge disposal, energy requirements, and plant economics. The collective arrangement of various treatment processes is called a flow scheme, flow diagram, a flow sheet or a process train [Qasim, 1985]. Choice of proper treatment processes and development of the flow scheme is not a simple task. It requires understanding of the unit operations and processes, operational capabilities and environmental effects of various treatment components that are arranged to develop the flow scheme for a desired application. Laboratory and pilot plant studies are often necessary to develop design parameters for physical, chemical, and biological

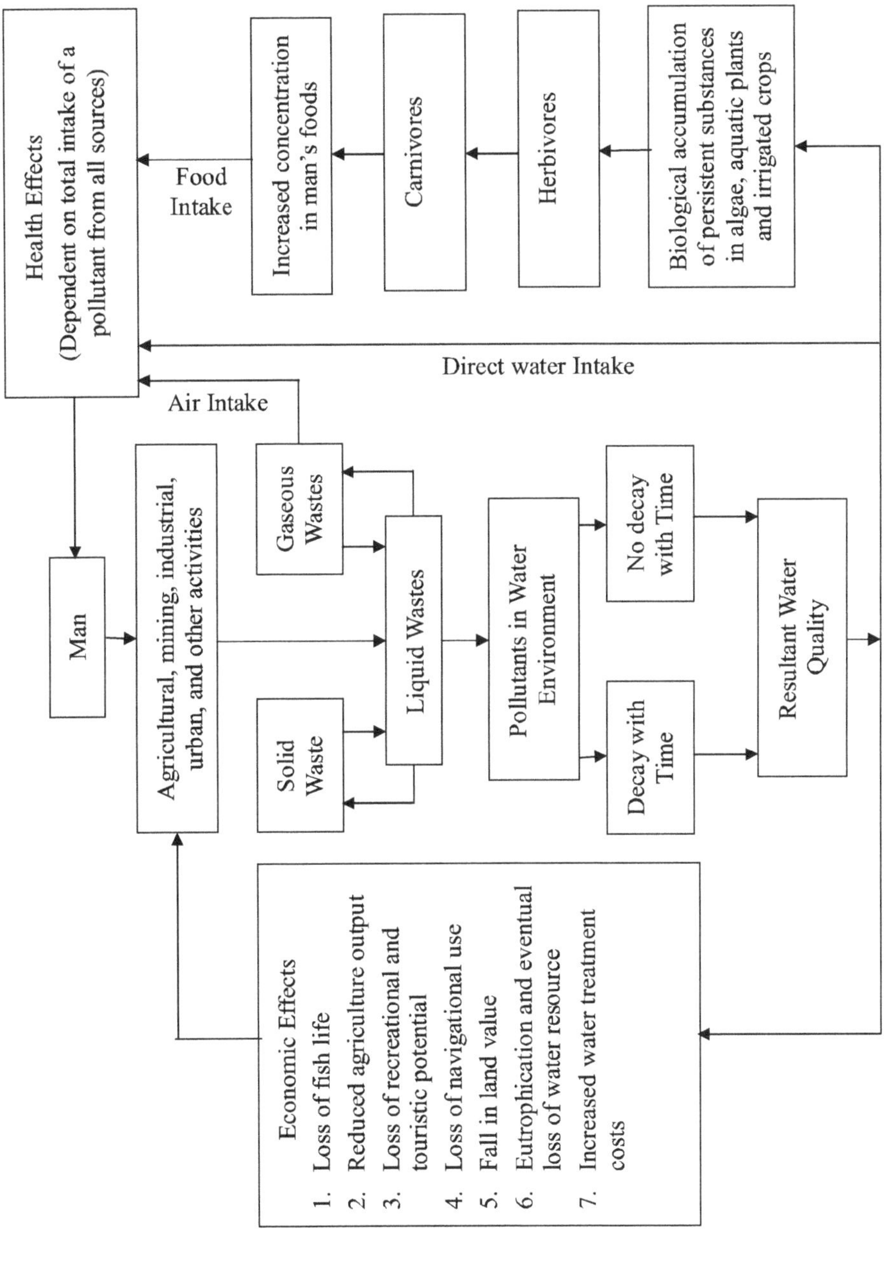

Figure 12.1: Some Health and Economic Effects likely to Result from Discharge of Solid, Liquid and Gaseous Wastes into Water.

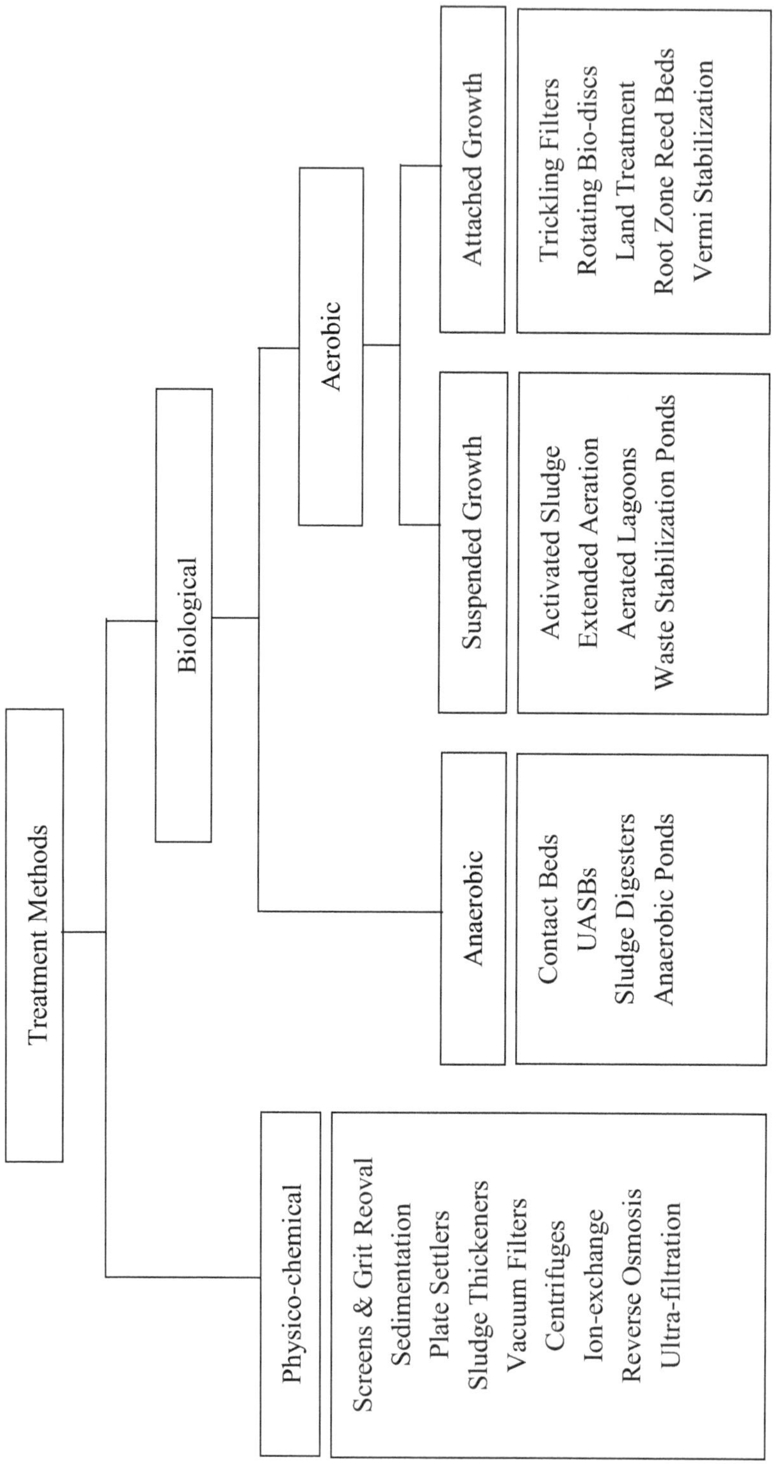

Figure 12.2: Some Methods of Wastewater Treatment (Ref: Arceivala, 1998).

treatment processes used for treatment of wastewater. Figure 12..2 lists some of the available methods of wastewater treatment [Arceivala, 1998].

Need for WWTP Design Software

The wastewater scenario and the situation alluded to earlier have led to the advent of methods for treating the waste so as to minimize the impact of wastewater discharges on receiving waters. Over a period of time several techniques and processes have been developed for wastewater treatment. These methods encompass from low-cost wastewater treatment methods such as stabilization ponds, to advanced wastewater treatment methods such as activated carbon adsorption, reverse osmosis, etc.

Design of wastewater treatment plant is the solution of a very complex problem for several reasons [Kroiss, 1994]. Some important ones are:

- ✰ Probabilistic nature of wastewater flow and composition.
- ✰ Restricted knowledge of the behavior of biological system in regard to adaptation.
- ✰ Strong influences of future developments during long lifetime of equipment and construction.
- ✰ Different definitions of effluent quality standards.
- ✰ Restricted knowledge of the influences of probabilistic effluent quality on the receiving waters.
- ✰ Restricted knowledge of the technical reliability of different process configurations.
- ✰ The influences of operational decisions on the treatment efficiency.
- ✰ Different cost calculation procedures (investment, capital, operation costs).

The design is never the solution of a series of mathematical equation only but always comparable to the work of a composer creating a symphony too. Engineering is basically more than the application of scientific knowledge to the solution of a specific local problem. Consequently, more than one optimal solution is possible. Nevertheless design engineers will have the task to find the least coat solution for any specific local case [Kroiss, 1994]. Design procedure for biological wastewater treatment plants as proposed by Kroiss [1994] is depicted in Figure 12.3. Starting points are the required treatment efficiency, the minimum requirements being contained in the effluent standards. The basis for a proper design is a standardized database consisting of a series of relevant loading conditions (scenarios). As there will always be a lack of information this database will contain 'safety factors'. For each specific case, the design engineer has to look for adequate models in order to calculate the basic design values as reactor volumes, oxygenation capacities, excess sludge production, surface of settlers or filters, etc. [Kroiss, 1994]. However, such an approach to incorporate reliability considerations in the design procedures requires large volume of data and in the absence of such large databases, the design procedures followed hitherto as reported in the literature and standard texts are being followed.

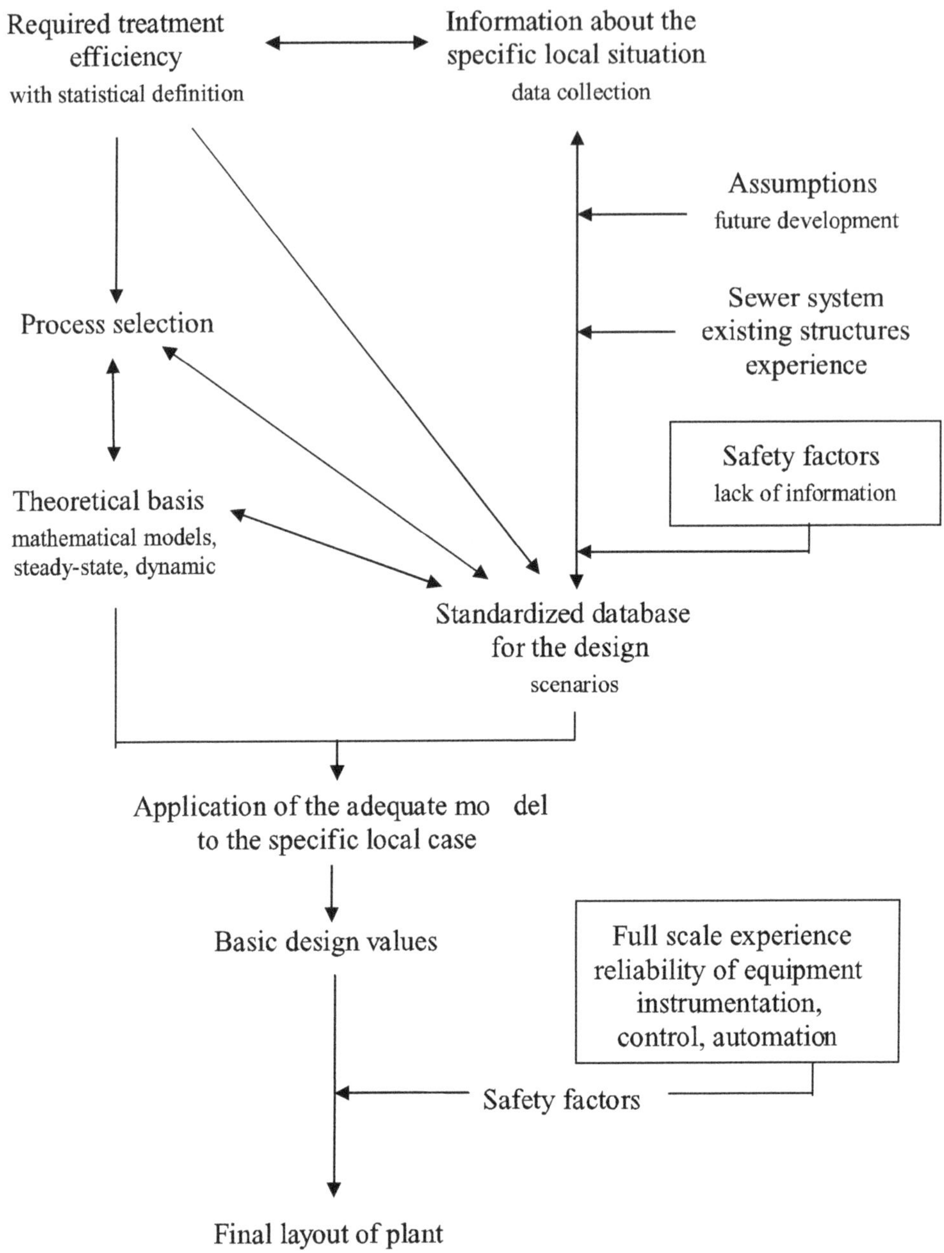

Figure 12.3: Design Procedure for Biological Wastewater Treatment Plants [*Source*: Kroiss, 1994].

Thus, in order to ensure that the quality of our watercourses is maintained at desirable levels, it is necessary to treat the wastewater before discharging into water bodies. It is only in last few years that the design approach for these plants has changed from empirical to sound engineering basis.

The design of wastewater treatment plant has to go through a number of steps, *viz.* conceptual design, preliminary design, special studies, and final design [Metcalf and Eddy, 1997].Preliminary literature review indicated that while much work has been reported on the development of mathematical models for treatment processes such as widely used activated sludge process, no or little work has been devoted to the development of computer based efficient tools for design of wastewater treatment plants. It was towards this end that development of a computer software for use by the environmental design engineers entrusted with the task of wastewater treatment designs was thought as a timely and apt action. It was envisaged to develop a computer software which would help in the design of a new treatment plant, in the augmentation of an existing treatment plant, and in the operation of treatment plant through troubleshooting guide.

Basic Considerations

Nature of Pollutants

The various pollutants entering an ecosystem may belong to one of the following three groups:

1. Degradable
2. Non-degradable (conservative)
3. Biologically accumulative (persistent)

Degradable pollutants are all complex organic substances (*e.g.*, sewage) and dead organisms, which can undergo gradual microbial decomposition. Degradable substances also include those that undergo physical degradation or decay (*e.g.*, radioactive isotopes, though some have such a long half-life that they may be considered as practically non-degradable).

Degradation is a time-rate process, the rate generally following first-order kinetics and can often be described by a typical equation such as: $C_t = C_o e^{-k.t}$ in which, C_o = concentration of pollutant initially at time zero; C_t = concentration of pollutant at time t; t = time over which degradation occurs; k = degradation rate constant per unit time for the substance in question.

Readily degradable substances show a high value of k while those more difficult to degrade show lower values. Microbial degradation of organic matter results in oxygen consumption and thus the latter can become an indirect method of estimating the amount of organic matter present. This is done in the biochemical oxygen demand (BOD) test.

Non-degradable (conservative) substances are those which are inert to biological action. These include inorganic chemicals (*e.g.*, Polychloride biphenyles).

Biologically accumulative (persistent) substances are those which tend to accumulate in the food chain. They include mercury, cadmium, arsenic, lead, manganese, pesticides and radioactive isotopes [Arceivala, 1981 and 1998].

Wastewater Characteristics

Municipal Wastewater

Domestic wastewater contains organic and inorganic matter in suspended, colloidal and dissolved form. The concentration in the wastewater depends on the original concentration in the water supply, and the uses to which the water has been put. The climate and the wealth and habits of the people have a marked effect on the wastewater characteristics. The presence of industrial wastes in public sewers can substantially alter the nature of the wastewater. Concentrations are also affected by the amount of water used per person since in many communities the amount of solids added per person vary within relatively narrow limits. Thus, wastewater characteristics vary not only from city to city but also from season to season and even hour to hour within a given city.

Raw domestic wastewater characteristics are shown in Table 12.1. The range of values given is typical for municipal wastewaters, which are predominantly domestic in character. The wastes are mainly organic in nature, containing "nutrients" like carbon, nitrogen, and phosphorus among others, with relatively high concentrations of microorganisms. They are readily putrescible, and the biological degradation of organic matter proceeds even as the wastes flow through sewers, thereby altering some of its characteristics as time passes. All values given in Table 12.1 are stated in terms of grams per capita to facilitate their use for design purposes and for comparison between different communities. Values given in terms of milligrams per litre are obscured by the fact that water usage varies markedly between communities, especially between those in developed and developing countries [Arceivala, 1998].

Values of biochemical oxygen demand (BOD) generally average around 54 g per person per day where the sewage collection system is reasonable efficient. In some developing areas, the BOD values may be only 30 to 40 g per person per day as all the sewage produced in the area may not enter the sewer system. Where combined sewers are used, the BOD values may be about 40 per cent higher, namely 77 g per person day. In the case of offices, factories, schools, etc., where there is part-time occupancy, the BOD values are generally taken as half of 54 g per person day or even less. For restaurants and cafeterias each meal served may be taken to contribute one-fourth of 54 g of BOD. Theaters and cinemas may be taken to contribute one-sixth of 54 g of BOD per seat. Hotels and hospitals, on the other hand, may contribute as much as 1.5 to 2.5 times more than the usual 54 g of BOD per person day. Domestic sewage is the primary source of nutrients such as nitrogen and phosphorus; most industrial wastes (except the food and fertilizer industries) are relatively minor sources of these nutrients in municipal wastewaters. In many developing countries the amount of nitrogen and phosphorus in wastewaters may lie at the lower end of ranges shown in Table 12.1.

Daily water usage ranges from 180 to 300 l person in most sewered communities although consumption in some large cities in developed countries can be even higher than 500 l person. At least 80-85 per cent of the water supplied returns as wastewater. Thus, if water supply is 180 l/person-day, the wastewater flow Q-at 80 per cent of the supply–is given by

Table 12.1: Domestic Wastewater Characteristics

Item	*Range of Values Contributed in Wastes (g/capita-day)*
Biochemical oxygen demand, 5 days, 20°C (BOD_5)	45-54
Chemical oxygen demand	1.6-1.9 x BOD_5
Total organic carbon	0.6-1.0 x BOD_5
Total solids	170-220
Suspended solids	70-145
Grit (inorganic, 0.2 mm and above)	5-15
Grease	10-30
Alkalinity (as calcium carbonate, $CaCO_3$)	20-30
Chlorides	4-8
Total nitrogen N	6-12
Organic nitrogen	~0.4 x total N
Free ammonia[2]	~0.6 x total N
Nitrite	–
Nitrate	~0.0-0.5 x total N
Total phosphorus, P	0.6-4.5
Organic phosphorus	~0.3 x total P
Inorganic (ortho-and polyphosphates)	–0.7 x total P
Potassium (as potassium oxide K_2O)	2.0-6.0
Microorganisms present in wastewater	(per 100 ml wastewater)
Total bacteria	10^9-10^{10}
Coliforms	10^9-10^{10}
Faecal Streptococci	10^5-10^6
Salmonella typhosa	10^1-10^4
Protozoan cysts	up to 10^3
Helminthic eggs	up to 10^3
Virus (plaque forming units)	10^2-10^4

1. Values given in the table represent approximate daily per capita contributions. To these must be added the contents in the original water supply. Per capita daily water usage ranges from 180 to 300 l in most sewered communities. The pH of wastewater generally ranges from 6.8 to 8.0, again, depending on raw water quality.

2. The major nitrogen compound in domestic waste is urea $CO(NH_2)_2$ which is readily hydrolyzed to ammonia (NH_3) and carbon dioxide (CO_2) by the enzyme urease present in sewage. Hence, NH_3 constitutes the major fraction of total nitrogen in domestic sewage.

$$Q = (180)\,(0.8) = 144 \cong 150\ \text{l/person-day}$$

$$BOD_5 = \frac{54\text{g/person} - \text{day} \times 10^3}{150\text{l/person} - \text{day}} = 360\text{mg/l}$$

Thus, concentration depends on both the wastewater flow and the biochemical oxygen demand, 5 days, 20°C (BOD_5) contribution per capita.

Industrial Wastewater

Characteristics of industrial wastewaters vary widely from industry to industry, and even within the same industry depending on raw materials used, process employed, and various other factors. It is, therefore, difficult to generalize, although a few typical values are shown in Table 12.2. One characteristic of most industrial wastes that must be mentioned at this stage is the likelihood of wide fluctuations in both flow and constituents of a waste [Arceivala, 1998].

Often flow measurement, as well as sampling, has to be done in industries in such a manner that individual streams from different departments, or even different machines, can be evaluated in order to assess possibilities for reuse of water, waste reduction, or segregation of wastes to facilitate treatment and/or disposal. Flow measurement if often done using a V-notch weir. The duration of sampling for industrial purposes is very important if a composite and representative sample is to be obtained. Generally, samples can be taken every 15 or 30 minutes, depending on variability, and composited over a total period long enough to cover one shift of operation in the industry or one clear cycle of operation of a unit. Data so collected can be used to prepare a material balance and flow diagram.

Some Wastewater Treatment Methods and their Characteristics

The methods range from physico-chemical to biological and in the latter group from aerobic to anaerobic (Figure 12.2). Typical performance characteristics of some of the methods are given in Table 12.3.

- ☆ Activated Sludge Process (ASP)
- ☆ Extended Aeration Process (EAP)
- ☆ Moving Bed Biological Reactor (MBBR)
- ☆ Membrane Bio-Reactor (MBR)
- ☆ Contact Stabilization Process (CSP)
- ☆ Trickling Filters and Rotating Bio-Discs (TF and RBC)
- ☆ Waste Stabilization Ponds (WSP)
- ☆ Mechanically Aerated Lagoons (MAL)
- ☆ Upflow Anaerobic Sludge Blanket Process (UASB)

Typical design factors for bar racks, grit chambers and settling tanks are presented in Table 12.4 to 12.7

Discharge Standards

The standards applicable for discharge of industrial wastewater are presented in the Schedule VI of the Environment (Protection) Third Amendment Rules, 1993. Different standards apply for different water bodies where the treated wastewater is discharged, *viz.* Inland surface water, Public sewers, Land for irrigation, and Marine coastal areas. Applicable standard for discharge of treated sewage in surface water is

Table 12.2: Typical Performance Characteristics of Some of the Wastewater Treatment Methods

Item	*Extended Aeration*	*Conventional Activated Sludge*	*Conventional Trickling Filters*	*Facultative Aerated Lagoons*
Performance (typical)				
BOD removal, per cent	95-98	85-92	80-90	75-85
Nutrient removal, per cent N	15-30 (b)	30-40 (b)	15-20	–
Nutrient removal, per cent P	10-20 (b)	30-45 (b)	10-20	–
Coliform removal,	60-90	60-90	60-90	60-90
Helminth removal, per cent	–	–	–	–
Land requirement (m2/person) (c)	0.15-0.20	0.20-0.25	0.20-0.30	0.30-0.40
Process power requirement (kWh/person-year) (d)	16-19	12-15	7-11	12-15
Sludge handling	No digestion, dry on sand beds or use mechanical dewatering devices.	First digest then dry on beds or use mechanical devices	First digest then dry on beds or use mechanical devices	Manual desludging once in 5-10 years
Equipment requirement (excluding screening and grit removal which are required in all cases)	Aerators, recycle pumps, sludge scrapers (for large settlers)	Aerators, recycle pumps, scrapers, thickeners, digesters, dryers, gas equipment	Trickling filter arms, recycle pumps, sludge scrapers, thickeners, digesters, gas equipment	Aerators only
Operational characteristics	Simpler than ASP	Skilled operation required	Skilled operation required	Simple
Effect of population size on cost	Relatively little	Considerable	Considerable	Slight
Special features	BOD removal highest; effluent nitrified; relatively high power requirement favoured for small and medium sized plants	Considerable equipment and skilled operation required, especially if gas collection and usage involved considered mainly for large sized plants	Considerable skilled operation required, especially if gas collection and usage involved often been preferred in India for large sized plants owing to lower power requirement compared to activated sludge	Power similar to activated sludge but construction and operation very simple; easy to enlarge or relocate if necessary

Contd...

Table 12.2–*Contd...*

Item	*UASBs*	*Waste Stabilization Ponds*	*Land Treatment/Irrigation*
Performance (typical)			
BOD removal, per cent	75-85	75-85	80-90
Nutrient removal, per cent N	–	40-50	80-90
Nutrient removal, per cent P	–	20-60	90-99
Coliform removal,	–	60-99.9	90-99
Helminth removal, per cent	Yes	Yes	–
Land requirement (m2/person) (c)	0.15-0.20 (excluding post treatment)	1.0-2.8	10-20
Process power requirement (kWh/person-year) (d)	Nil	Nil	Nil
Sludge handling	Directly dry on sand beds or use mechanical devices	Manual desludging once in 5-10 years	Nil
Equipment requirement (excluding screening and grit removal which are required in all cases)	Nil (except gas collection and flaring; gas conversion to electricity is optional)	Nil	Sprinklers or drip irrigation (optional)
Operational characteristics	Simpler than activated sludge	Simplest	
Effect of population size on unit cost	Relatively little	Slight	Slight
Special features	Minimal to negligible power requirement of the system makes it an economical alternative even if gas revenue is neglected; land requirement is also relatively small but depends on type of post-treatment adopted	Simplest treatment method and nil power requirement but this advantage may be offset by high land requirement which may be unavailable or expensive near urban areas	Adequate pretreatment required depending on type of crops and distribution system used. Some cost is offset by revenue from crops

IS 4764: 1973 (BOD 20mg/L and SS 30mg/L). In addition, provision has been made in the software to enter user-defined limits for relevant parameters corresponding to other applicable standards.

Details of Activated Sludge Process

Activated sludge process (ASP) has been utilized for treatment of both domestic and industrial wastewater for more than half a century. Design of ASP was carried out to a large extent in an empirical manner. It was only after 1960s that a rational approach was developed. The observation that whenever wastewater is aerated for a period of time, the content of organic matter is reduced and at the same time a flocculent sludge is formed, led to the origin of ASP. ASP has been developed as a continuous operation by recycling the sludge. A flow diagram of ASP is shown in Figure 12.3 and all important process variables are indicated in the figure and defined in Table 12.8.

Fresh feed is combined with recycled sludge and enters the reactor. Two fundamental assumptions for the CSTR are (i) steady state and (ii) complete mixing conditions. Reactor effluent enters the secondary clarifier and clarifier underflow splits into two streams: wastage and recycled sludge. The recycled sludge is then combined with fresh feed to form combined feed. A constant MLSS is maintained in the reactor while oxygen is supplied through mechanical devices.

Modifications of ASP

Several modifications of activated sludge process have been suggested in the literature optimizing one or the other parameter. A few modifications are:

- Conventional ASP
- Tapered aeration
- Step aeration
- High rate ASP
- Complete mix ASP
- Extended aeration
- Contact stabilization
- Sludge reaeration
- Pure oxygen ASP
- Oxidation ditches

Design of ASP

It is desirable to portray ASP by mathematical models from experimental data obtained through a series of bench-scale lab. reactors. Relationships that are pertinent fall into three groups: (1) kinetic relationships (2) material balance relationships for O_2 utilization and net yield of MLSS; and (3) relationships for optimum settling conditions of sludge.

Study of kinetics of aerobic biological treatment yields the rate at which microorganisms degrade a specific waste and provide basic information required for sizing the reactors.

Table 12.3: Degree of Treatment Achieved by Unit Operations and Processes

Treatment Units or Combinations	*Removal Efficiency (Per cent) across the Units*					
	BOD_5	*COD*	*TSS*	*TP*	*ON*	*NH_3-N*
Preliminary treatment (PT)	small	small	small	small	small	small
Primary sedimentation	30-40	30-40	50-65	10-20	10-20	0
Activated sludge (conventional)	80-85	80-85	80-90	10-25	15-50	8-15
Trickling filter (high rate)	60-80	60-80	60-85	8-12	15-50	8-15
Disinfection	small	small	small	small	small	small
Coagulation and sedimentation after preliminary or secondary treatment	40-70	40-70	50-80	70-90	50-90	0
Coagulation in biological treatment process	80-90	80-90	70-90	75-85	60-90	0
Single-stage lime addition after prelimanary treatment, or after secondary treatment	50-70	50-70	60-80	70-90	60-90	0
Single-stage lime addition in biological treatment	80-90	80-90	70-80	75-85	60-90	0
Two-stage lime addition after preliminary treatment, or after biological treatment	50-85	50-85	50-90	85-95	70-90	0
Nitrification single stage with carbonaceous BOD removal	80-95	80-90	70-90	10-15	75-85	85-95
Nitrification separate stages suspended or attached growth	50-70	50-60	small	–	40-50	90-96
Denitrification separate stages suspended or attached growth	small	small	small	small	small	small
Ammonia stripping	0	–	0	0	–	60-95
Breakpoint chlorination	–	–	0	0	60-85	80-90
Ion exchange	0	0	0	0	0	90-95
Filtration	20-50	20-50	60-80	20-50	50-70	0
Carbon adsorption	50-85	50-85	50-80	10-30	30-50	–
Reverse osmosis	90-100	90-100	–	90-100	90-100	60-90
Electrodialysis	20-60	20-60	–	–	80-95	30-50

Source: Qasim, 1985.

The parameters are:

- ✰ Substrate removal rate (k)
- ✰ Half saturation coefficient (K_s) (Michaelis-Menten constant)
- ✰ Biomass yield (Y)
- ✰ Endogenous respiration rate (k_d)
- ✰ O_2 utilization parameters a and b

Table 12.4: Design Factors for Manually Cleaned and Mechanically Cleaned Bar Racks

Design Factor	*Manually Cleaned*	*Mechanically Cleaned*
Velocity through rack (m/s)	0.3-0.6	0.6-1.0
Bar size		
Width (mm)	4-8	8-10
Depth (mm)	25-50	50-75
Clear spacing between bars (mm)	25-75	10-50
Slope from horizontal (degrees)	45-60	75-85
Allowable head loss, clogged screen (mm)	150	150
Maximum head loss, clogged screen (mm)	800	800

Source: Qasim, 1985.

Table 12.5: Typical Design Information for Horizontal-Flow Grit Chambers

Item	*Value*	
	Range	*Typical*
Detention time, s	45–90	60
Horizontal velocity, ft/s	0.8–1.3	1.0
Settling velocity for removal of:		
65-mesh material, ft/min[a]	3.2 – 4.2	3.8
100-mesh material, ft/min[a]	2.0 – 3.0	2.5
Headloss in a control section as per cent of depth in channel, per cent	30–40	36[b]
Allowance for inlet and outlet turbulence	$2D_m$ – 0.5L[c]	

a: If the specific gravity of the grit is significantly less than 2.65, lower velocities should be used.

b: For Parshall flume control.

c: D_m = maximum depth in grit chamber; L = theoretical length of grit chamber.

Source: Metcalf and Eddy, 1997

These are several methods to determine these parameters experimentally as described in standard texts [Metcalf and Eddy, 1997; Ramalho, 1983; Eckenfelder,

1989; Arceivala, 1981]. Typical values of kinetic coefficients for ASP are provided in Table 12.9. Performance evaluation of several wastewater treatment plants in India has been presented in a report by NEERI [1994]. The ranges of values of various kinetic coefficients presented in the report have been incorporated in the WWTP software to guide the user to select appropriate values in the absence of data on laboratory studies. Table 12.10 presents the allowable concentration of heavy metals for the growth of micro-organisms in the ASF.

Table 12.6: Design Factors and Typical Values for Aerated Grit Chambers

Design Factor	*Range of Value*	*Comment*
Dimensions		The width of the basin is limited to provide roll action in the tank.
Depth, m	2-5	
Length, m	7.5-20	
Width, m	2.5-7.0	
Width/depth ratio	1: 1-5: 1	
Length/width ratio	2.5: 1-5: 1	
Transverse velocity at surface	0.6-0.8 m/s	
Detention time at peak flow, min	2 to 5	If grit chamber is used for pre-aeration or to remove grit less than 65-mesh (0.21 mm), longer detention time may be provided.
Air supply	4.6 to 12.4 l/s.m of tank length (3-8 cfm/ft)	Higher air rate should be used for wider and deeper tanks with provision to vary air flow. An air flow rate of 4.6-8 l/s.m in a 3.5–5 m wide and 4.5 m deep tank give surface velocity of approximately 0.5-0.7 m/d. The velocity at the floor of the tank is 75 per cent of surface velocity. A velocity of 0.23 m/s is required to move a 0.2 mm sand particle along the tank bottom.
Inlet and outlet structures	–	Inlet and outlet structures must be such as to prevent short-circuiting and turbulence. Inlet to the chamber should introduce the influent into circulation pattern. Outlet should be at a right angle to the inlet. The inlet and outlet are sized and build in such a way that the velocity exceeds 0.3 m/s under all flow conditions to minimize the deposits.
Baffles	–	Longitudinal and transverse baffles improve grit removal efficiency. If grit chamber is much longer than the width, a transverse baffle should be considered.
Chamber geometry		Location of air diffusers, sloping tank bottom, grit hopper, and accommodation of grit collection and removal equipment should all be given consideration in chamber geometry. The diffusers are normally located 0.6 m above the sloping tank bottom. Beneath the air diffusers would be the hopper for grit collection.

Source: Qasim, 1985.

Table 12.7: Design Parameters for Settling Tanks

	Type of Settling	Overflow Rate, m^3/m^2 d		Solid Loading, kg/m^2, d		Depth (m)	Detention Time (hr)
		Average	Peak	Average	Peak		
A.	PRIMARY SETTLING						
1.	Primary Settling only	25-30	50-60			2.5-3.5	2.0-2.5
2.	Primary settling followed by secondary treatment	35-50	80-120			2.5-3.5	2.0-2.5
3.	Primary settling with sctivated sludge return	25-35	50-60			3.5-4.5	2.0-2.5
B.	SECONDARY SETTLING						
1.	Secondary settling for trickling filter	15-25	40-50	70-120	190	2.5-3.5	1.5-2.0
2.	Secondary settling for activated sludge (excluding extended aeration)	15-35	40-50	70-140	210	3.5-4.5	1.5-2.0
3.	Secondary settling for extended aeration	8-15	25-35	25-120	170	3.5-4.5	1.5-2.0

Source: CPHEEO, 1993.

Table 12.8: Definition of Symbols Used in ASP Schematic Diagram

Key

For suspended solids double subscripts are utilized, *e.g.*, $X_{v,i}$, $X_{nv,i}$. The first subscript (v or nv) designates volatile and nonvolatile suspended solids, respectively. The second subscript (i) refers to the specific stream in question:

F–fresh feed [Stream 1]

o–combined feed [Stream 2]

a–reactor effluent [Stream 3]

e–net effluent [Stream 4]

u–underflow from secondary clarifier [Stream 5]

Symbols

1. Flow rates

   Q_f - fresh feed; (million gallons per day) [stream 1]

   Q_r - recycle; [stream 7]

   r - recycle ratio; dimensionless ($r = Q_r/Q_f$)

   Q_o - combined feed; $Q_o = Q_f + Q_r = Q_f(1 + r)$ [stream 2]

   Q_e - net effluent; [stream 4]

   Q_w - wastage; [stream 6] (notice that $Q_f = Q_e + Q_w$)

   Q_u - clarifier underflow; $Q_u = Q_w + Q_r = Q_w + rQ_f$ [stream 5]

2. Concentrations (mg/litre) of soluble BOD

   S_f - soluble BOD of fresh feed

   S_o - soluble BOD of combined feed

   S_e - soluble BOD of effluent

3. Concentrations (mg/liter) of volatile suspended solids (VSS)

   $X_{v,f}$ - VSS in fresh feed

   $X_{v,o}$ - VSS in combined feed

   $X_{v,a}$ - VSS in reactor. This also is equal to concentration of VSS in reactor effluent (complete mix reactor at steady state)

   $X_{v,u}$ - VSS in secondary clarifier underflow

   $X_{v,e}$ - VSS in net effluent

4. Concentrations (mg/litre) of nonvolatile suspended solids (NVSS)

   $X_{nv,f}$ - NVSS in fresh feed

   $X_{nv,o}$ - NVSS in combined feed

   $X_{nv,a}$ - NVSS in reactor. This also is equal to concentration of VSS in reactor effluent (complete mix reactor at steady state)

   $X_{nv,u}$ - NVSS in secondary clarifier underflow

   $X_{nv,e}$ - NVSS in net effluent

5. Wastage

   $(VSS)_w$ - kg/day of VSS in wastage

   $(NVSS)_w$ - kg/day of NVSS in wastage

   $(TSS)_w$ - kg/day of TSS in wastage

6. Reactor volume - V

7. Sludge production - ΔX_v

Table 12.9: Typical Values of Kinetic Coefficients for Activated Sludge Process

Coefficient	*Basis*	*Values*	
		Range	*Typical*
k	d^{-1}	2-8	4
k_d	d^{-1}	0.03-0.07	0.05
K_s	mg/l, BOD_5	40-120	80
	mg/l, COD	20-80	40
Y	VSS/BOD_5	0.3-0.7	0.5
	VSS/COD	0.2-0.5	0.4

Source: Qasim, 1985.

Table 12.10: Allowable Concentrations of Heavy Metals that can Usually be Tolerated by Activated Sludge Microorganisms

Constituent	*Concentration at Which Damage to Activated Sludge Might Occur*	
	Continuous loading	*Slug loading*
Cadmium	1 mg/l	10 mg/l
Chromium	2 mg/l	2 mg/l
Copper	1 mg/l	1.5 mg/l
Iron	35 mg/l	100 mg/l
Lead	1 mg/l	–
Manganese	1 mg/l	–
Mercury	0.002 mg/l	0.5 mg/l
Nickel	1 mg/l	5 mg/l
Silver	0.03 mg/l	0.25 mg/l
Zinc	1 to 5 mg/l	25 mg/l
Cobalt	> 1 mg/l	–
Cyanide	1 mg/l	1 to 10 mg/l
Arsenic	0.7 mg/l	

Knowledge of production of MLVSS and of oxygen utilization is needed for design of aerobic biological reactors. In order to develop mathematical models, which yield these two values, the mass balance approach if followed as described in standard literature.

Knowledge of oxygen requirements to effect a specified BOD removal is necessary for specification of aeration equipment. Oxygen is required for two purposes: (1) to oxidize substrate in order to provide energy requirements for cells, and (2) for the endogenous respiration process.

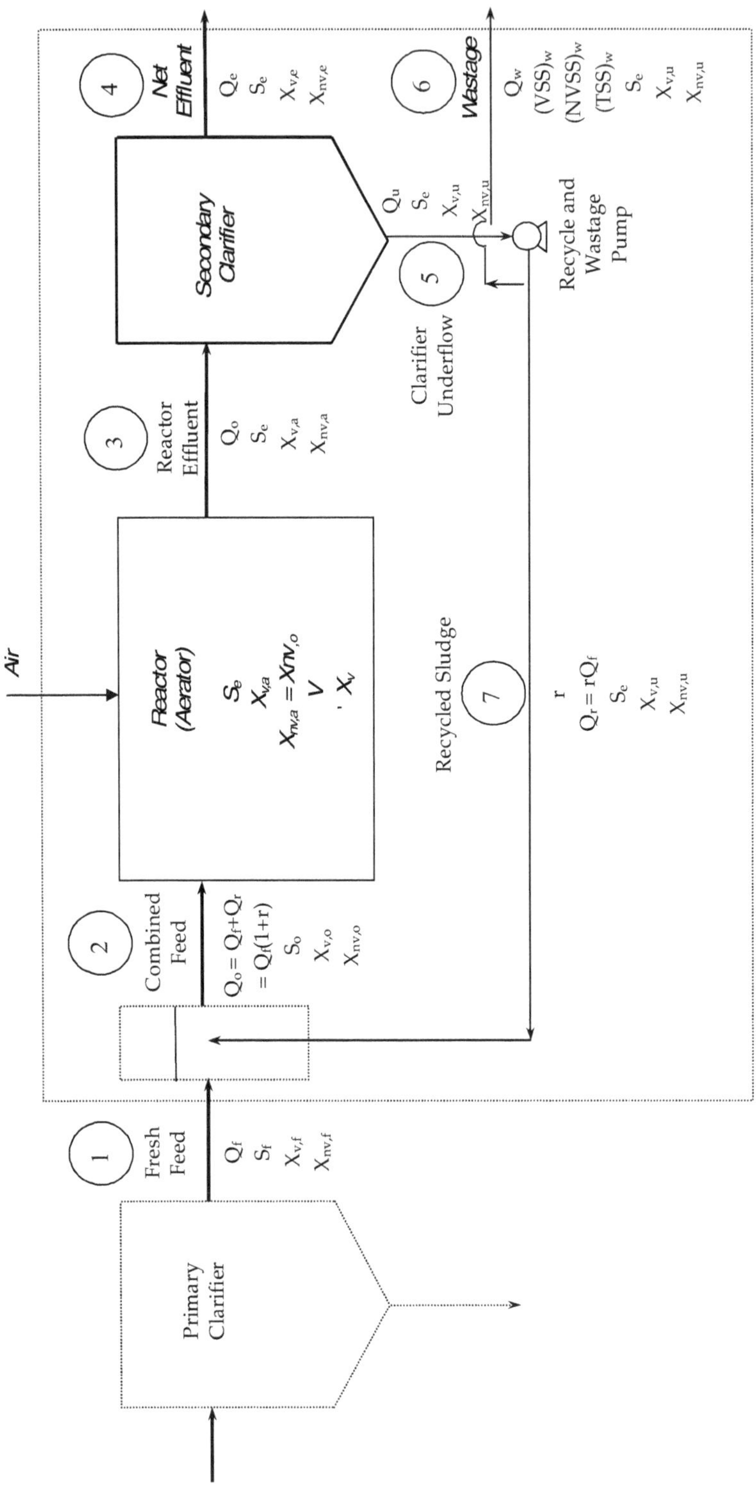

Figure 12.3: Flow Diagram of Activated Sludge Process.

For adequate operation of the ASP, MLVSS in the reactor effluent should be readily separated in the secondary clarifier. Settling characteristics of sludge are evaluated from sedimentation tests performed in laboratory. Two commonly used parameters are: zone settling velocity, and sludge volume index.

Basic Relationships

The true mean solids residence time cell residence time q_m is the ratio of mass of VSS in the system to the total input of VSS into the system by biomass synthesis and hydraulic transport along with fresh feed and is given by:

$$\theta_m = \frac{X_{v.a}.V}{Y(S_F - S_e)Q_F + Q_F.X_{v,f}}$$

A more useful concept is that of sludge age q_c, which is defined as the ratio of the mass of VSS in the system to the net yield of biomass ΔX_v:

$$\theta_c = \frac{X_{v.a}.V}{\Delta X_v}$$

This fundamental relationships, for the case when first order kinetics for substrate removal is assumed, becomes:

$$\frac{1}{\theta_c} = YkS_e - k_d \text{ and } Y_n = \frac{Y}{1 + k_d\theta_c}$$

Mass concentration of microorganism in the reactor can be obtained by

$$X = \frac{\theta_c.Y.(S_o - S_e)}{\theta(1 + k_d\theta_c)}$$

and effluent substrate concentration is found to be equal to

$$S_e = \frac{Ks(1 + \theta_c k_d)}{\theta_c(Yk - k_d) - 1}$$

A term closely related to the specific utilization rate given by kS_e in above equation and commonly used in practice as a design and control parameter is known as food–microorganism ratio (F/M), which is defined as:

$F/M = S_o/\theta X$

and $kS_e = (F/M)\, E/100$

where, $E = \frac{S_o - S_e}{S_o o} x\, 100$ is efficiency of process.

Aeration in Wastewater Treatment

Requisite oxygen must be transferred from air to a wastewater subjected to biological aerobic treatment. Knowledge of the rate of oxygen transfer is essential for design of aeration system.

The simplest explanation for mechanism of gas transfer is given by the two-film theory [Ramalho, 1983]. The rate of transfer is given by Fick's diffusion equation applied across the liquid film.

$$\frac{\partial M}{\partial t} = -DA\frac{\partial C}{\partial x}$$

which can be reduced to

$$\frac{dc}{dt} = K_L.a.(C_{sw} - C) = OC$$

For the case of aeration of activated sludge liquor, it is possible to utilize a steady state technique for determination of overall oxygen transfer coefficient K_La.

For actual field condition K_La may need to be corrected for temperature, dissolved solids and pressure. Thus,

$$\frac{(OC)_{actual}}{(OC)_{stan\,dad}} = \alpha_{20\circ C}x1.024^{(T-20)}x\frac{\beta C_{s,s} - C_L}{9.2}$$

where,

$C_{s,s} = (C_{s,s})_{760}\,(P-P^v)/\,(760\text{-}P^v)$

$\beta = C_{sw}/C_s$

C_{sw}, C_s = DO Saturation concentration in wastewater and tap water

$\alpha = K_La$ of wastewater/K_La of tap water

Design Procedure for Aeration System using Surface Aerators

Fundamental information required is as follows:

- ☆ Volume of aeration tank calculated from biological reactor requirements
- ☆ Oxygen requirements (kg-O_2/hr), also calculated from biological reactor requirements
- ☆ Operating temperature
- ☆ Operating steady state DO (mg/l), usually 0.5 to 1.5 mg/L
- ☆ Oxygen transfer coefficient α and solubility coefficient β
- ☆ Characteristics of aerator, including correlation between standard transfer efficiency versus power required per 1000 m^3 and correlation between unit horsepower and diameter of influence for solids in suspension.

Design Steps

Step 1

$$K = \alpha_{20\circ C}x1.024^{(T-20)}x\frac{\beta C_{s,s} - C_L}{9.2}$$

Range of values of $\alpha_{20°C}$ and β are 0.3 to 1.2 and 0.7 to 0.98, respectively depending upon the type of wastewater, aeration device and tank geometry [Metcalf and Eddy, 1997].

$$\alpha = \frac{K_a \text{a of wastewater}}{K_a \text{a of tapwater}}$$; the value of α is usually < 1

$$OC_{standard} = (K_L a)_{20°C} \cdot C_s$$

$$OC_{actual} = (K_L a)_{W,T} (C_{sw} - C_L)$$

where,

$K_L a$ = overall oxygen transfer coefficient

OC = oxygenation capacity

Calculate the transfer efficiency, TE_{actual} for summer and winter temperature. The lower K corresponds to controlling condition. In general $K_{summer} < K_{winter}$

Step 2

Using the aerator characteristic curves calculate power requirement for assumed power level as:

$$\text{Power requirement} = \frac{O_2 \text{ required (Kg/}O_2\text{/h)}}{(TE)_{actual} \text{ (Kg/}O_2\text{/hp.h)}} = \text{hp}$$

where,

O_2 required is already calculated from reactor requirements.

Compare the two power levels per unit volume for close agreement.

If need be, repeat the above till an agreement within 5 per cent is achieved.

Select hp per unit and calculate number of units.

Step 3

Spacing between aerators is decided based on the manufactures' data on particular aerators employed. The data on zones of influence–mixing and aeration, is essential so as to reduce the interference of aerators with each other in case multiple aerators are used in a single tank.

Surface aerators – floating and fixed type, are most widely used and appropriate data on such aerators as provided in literature [Metcalf and Eddy, 1997] has been stored in database for use in the WWTP programme.

The WWTP Software

General Considerations

Choice of Development Platform – Hardware and Software

Starting from the large main frame computer systems that were prone to frequent breakdowns and requiring great amount of operating and programming skill, the computer has today invaded not only big business houses and consultants but also the common man's household. This computer known as PC (Personal Computer) is ubiquitous and the impact is so great that any scientific design/development work today is incomprehensible without a PC. With the advent of powerful Pentium family of processors, the PC power has been increasing at a very fast pace and today's PC is much faster with very large core memory and external storage space. The PC selected for development of WWTP software belongs to this category with the following base specifications:

- ☆ Pentium Pro processor @ 266 MHz clock speed
- ☆ 64 MB RAM
- ☆ 6 GB Hard disk space
- ☆ 14" VGA color monitor

In the beginning the PC was operating under an operating system (OS) called MS-DOS and most often used programming language for scientific work was FORTRAN and PASCAL. Although FORTRAN was quite successful (more than PASCAL), both – the OS and FORTRAN had limitations, which would not allow development of an easy to use user-interface. The programs developed under these environments required considerable amount of knowledge and experience on the part of the user for using these design tools. The user interface is perhaps the most important part of an application; it's certainly the most visible. To users, interface is the application. No matter how much time and effort is put in writing and optimizing the code, the usability of the application depends on the interface [Visual BASIC 5.0 Books Online Help].

With the introduction of better OS, *viz.* Microsoft WINDOWS, newer computing techniques and software platforms, that are more efficient and offer faster development by way of object oriented programming have been developed. These techniques usually offer an integrated development environment that would relieve the developer of most of the routine chores, which earlier required the user to have mandatory knowledge of computer systems. There are several reasons why Windows has become so popular in a relatively short time. A few are [Walton, 1995]:

- ☆ *Device Independent Programs*: Windows lets one write device independent programs. This means that while writing the application, one does not have to concern himself with what type of printer, mouse, monitor, keyboard, sound card, CD-ROM drive or devices the target users own.
- ☆ *Pre-installed Code*: A lot of code is already installed in the PC. Once Windows is installed the PC contains much Windows related software. This means that before one even starts writing the first line of code, more than half the program is already there in the PC.
- ☆ *Standard User Interface*: The user interface mechanism is the same for all Windows applications. Without reading the application's documentation, users know how to execute the application.

Amongst the several programming languages, the choice of a particular language depends solely on the developer and particular application being developed. Microsoft

Visual Basic (VB) is an Object Oriented Programming language that offers most of the features required for development of good application software (Objects are central to VB programming; *e.g.* Forms and Controls as also Database are objects and their use simplifies coding and increases code reuse). These features include an integrated development environment that reduces overall development time; powerful tools for creating graphical user interfaces (GUI) help in developing better and uniform user interfaces; the application can be developed as a standalone executable program (a VB project can be compiled to native code for faster execution); Crystal Reports can be used to create customized reports, lists, form letters, etc. using data from an existing database.

Owing to the its distinct advantages as alluded to earlier, Wastewater Treatment System Design (WWTP) Software has been developed using Microsoft Visual BASIC. The hardware platform used, *i.e.* a Pentium Pro PC running Microsoft Windows-95 OS, is a normal configuration available with the environmental consultants and design offices, that are the target users of this software.

Objective of Software

The treatment objectives are concerned with (1) the removal of suspended and floatable material, (2) the treatment of biodegradable organics, and (3) the elimination of pathogenic organisms. Wastewater treatment plants utilize a number of treatment processes to achieve the desired degree of treatment. In addition, the environmental design engineer must evaluate numerous other important factors in selection of the treatment processes. These include constituents treated, effluent limitations, proximity to build-up areas, hydraulic requirements, sludge disposal, energy requirements, and plant economics. The collective arrangement of various treatment processes is called a flow scheme, flow diagram, a flow sheet or a process train [Qasim, 1985].

The contaminants in wastewater (unwanted substances termed as pollutants) are removed by physical, chemical, biological or other means and the treated wastewater can be used for consumptive purposes or can be discharged into the water bodies without causing quality degradation beyond those permitted.

Choice of proper treatment processes and development of the flow scheme is not a simple task. It requires understanding of the unit operations and processes, operational capabilities and environmental effects of various treatment components that are arranged to develop the flow scheme for a desired application. Laboratory and pilot plant studies are often necessary to develop design parameters for physical, chemical, and biological treatment processes used for treatment of wastewater.

Thus, decision for a suitable treatment alternative for a particular wastewater under certain conditions, warrants evaluation of several possible treatment alternatives. The evaluation includes:

- ☆ Selection of appropriate unit processes and development of process flow sheet incorporating various conventional low-cost and state-of-art treatment technologies based on mode of disposal so as to conform to the prescribed effluent standards.

- ☆ Develop design parameters for physical, chemical, and biological treatment processes used for treatment of wastewater.
- ☆ Design of individual treatment units.
- ☆ Estimation of capital as well as operation and maintenance costs of each process, and
- ☆ Ranking of these alternatives with a view to selecting the most appropriate treatment process.

Thus, the process of selection of a suitable treatment alternative is not only tedious but also time consuming. With a view to simplifying the task of preparing designs of effluent treatment plant, development of computer software was undertaken in this dissertation.

It was, therefore, envisaged that the software developed under this research would not only help in designing the treatment units and estimation of various costs but would also assist in evaluation of several alternative treatment flow-sheets in minimal time. Since an existing treatment plant may need augmentation/upgradation over a period of time due to increased organic and/or hydraulic loading, a module on augmentation of treatment plant was also envisaged in the software. Further, since proper operation of treatment plant is an important aspect in the overall objective of wastewater treatment, troubleshooting guidelines have been included in the software to aid in day-to-day operation. Moreover, the use of such a tool for designing treatment plants would ensure selection of the most suitable alternative in a particular situation.

Framework for WWTP Software

Commensurate with the objective of the research, the WWTP Design software has been conceptualized to comprise the following three modules:

- ☆ Design Module
- ☆ Augmentation Module
- ☆ Troubleshooting Module

The design procedures for various unit operations/processes as described in the earlier chapters have been appropriately coded in the software with a view to obtain the preliminary design and flow sheet of the treatment plant. The entire software comprises a number of VB Forms (a form in Visual Basic is a specifically designed screen window for user interaction –both input as well as output); data files- either ASCII data file or MDB database; and report formats.

Visual Basic form files are created and saved in ASCII format. The structure of a form consists of:

- ☆ The version number of the file format
- ☆ A block of text containing the form description
- ☆ A set of form attributes
- ☆ The Basic code for the form.

The form description contains the property settings of the form. Blocks of text that define the properties of controls on the form are nested within the form. Controls contained within other controls have their properties nested within the text of the container. Given below is the structure of the form description [Microsoft, 1997].

```
Version 5.00
Begin VB.Form formname
        Form properties
        Begin VB.controltype controlname
                Control properties
                Begin VB.controltype controlname
                        Control properties
                End
                Begin VB.controltype controlname
                        Control properties
                End
        End
End
Attributes
Basic code for this form begins here.
```

Several fixed data are required to arrive at a design, such as information on mechanical aerator parameters *i.e.* their zones of influence, aeration capacity, etc.; properties of water as a function of temperature *i.e.* viscosity, Saturation DO as a function of salinity, temperature, altitude/atmospheric pressure. All such data are compiled from literature and stored in ASCII data file/MDB database for use in the software as and when required. The dynamic data specific for a particular unit operation, however, has been restricted to the forms only instead of hard coding.

One set of linked forms is dedicated to each task to be performed, *e.g.* for obtaining Input Data from user, presenting the design results on screen, carrying out the design in an interactive mode, etc.

Input data and Design details are stored in an Access Database (WWOP.mdb) consisting of Tables corresponding to the respective units. Summary reports for each of the design units have been developed using 'Crystal Reports' instead of creating plain ASCII files. This approach offers the advantage of pre-designed report format and efficient linking with the database. A menu of reports is displayed to select the report to be printed/saved.

Design Module

The wastewater treatment plant design, as has already been explained in earlier chapters, comprises a number of treatment unit processes and/or unit operations connected in series to achieve the overall objective of reducing the strength of the waste so that it is acceptable for disposal. The choice of specific units depends on several factors, including influent flow and characteristics (diurnal flow variations, BOD loads, etc.), type of influent (industrial/domestic/combined/mixed, etc.), availability of skilled labor/manpower, process reliability, land area available for construction, mode of final disposal, governing rules and local laws, etc.

In view of the foregoing, it is not practicable and advisable to pre-decide the treatment flow-sheet and go ahead with the design. Rather, a number of flow-sheets must be developed, evaluated and ranked for delineation of the most cost effective and appropriate flow-sheet for specific cases. In general, the treatment scheme can be divided into the following sub-categories: *Preliminary Treatment, Primary Treatment, Secondary Treatment, and Tertiary Treatment*. Therefore, the software has been designed keeping in view the above sub-categories. The choice of individual units is selectable by the user rather than fixing the choice in the code itself. The entire wastewater treatment plant design procedure is thus modularized.

The design module comprises a total of 26 Visual BASIC Forms designed and developed for various operations in the software. These forms can be linked together in the desired sequence. Except the following four forms, *viz.* Splash Form, Main Menu Form, Input Data Form and Unit Selection Form, all Forms must be displayed and utilized in a particular sequence. The sequence of displaying the forms corresponding to individual units can not be pre-decided at design-time (due of dynamic choice of units at run-time). Therefore, an elaborate scheme for chain-linking the desired forms has been designed and an algorithm has been developed and implemented in Visual Basic.

Apart from common input data required for the design (such as, wastewater flow, peaking factor, influent BOD, COD, SS, etc.), specific input data has been restricted to individual forms requiring it for design of that particular unit. Wherever required (when larger input, output and graphical depiction is necessary) two or more forms constitute the design of a single unit.

Besides being modular, this framework offers enough flexibility for future expansion and easy software maintainability.

Augmentation Module

An existing system needs augmentation over a period of time for two reasons: (1) Increase in the wastewater flow, and (2) Changes in wastewater characteristics. In the modular design of treatment plant, duplication of modules would take care of augmentation needs due to the first reason. Second reason, however, requires a careful analysis if the organic mass loading increases significantly.

The application of today's technology in upgrading the plant includes (a) techniques that can be used to maintain the original treatment plant efficiency under increasing organic and/or hydraulic loading, (b) the addition of processes that can be used to increase overall plant removal efficiencies, and (c) process addition or modifications for specific contaminant removal. Physically these modifications can be applied ahead of the plant; as modifications of the treatment process itself; or as effluent polishing techniques [Smith, 1972].

The design for augmentation is thus specific to individual cases. However, with a view to create a framework for further development, it was though appropriate to include the design of a specific treatment plant augmentation. In the case study, the wastewater flow has not increased significantly and is within the capacity of treatment facility but the BOD has increased significantly.

In view of the foregoing, a rudimentary augmentation scheme for an existing sewage treatment plant (STP) based on Extended Aeration system has only been planned. In this particular case of STP, due to increase in influent BOD, it has become necessary to augment the system. It is proposed to reduce the BOD load on the existing STP by providing anaerobic pre-treatment before the extended aeration system. The same scheme has been coded in the software.

Troubleshooting Module

An important aspect of the overall objective of wastewater treatment facility is proper operation and maintenance of wastewater treatment plants. A guide to troubleshooting the problems encountered during the operation of a treatment plant provides an effective tool for proper operation of the facility.

The field observations in the condition of various unit operations/processes can provide important clues to the causes and remedies for the problems as also the laboratory tests like sludge settleability test. The common operating problems encountered in the activated sludge process have been collected from various references and compiled in the form of a database. Probable causes, checks to determine the cause, and remedial/corrective measures for each observation have been compiled in the form of VB Forms.

In light of the detailed description (in earlier chapter) of the problems associated with the operation of activated sludge process, their causes, and remedies, there are seven problems that frequently occur in operating the ASP plant, which are considered in the software under troubleshooting guide.

The problems, categorized between A) *aeration tank* and B) *secondary clarifier* – two in aeration tank and five in the secondary clarifier, are as follows:

Aeration Tank:

- ☆ Aeration System Problems
- ☆ Foaming Problems

Secondary Clarifier

- ☆ Solids Washout
- ☆ Bulking Sludge
- ☆ Clumping/Rising Sludge
- ☆ Clouding Secondary Effluent
- ☆ Ashing

☆ Pinpoint Floc
☆ Stragglers/Billowing Solids

Detailed Description of Software

The following describes the detailed description of various modules of the WWTP software, the available options for design, augmentation, and troubleshooting. System flow chart of WWTP software is depicted in Figure 12.4.

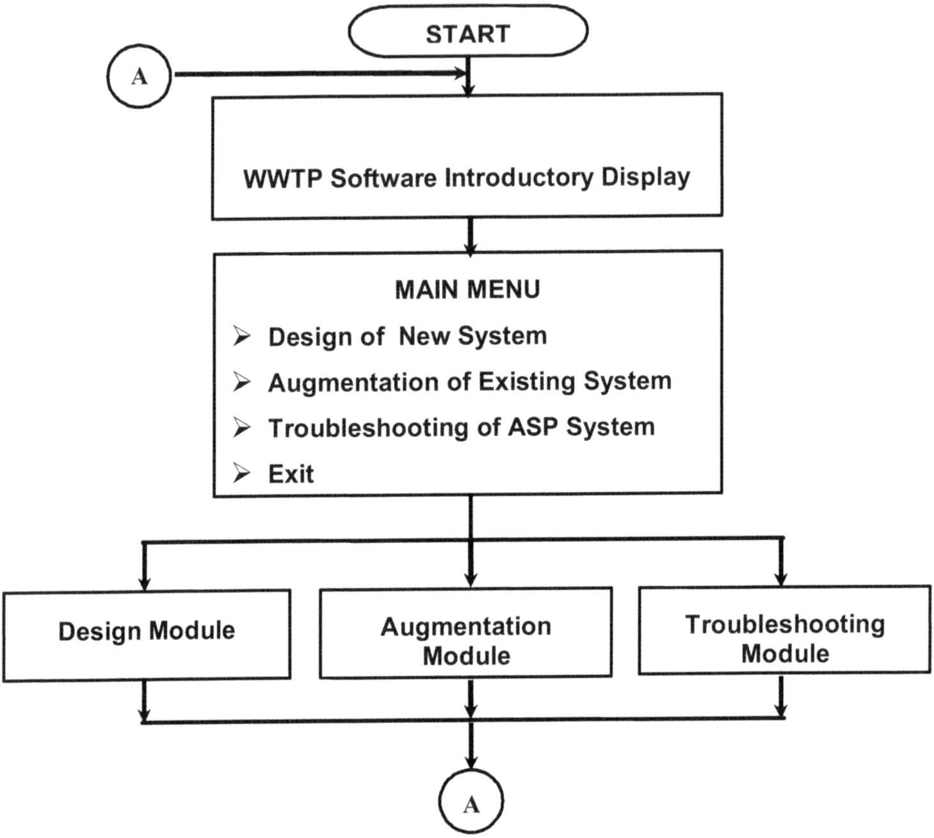

Figure 12.4: WWTP System Flow Chart.

Instead of describing each form in detail, the approach followed is to describe the details of the software during execution. Beginning with the self-introduction, the Main Menu is displayed for selection of Module to execute and thereafter the execution control transfers from form to form based on the selections of units. The image printouts of the VB Forms comprising the Design and Augmentation Modules are given in ***Annexure-I***. Sample design calculations for a typical wastewater treatment plant, which form the basis of designing various treatment units are given in ***Annexure-II***.

Self-Introduction Window

The introductory form depicts a typical treatment plant pictorially and includes general information about the software such as name of the institute where the software has been developed, author's name and affiliations, software version, etc. Typical 'Windows Help' facility is provided here, which tell about the program, the Institute and the author. The Form provides two Click Buttons named 'Run' and 'Exit'.

Main Menu

Over the backdrop of a picture, this form presents three options, *viz. Design, Augmentation and Troubleshooting* are listed in the menu apart from EXIT. These are Click Buttons for user selection and directing the further action.

Design Module

The system flowchart of *WWTP Design Module* is depicted in Figure 12.5. A new treatment plant can be designed using this option. On invoking this option the programme automatically follows the sequence of other forms required to accomplish the task of designing the plant. The design option comprises the following steps:

Input Data Entry

Design input data such as flow, site conditions, wastewater characteristics, effluent standards etc. are provided by the user in this Form. The data entered by the user is stored in VB global variables for later reference in the other forms.

The values of following parameters are required -

Flow: Average flow, peak factor

Site Conditions: Height above MSL, ambient temperature

Wastewater Characteristics: Temperature, Solids, pH, BOD, COD, Nutrients, and alkalinity.

Treated Effluent Characteristics: Discharge limits for BOD, COD and SS under various conditions. Four options are programmed in keeping with the standards stipulated by regulatory body, *viz.* Discharge into inland surface waters, discharge into sea waters, reuse as irrigation water, and user defined values for other Standards.

The input is checked for obvious erroneous entries only since an elaborate check on the values of each parameter is a very tedious task and not essential also. On verification, the program executes further for interactive selection of various units to be included in the treatment scheme.

Unit Selection

Different options for Preliminary, Primary, Secondary and Tertiary treatment units are included. Figure 4.3 describes the unit/process options available for design. Desired units are selected by checking the boxes provided with each option (Refer Unit Selection Form image in *Annexure-I*). Selecting an option activates the options-list for types of individual treatment unit for that option where applicable. Clicking

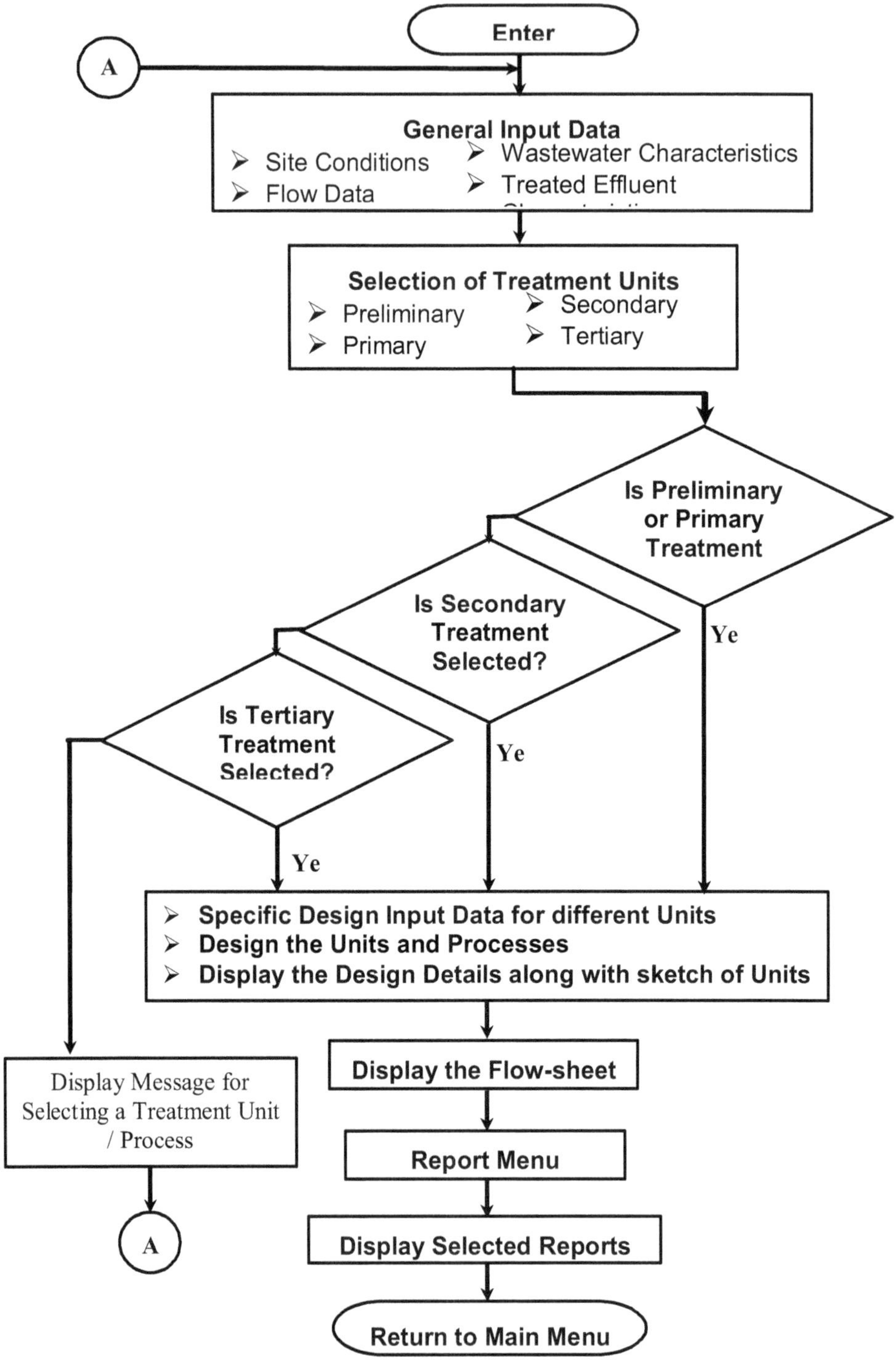

Figure 12.5: System Flow Chart of Design Module.

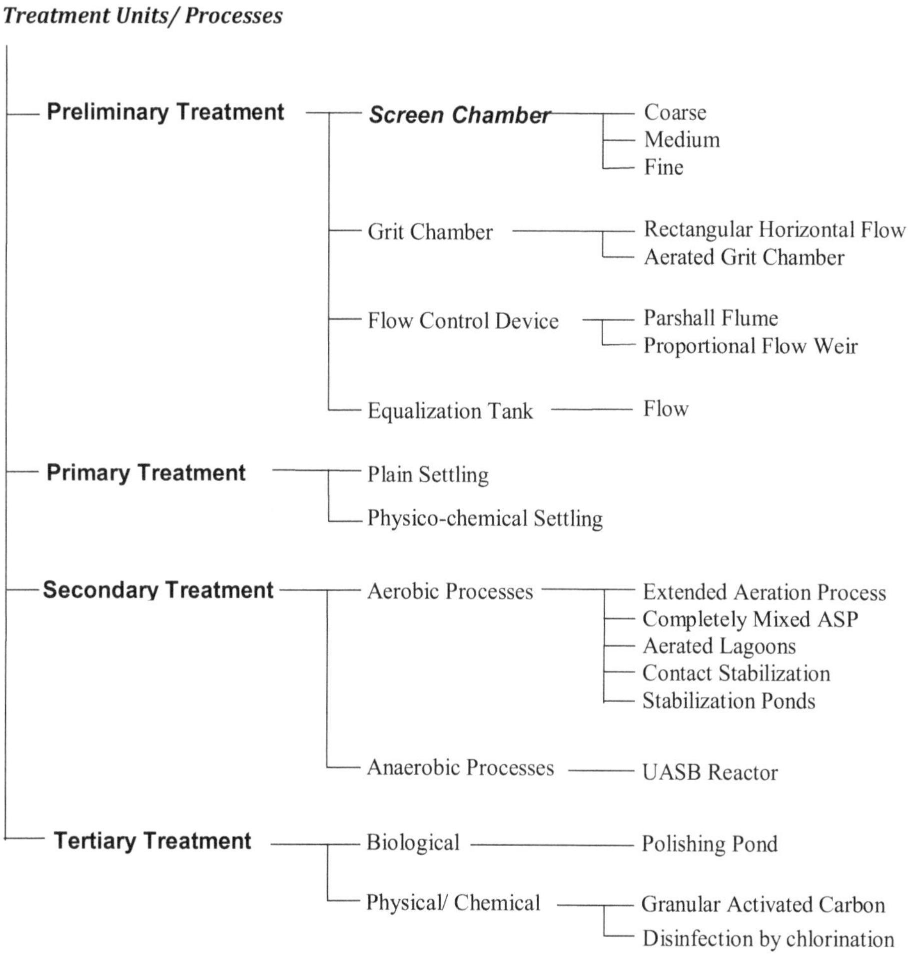

Figure 12.6: WWTP Design Module Options Tree.

in the list allows selection of a particular option, *e.g.* when 'Screen Chamber' is selected, the options-list for selection of Coarse/Medium/Fine Screen pops up. Similarly, options-list pops up for other units such as when 'Biological Treatment' is selected, the options-list of available biological treatment methods is activated and the desired option can be selected from the list. Figure 12.6 gives the detailed list of units/processes available for design.

Appropriate checks are built-in so that a proper selection of units is made. Appropriate error messages are generated in case 'Next' Button is clicked without making a selection. Specific design data required for each unit wherever needed is entered next.

Unit Design Data and Design

On completion of selection of units, the order of displaying rest of the Forms based on the units selected is determined and appropriate windows are displayed sequentially. If any of the preliminary or primary units is selected, the window for design input data for these units is displayed. Then the individual windows are displayed along with the design of the unit and schematic diagram of the respective unit.

The Equalization Tank design includes the electronic spreadsheet type data grid for entry of hourly flow measurement data over a day in order to calculate the volume of equalization tank required. The corresponding Mass diagram is also drawn along with the design of the unit. Selection of type of tank and rating of aerators for mixing is interactive, selection of different ratings modifies the design and displays the tank dimensions and available power per unit volume of tank.

In case of biological and tertiary treatment units, Tabbed Form Interface has been chosen and the design-input data and the design details of the unit are included in the same Form on different Tabs. Selection of rating of aerator wherever needed has an interactive interface which helps in decision making during the process of design itself. The configuration of aeration tank and aerators as also optimization of power provided through a combination of aerator rating, number of aerators in each tank and number of aeration tanks to be provided can be decided judiciously. Wherever applicable, default values for design parameters have been given in the form, which can be changed as needed. The design details, along with a schematic diagram of the process units is given on individual Tabs of the Form.

Figure 12.7 to Figure 12.9 depict the system flow charts for design of Extended Aeration Process, UASB Process, and Polishing Pond, respectively.

Summary of Design

A Menu of Design Summary Sheets is given in a separate window. Facility for saving the desired design summaries is provided in the Report Forms. The treatment flow sheet formed from the selected treatment units can be visualized by pressing a button. A printout of the individual design summaries can be also taken from within the Report Form.

The design module execution ends at this point. The entire treatment plant design is, however, retained in memory till Exit from the programme and can be reviewed by going back through the Design Forms. New designs can be obtained just by changing the necessary input/design-parameter data as required, *e.g.* the design can be modified to see the effect of changes in MLSS concentration or F/M ratio in the aeration tank. This helps in conducting rudimentary sensitivity analyses to observe the design changes by varying the design parameters for various processes.

Augmentation Option

An existing system needs augmentations for two reasons: (1) Increase in the wastewater flow over a period of time, and (2) Changes in wastewater characteristics.

Presently, a rudimentary augmentation scheme for an existing sewage treatment plant (STP) based on Extended Aeration system has only been planned. In this particular case of STP, due to increase in influent BOD, it has become necessary to augment the system. It is proposed to reduce the BOD load on the existing STP by providing anaerobic pre-treatment before the extended aeration system.

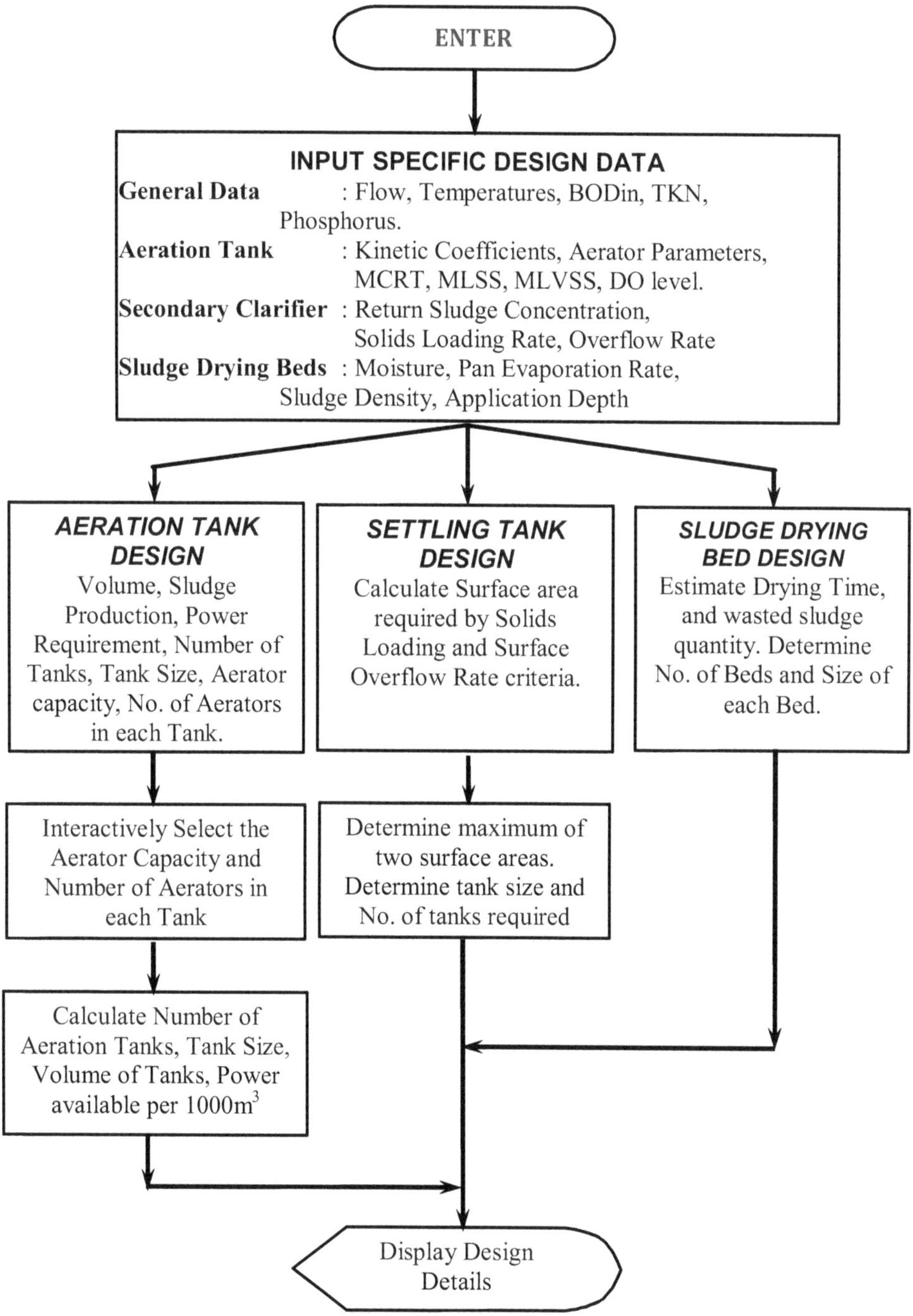

Figure 12.7: System Flow Chart of Extended Aeration System Design.

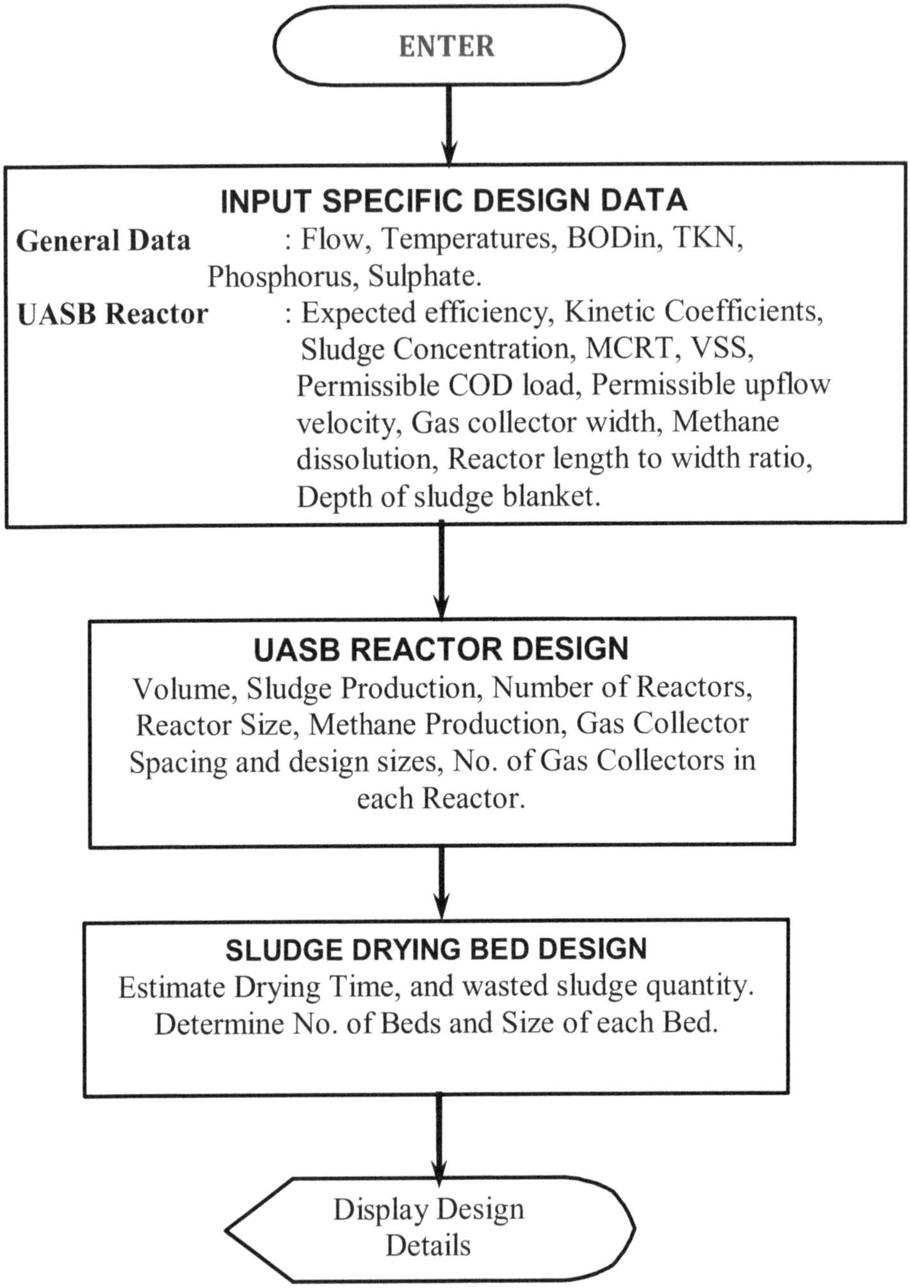

Figure 12.8: System Flow Chart of Upflow Anaerobic Sludge Blanket Reactor Design.

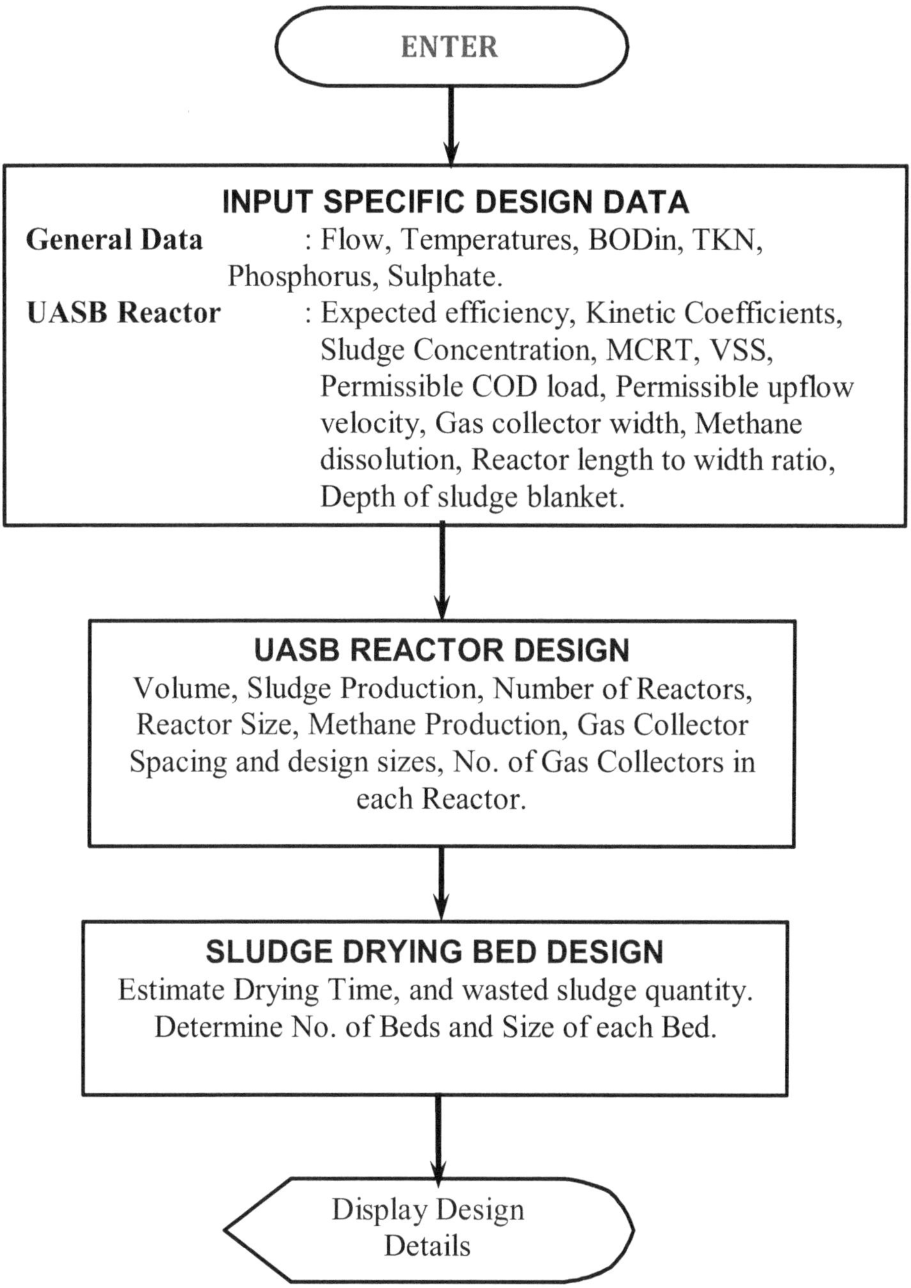

Figure 12.9: System Flow Chart for Design of Polishing Pond.

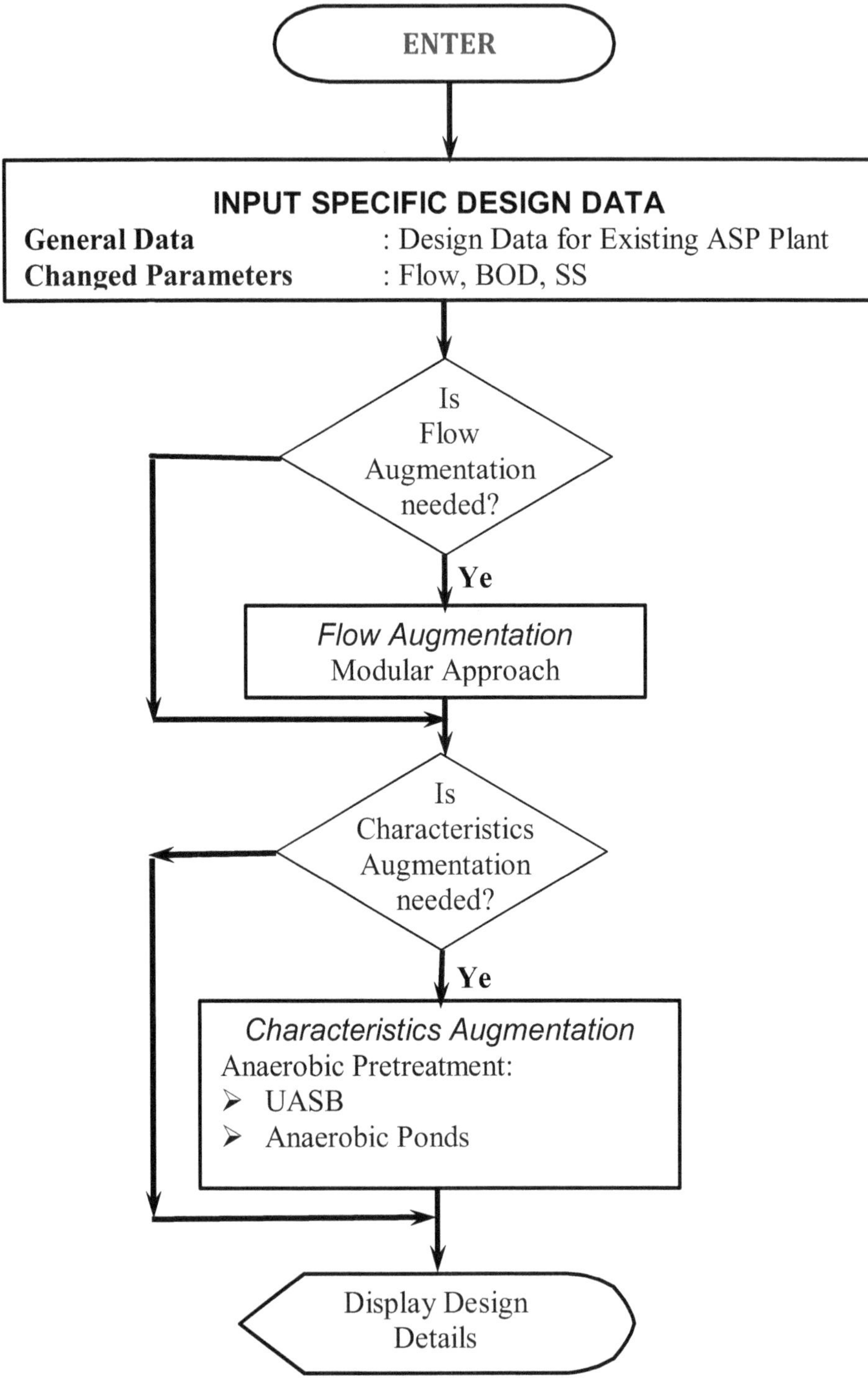

Figure 12.10: System Flow Chart for Design of Augmentation Module.

A framework for augmentation/upgradation of treatment plant by providing pre-treatment employing UASB or Anaerobic pond prior to the existing treatment plant has been planned. The design procedure for UASB and Anaerobic pond already included in the software form the basis for designing the pre-treatment units. The differential of the design organic loading and increased organic loading forms the basis for designing and sizing the pretreatment units.

System flow chart for the Augmentation Module is depicted in Figure 12.10.

Troubleshooting Option

On invoking this option, a Form containing a listing of the seven categorized problems is presented. Each problem has a check box attached to it checking which extracts from the 'Troubleshoot Database' the available Observations under that particular problem type and displays the list of observations in another window on the Form. Each of the Observations in turn has a check box attached to it checking which will display the appropriate set of Forms depicting the Probable Cause, Necessary Checks, and Suggested Remedies for solving that problem. The problems, their causes, checks and remedies have been segregated for easy handling and in all 22 Forms have been designed and developed to portray the cause and remedies for these observable problems indexed by the observations. Figure 12.11 presents the system flow chart of Troubleshooting Module.

The list of the observations and brief description is given below:

(i) Aeration Tank

Aeration System Problems

- ☆ Boiling action, violent turbulence throughout aeration tank surface. Large air bubbles, ½″ or greater, apparent.
- ☆ Uneven surface aeration pattern. Dead spots or inadequate mixing in some areas of tank.
- ☆ Excessive air rates being used with no apparent change in organic or hydraulic loading. Difficult to maintain adequate D.O. level.

Foaming Problems

- ☆ White, thick, billowing or sudsy foam on aeration tank surface.
- ☆ Shiny, dark-tan foam on aeration tank surface.
- ☆ Thick, scummy dark-tan foam on aeration tank surface.
- ☆ Dark-brown, almost blackish sudsy foam on aeration tank surface. Mixed liquor color is very dark-brown to almost black. Detection of septic or sour odour from aeration tank.

(ii) Secondary Clarifier

Solids Washout/Billowing Solids

- ☆ Localized clouds of homogenous sludge solids rising in certain areas of the clarifier. Mixed liquor in settleability test settles fairly well with a clear supernatant.

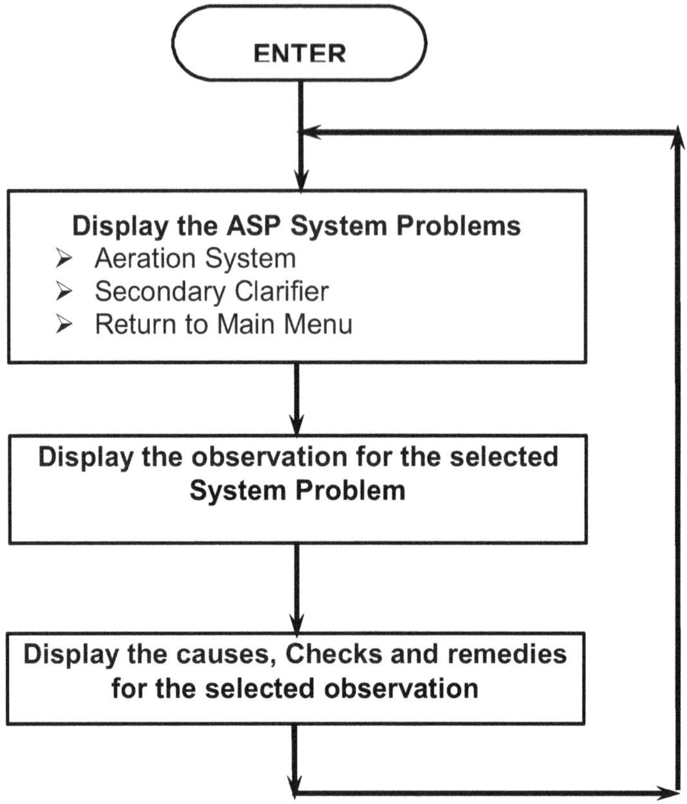

Figure 12.11: System Flow Chart of Troubleshooting Module.

☆ Localized clouds of fluffy homogenous sludge rising in certain areas of the clarifier. Mixed liquor in settleability test settles slowly, leaving stragglers in supernatant.

Bulking Sludge

☆ Clouds of billowing homogenous sludge rising and extending throughout the clarifier tank. Mixed liquor settles slowly and compacts poorly in settleability test, but supernatant is fairly clear.

Sludge Clumping

☆ Sludge clumps (from size of a golf ball or as large as a basketball) rising to and dispersing on clarifier surface. Bubbles noticed on clarifier surface. Mixed liquor settleability test settles fairly well, however a portion of and/or all of the settled sludge rises to the surface within four hours after test is started.

Cloudy Secondary Effluent

☆ Secondary effluent from clarifier is cloudy and contains suspended matter. Mixed liquor in settleability test settles poorly, leaving a cloudy supernatant.

Ashing and Pinpoint/Straggler Floc

- ☆ Fine dispersed floc (about the size of a pinhead) extending throughout the clarifier with little islands of sludge accumulated on the surface and discharging over the weirs. Mixed liquor, in settleability test, settles fairly well. Sludge is dense at bottom with fine particles of floc suspended in fairly clear supernatant.
- ☆ Small particles of ash-like material floating on clarifier surface.
- ☆ Particles of straggler floc about ¼" or larger, extending throughout the clarifier and discharging over the weirs. Mixed liquor, in settleability test, settles fairly well. Sludge does not compact well at the bottom with chunks of floc suspended in fairly clear supernatant.

Many of the above mentioned problems have multiple causes and/or remedies spread over multiple pages for which back and forth reference to other Forms is necessary as required. Therefore, links to appropriate Forms are programmed in the module using VB Button Controls pressing which displays additional information. Command Buttons captioned as 'Next>>', '<<Back', 'Okay', 'Additional Information', 'Click for More', etc. are provided on the Forms as necessary for negotiating various observations and associated display windows in a user friendly manner.

The WWTP software comprises a number of VB Forms. A list of forms is given in ***Appendix*** along with a brief description of the purpose and printouts of the important forms.

Conclusion

Today, both urbanization and industrialization are on the increase and as the growth increases, the pressures on the environment due to disposal of wastes also grow. Consequently, the standards for discharge of wastes stipulated by the regulatory agencies also become more and more stringent. The need for installation of pollution abatement thus becomes more urgent. Therefore, need for appropriate tools for designing the wastewater treatment plants and for proper diagnosis of operational problems encountered during the operation of these wastewater treatment facilities and remedies thereof has also gained importance. This in turn would ensure that the treatment plants designed and operated properly to achieve the best efficiencies at lesser cost. The software package – WWTP, developed under this dissertation thesis has the above objective in the fore amongst others.

Three major aspects of wastewater treatment, *viz. design of new facilities, augmentation of existing facilities*, and *troubleshooting* (diagnosis of operational problems and their causes and remedies thereof) have been addressed in this dissertation work. The governing design equations and criteria for the individual unit operations/ processes and the design criteria for the mechanical equipment such as aerators, are based on extensive literature review. A uniform user-interface, compatible with commercial Windows applications, has been designed for the software for ease of its execution without any knowledge of computer fundamentals or programming aspects.

A treatment scheme comprises a number of serially connected unit operations/ processes with feedback in some of the processes. The fact that the performance of each individual unit affects the performance of the units downstream, warrants the use of an integrated design approach wherein the system design is based on the overall system efficiencies with due consideration of the effect of performance variations of individual units on other units. Though, the entire design procedure should ideally be based on mathematical models describing the performance of individual units, it was not feasible in this case. This was due to the fact that still a number of empirical formulae and heuristic decisions are involved in the absence of well-defined performance models for some of the units. However, the heuristic approach used in the software for selection of aeration tank and aerator configuration affords a near optimal solution. The added advantage is that the configuration can be decided online with respect to the availability of land area for the treatment plant.

The integrated design approach, similar to the one adopted by Mhaisalkar *et al.*, [1993] for design of water treatment plant, would afford inclusion of cost of treatment units in the overall framework of a cost-optimization model. Joshi *et al.*, [1983] presented an approach to demonstrate the application of dynamic programming technique for optimization of an activated sludge process. Although, a cost-optimization problem for design of wastewater treatment plant can be formulated for a particular type of treatment, a generalized optimization model for several treatment options considered in the software is not easy to formulate. In view of the above and the time constraint, no attempt was made to formulate the problem as a mathematical programming problem employing the available optimization techniques, such as dynamic programming, non-linear programming, etc. Instead a heuristic approach has been adopted wherever required for optimization necessitated due to rounding off, *e.g.* volume of aeration tank actually provided may deviate substantially from the calculated volume depending on the configuration of aeration tanks and number of aerators in each tank and rating of each aerator. A near optimal solution is obtained heuristically using the commercially available aerator ratings and other mechanical equipment/devices.

The troubleshooting guidelines for a treatment plant based on activated sludge process only could be included in the software for want of time, although diagnostics for all the processes (considered in the software) through the implementation of an expert system/advisor was aimed at. Building an appropriate knowledge-base and database based on the analysis of operational problems encountered in a large number of treatment plants is a prerequisite to the development of expert system/advisor. Such information is not readily available in the literature.

It is hoped that the software even in its present form would prove to be a very useful tool for design of wastewater treatment plants as well as for providing decision support for operation and maintenance of an activated sludge process.

Annexure-I

WWSD-NEERI : Wastewater Treatment System Design, NEERI, Jaipur
Wastewater Treatment System Design
Version 1.0.0
Developed by J.K. Bassin, M.E. (Env) III-Semester
Guide: Dr. A.B.Gupta, Prof&Head, Civil Engg, MREC
Malaviya Regional Engineering College, Jaipur
RUN
DEVELOPED AT
National Environmental Engineering Research Institute, Jaipur
EXIT

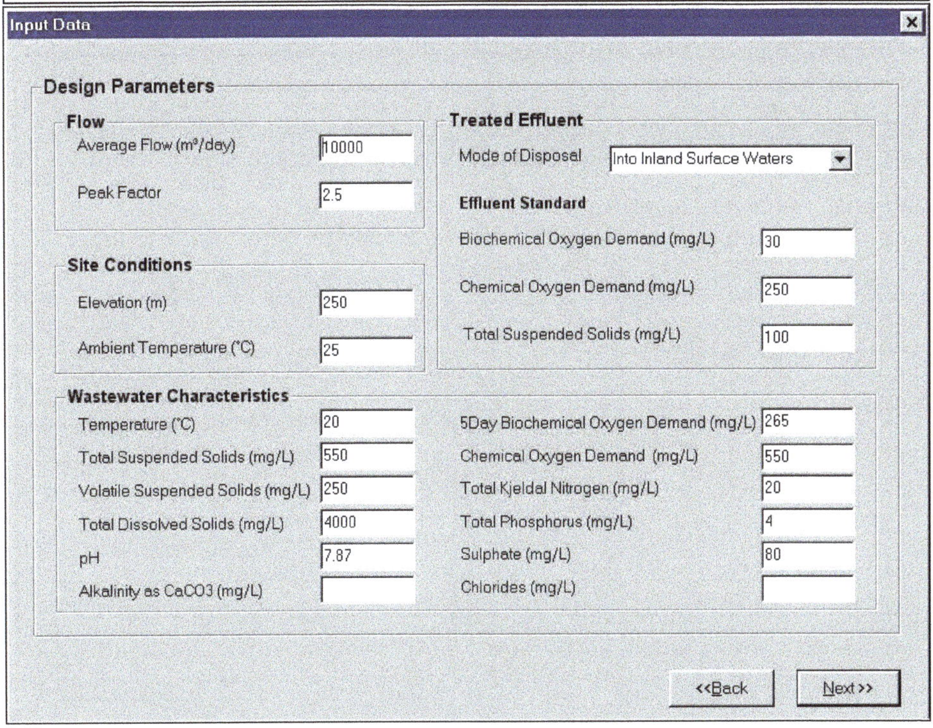
Input Data
Design Parameters
Flow
Average Flow (m³/day) 10000
Peak Factor 2.5
Site Conditions
Elevation (m) 250
Ambient Temperature (°C) 25
Treated Effluent
Mode of Disposal Into Inland Surface Waters
Effluent Standard
Biochemical Oxygen Demand (mg/L) 30
Chemical Oxygen Demand (mg/L) 250
Total Suspended Solids (mg/L) 100
Wastewater Characteristics
Temperature (°C) 20
Total Suspended Solids (mg/L) 550
Volatile Suspended Solids (mg/L) 250
Total Dissolved Solids (mg/L) 4000
pH 7.87
Alkalinity as CaCO3 (mg/L)
5Day Biochemical Oxygen Demand (mg/L) 265
Chemical Oxygen Demand (mg/L) 550
Total Kjeldal Nitrogen (mg/L) 20
Total Phosphorus (mg/L) 4
Sulphate (mg/L) 80
Chlorides (mg/L)
<<Back
Next>>

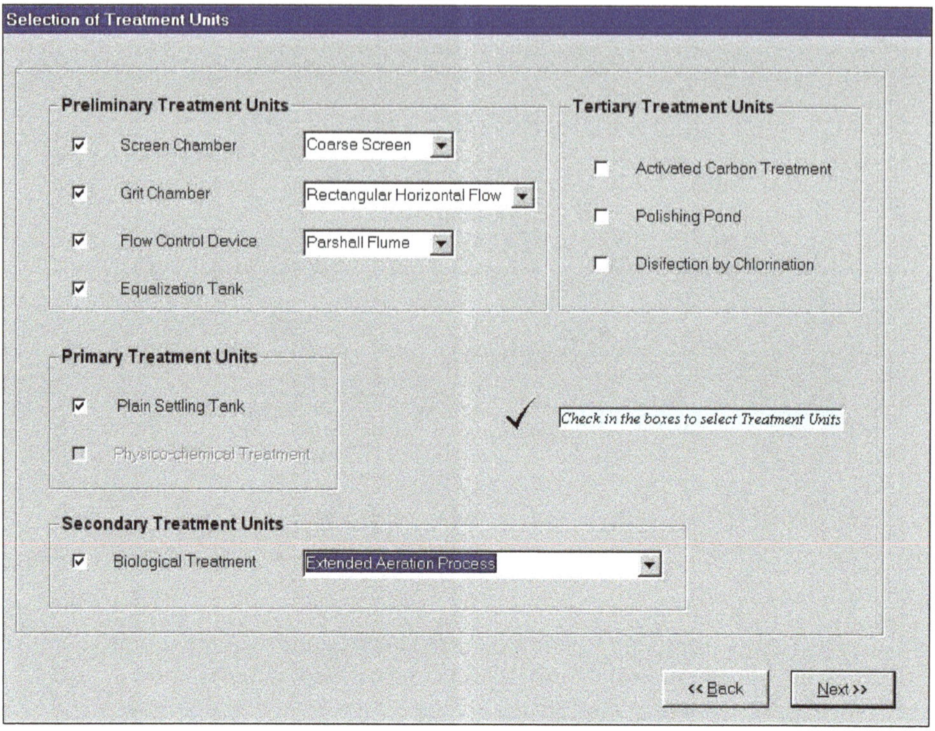
Selection of Treatment Units
Preliminary Treatment Units
Screen Chamber
Coarse Screen
Grit Chamber
Rectangular Horizontal Flow
Flow Control Device
Parshall Flume
Equalization Tank
Tertiary Treatment Units
Activated Carbon Treatment
Polishing Pond
Disifection by Chlorination
Primary Treatment Units
Plain Settling Tank
Physico-chemical Treatment
Check in the boxes to select Treatment Units
Secondary Treatment Units
Biological Treatment
Extended Aeration Process
<< Back
Next >>

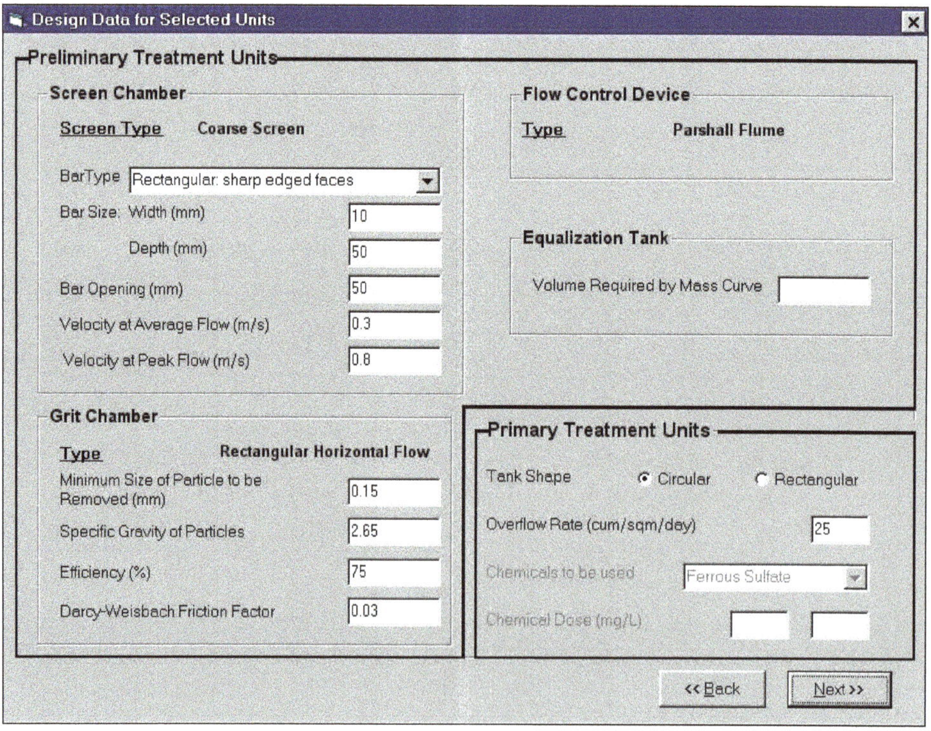
Design Data for Selected Units
Preliminary Treatment Units
Screen Chamber
Screen Type
Coarse Screen
BarType
Rectangular: sharp edged faces
Bar Size: Width (mm)
10
Depth (mm)
50
Bar Opening (mm)
50
Velocity at Average Flow (m/s)
0.3
Velocity at Peak Flow (m/s)
0.8
Flow Control Device
Type
Parshall Flume
Equalization Tank
Volume Required by Mass Curve
Grit Chamber
Type
Rectangular Horizontal Flow
Minimum Size of Particle to be Removed (mm)
0.15
Specific Gravity of Particles
2.65
Efficiency (%)
75
Darcy-Weisbach Friction Factor
0.03
Primary Treatment Units
Tank Shape
Circular
Rectangular
Overflow Rate (cum/sqm/day)
25
Chemicals to be used
Ferrous Sulfate
Chemical Dose (mg/L)
<< Back
Next >>

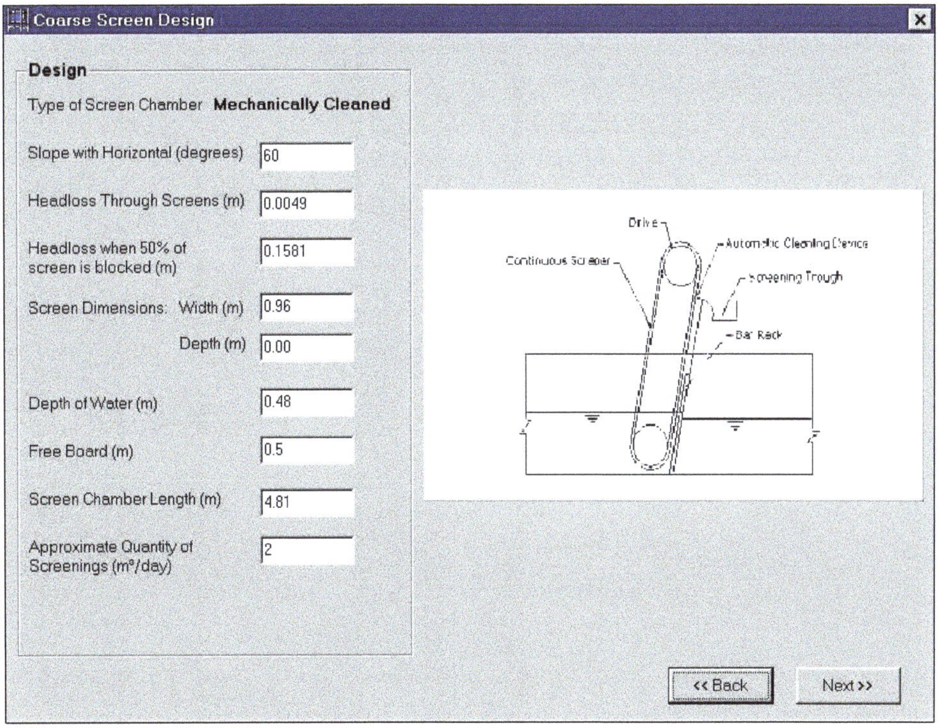
Coarse Screen Design
Design
Type of Screen Chamber **Mechanically Cleaned**
Slope with Horizontal (degrees) 60
Headloss Through Screens (m) 0.0049
Headloss when 50% of screen is blocked (m) 0.1581
Screen Dimensions: Width (m) 0.96
Depth (m) 0.00
Depth of Water (m) 0.48
Free Board (m) 0.5
Screen Chamber Length (m) 4.81
Approximate Quantity of Screenings (m³/day) 2
Drive
Automatic Cleaning Device
Continuous Scraper
Screening Trough
Bar Rack
<< Back
Next >>

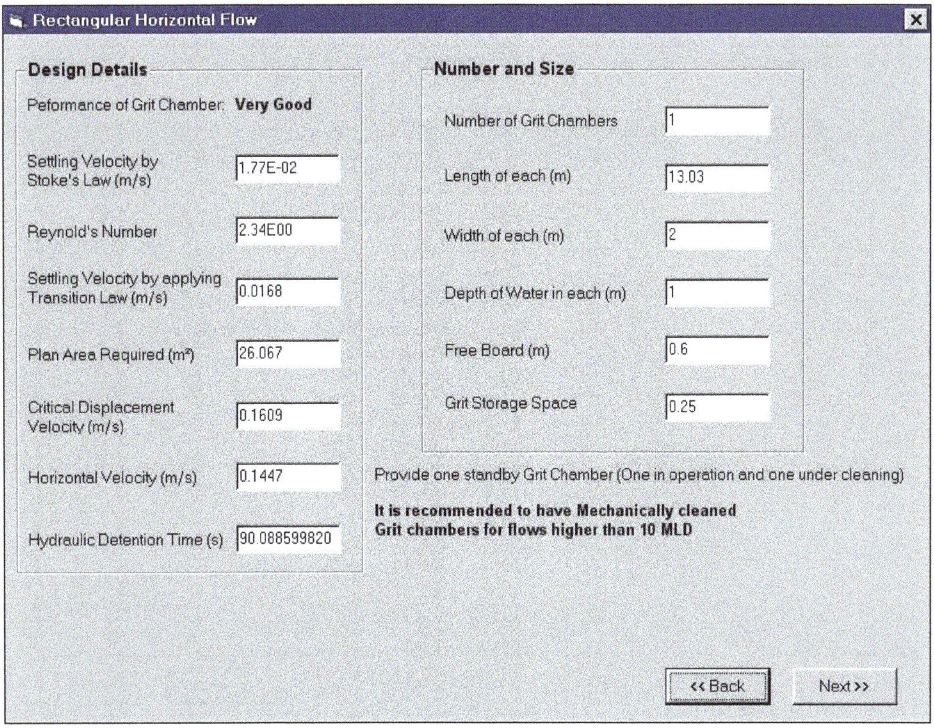
Rectangular Horizontal Flow
Design Details
Peformance of Grit Chamber: **Very Good**
Settling Velocity by Stoke's Law (m/s) 1.77E-02
Reynold's Number 2.34E00
Settling Velocity by applying Transition Law (m/s) 0.0168
Plan Area Required (m²) 26.067
Critical Displacement Velocity (m/s) 0.1609
Horizontal Velocity (m/s) 0.1447
Hydraulic Detention Time (s) 90.088599820
Number and Size
Number of Grit Chambers 1
Length of each (m) 13.03
Width of each (m) 2
Depth of Water in each (m) 1
Free Board (m) 0.6
Grit Storage Space 0.25
Provide one standby Grit Chamber (One in operation and one under cleaning)
It is recommended to have Mechanically cleaned Grit chambers for flows higher than 10 MLD
<< Back
Next >>

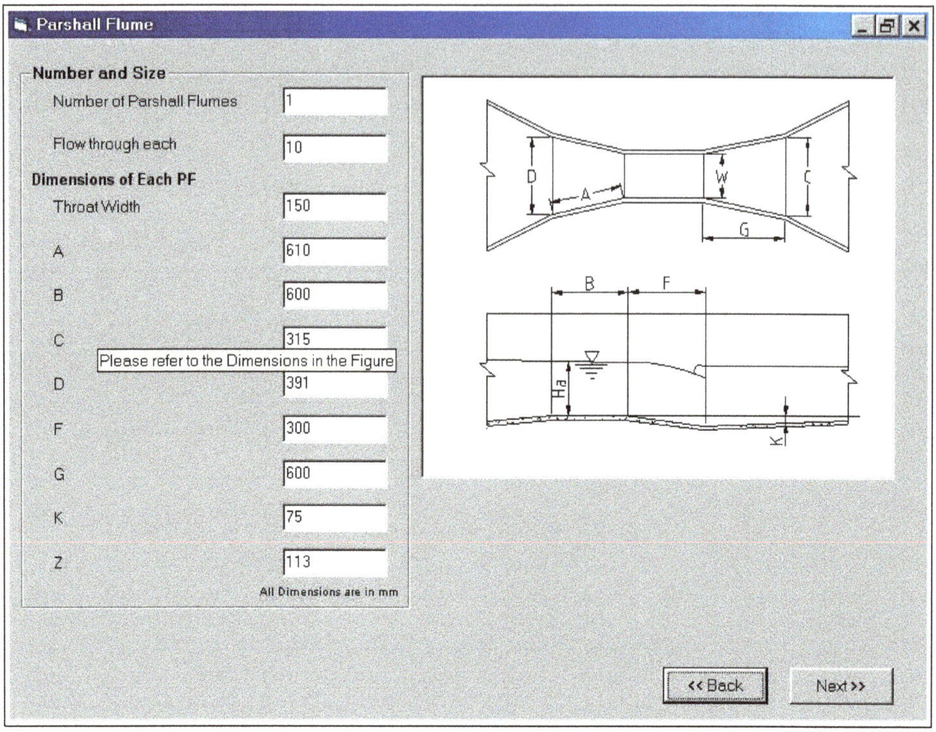

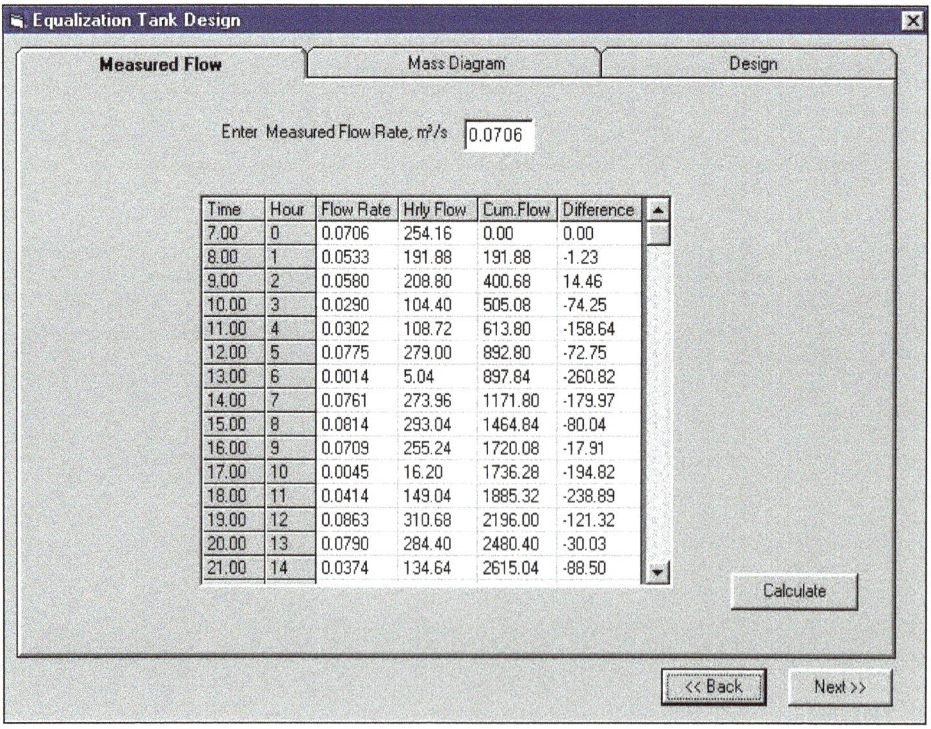

Time	Hour	Flow Rate	Hrly Flow	Cum.Flow	Difference
7.00	0	0.0706	254.16	0.00	0.00
8.00	1	0.0533	191.88	191.88	-1.23
9.00	2	0.0580	208.80	400.68	14.46
10.00	3	0.0290	104.40	505.08	-74.25
11.00	4	0.0302	108.72	613.80	-158.64
12.00	5	0.0775	279.00	892.80	-72.75
13.00	6	0.0014	5.04	897.84	-260.82
14.00	7	0.0761	273.96	1171.80	-179.97
15.00	8	0.0814	293.04	1464.84	-80.04
16.00	9	0.0709	255.24	1720.08	-17.91
17.00	10	0.0045	16.20	1736.28	-194.82
18.00	11	0.0414	149.04	1885.32	-238.89
19.00	12	0.0863	310.68	2196.00	-121.32
20.00	13	0.0790	284.40	2480.40	-30.03
21.00	14	0.0374	134.64	2615.04	-88.50

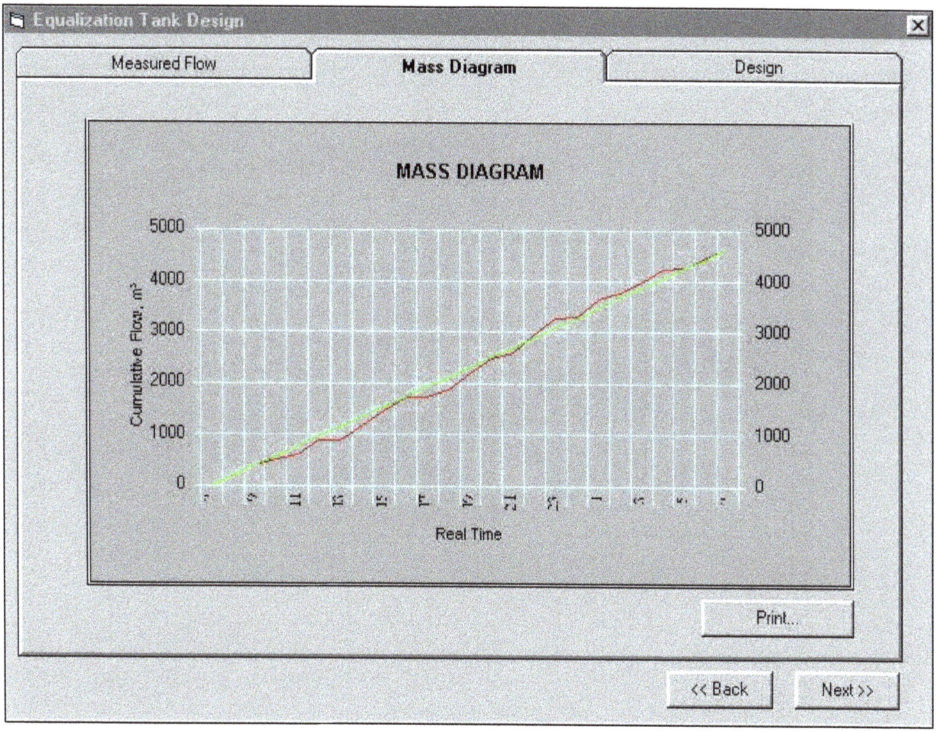
Equalization Tank Design
Measured Flow
Mass Diagram
Design
MASS DIAGRAM
Cumulative Flow, m³
5000
4000
3000
2000
1000
0
Real Time
Print...
<< Back
Next >>

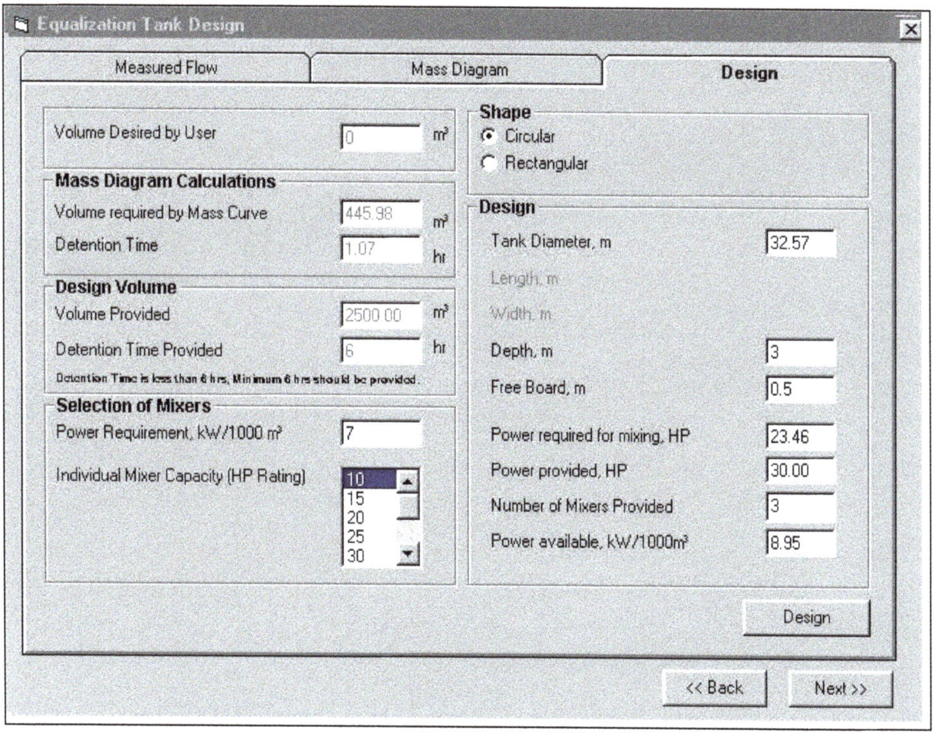
Equalization Tank Design
Measured Flow
Mass Diagram
Design
Volume Desired by User
0
m³
Mass Diagram Calculations
Volume required by Mass Curve
445.98
m³
Detention Time
1.07
hr
Design Volume
Volume Provided
2500.00
m³
Detention Time Provided
6
hr
Detention Time is less than 6 hrs, Minimum 6 hrs should be provided.
Selection of Mixers
Power Requirement, kW/1000 m³
7
Individual Mixer Capacity (HP Rating)
10
15
20
25
30
Shape
Circular
Rectangular
Design
Tank Diameter, m
32.57
Length, m
Width, m
Depth, m
3
Free Board, m
0.5
Power required for mixing, HP
23.46
Power provided, HP
30.00
Number of Mixers Provided
3
Power available, kW/1000m³
8.95
Design
<< Back
Next >>

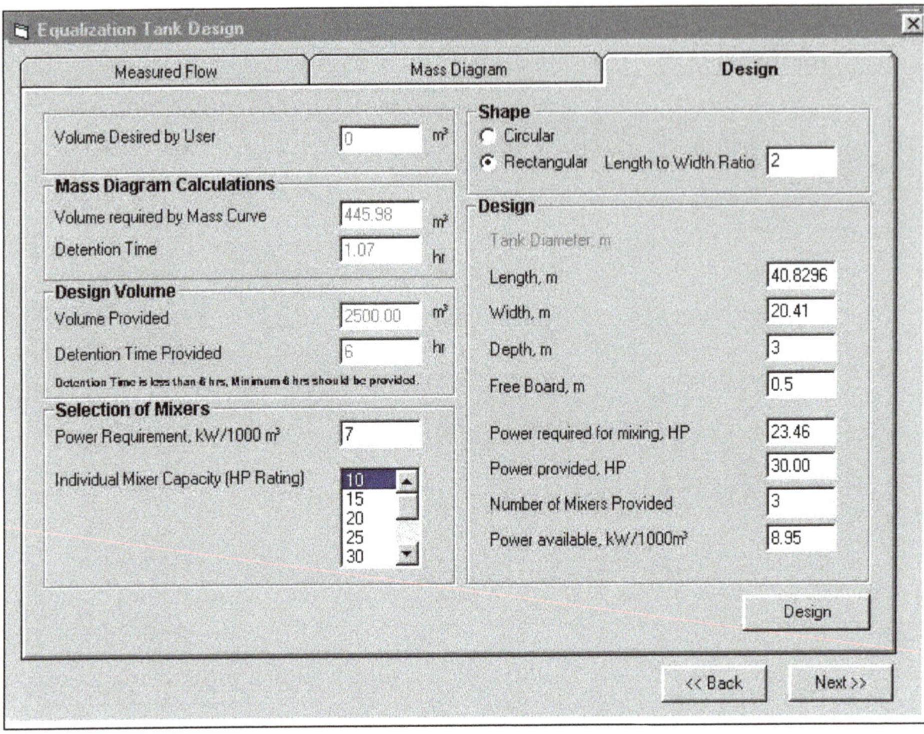

Equalization Tank Design
Measured Flow
Mass Diagram
Design
Volume Desired by User
0
m³
Mass Diagram Calculations
Volume required by Mass Curve
445.98
m³
Detention Time
1.07
hr
Design Volume
Volume Provided
2500.00
m³
Detention Time Provided
6
hr
Detention Time is less than 6 hrs. Minimum 6 hrs should be provided.
Selection of Mixers
Power Requirement, kW/1000 m³
7
Individual Mixer Capacity (HP Rating)
10
15
20
25
30
Shape
Circular
Rectangular
Length to Width Ratio
2
Design
Tank Diameter, m
Length, m
40.8296
Width, m
20.41
Depth, m
3
Free Board, m
0.5
Power required for mixing, HP
23.46
Power provided, HP
30.00
Number of Mixers Provided
3
Power available, kW/1000m³
8.95
Design
<< Back
Next >>

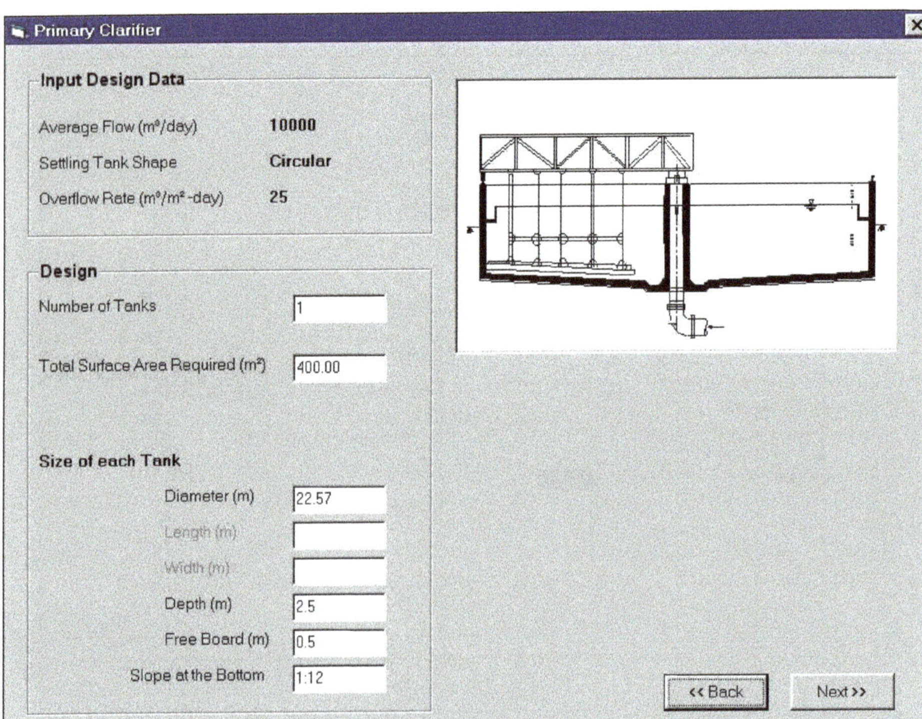

Primary Clarifier
Input Design Data
Average Flow (m³/day)
10000
Settling Tank Shape
Circular
Overflow Rate (m³/m²-day)
25
Design
Number of Tanks
1
Total Surface Area Required (m²)
400.00
Size of each Tank
Diameter (m)
22.57
Length (m)
Width (m)
Depth (m)
2.5
Free Board (m)
0.5
Slope at the Bottom
1:12
<< Back
Next >>

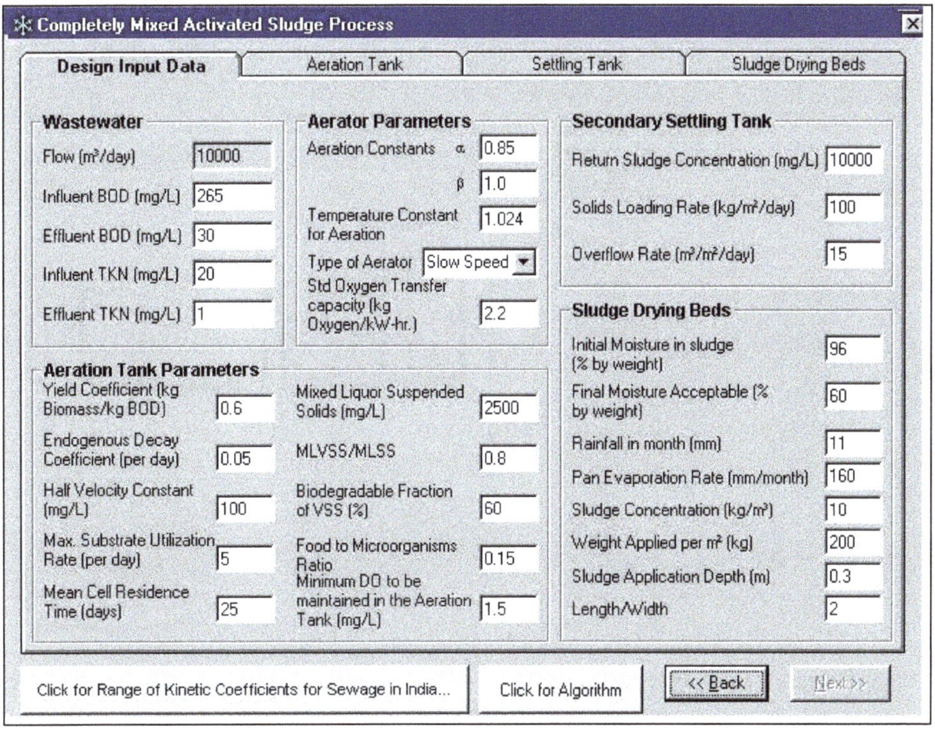
Completely Mixed Activated Sludge Process
Design Input Data
Aeration Tank
Settling Tank
Sludge Drying Beds
Wastewater
Flow (m³/day) 10000
Influent BOD (mg/L) 265
Effluent BOD (mg/L) 30
Influent TKN (mg/L) 20
Effluent TKN (mg/L) 1
Aerator Parameters
Aeration Constants α 0.85
β 1.0
Temperature Constant for Aeration 1.024
Type of Aerator Slow Speed
Std Oxygen Transfer capacity (kg Oxygen/kW-hr.) 2.2
Secondary Settling Tank
Return Sludge Concentration (mg/L) 10000
Solids Loading Rate (kg/m²/day) 100
Overflow Rate (m³/m²/day) 15
Sludge Drying Beds
Initial Moisture in sludge (% by weight) 96
Final Moisture Acceptable (% by weight) 60
Rainfall in month (mm) 11
Pan Evaporation Rate (mm/month) 160
Sludge Concentration (kg/m³) 10
Weight Applied per m² (kg) 200
Sludge Application Depth (m) 0.3
Length/Width 2
Aeration Tank Parameters
Yield Coefficient (kg Biomass/kg BOD) 0.6
Endogenous Decay Coefficient (per day) 0.05
Half Velocity Constant (mg/L) 100
Max. Substrate Utilization Rate (per day) 5
Mean Cell Residence Time (days) 25
Mixed Liquor Suspended Solids (mg/L) 2500
MLVSS/MLSS 0.8
Biodegradable Fraction of VSS (%) 60
Food to Microorganisms Ratio 0.15
Minimum DO to be maintained in the Aeration Tank (mg/L) 1.5
Click for Range of Kinetic Coefficients for Sewage in India...
Click for Algorithm
<< Back
Next >>

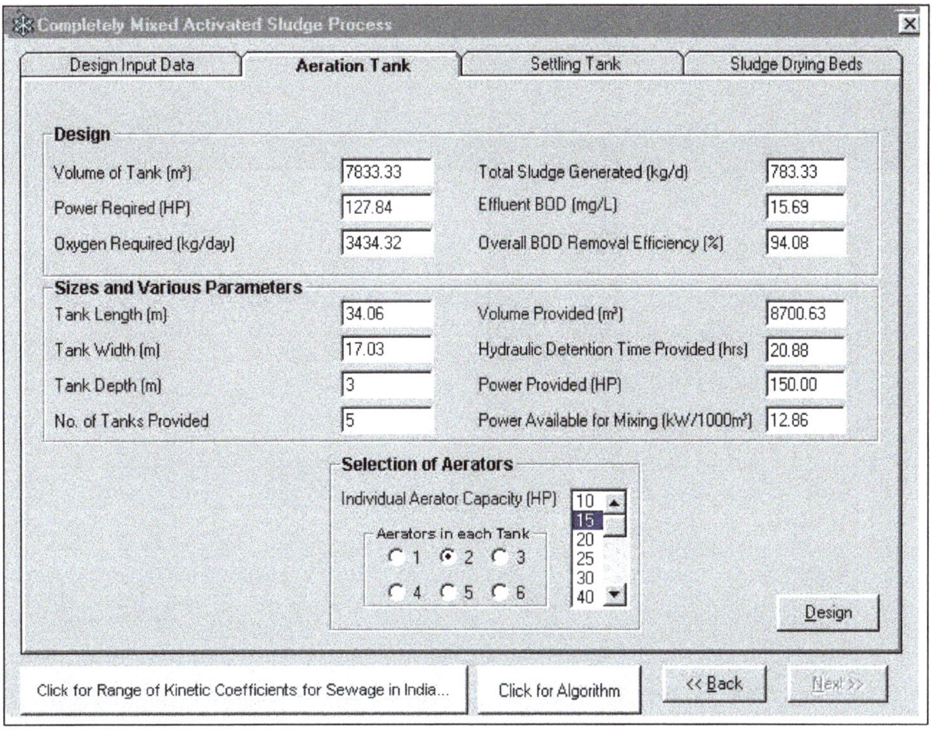
Completely Mixed Activated Sludge Process
Design Input Data
Aeration Tank
Settling Tank
Sludge Drying Beds
Design
Volume of Tank (m³) 7833.33
Power Reqired (HP) 127.84
Oxygen Required (kg/day) 3434.32
Total Sludge Generated (kg/d) 783.33
Effluent BOD (mg/L) 15.69
Overall BOD Removal Efficiency (%) 94.08
Sizes and Various Parameters
Tank Length (m) 34.06
Tank Width (m) 17.03
Tank Depth (m) 3
No. of Tanks Provided 5
Volume Provided (m³) 8700.63
Hydraulic Detention Time Provided (hrs) 20.88
Power Provided (HP) 150.00
Power Available for Mixing (kW/1000m³) 12.86
Selection of Aerators
Individual Aerator Capacity (HP)
10
15
20
25
30
40
Aerators in each Tank
1 2 3
4 5 6
Design
Click for Range of Kinetic Coefficients for Sewage in India...
Click for Algorithm
<< Back
Next >>

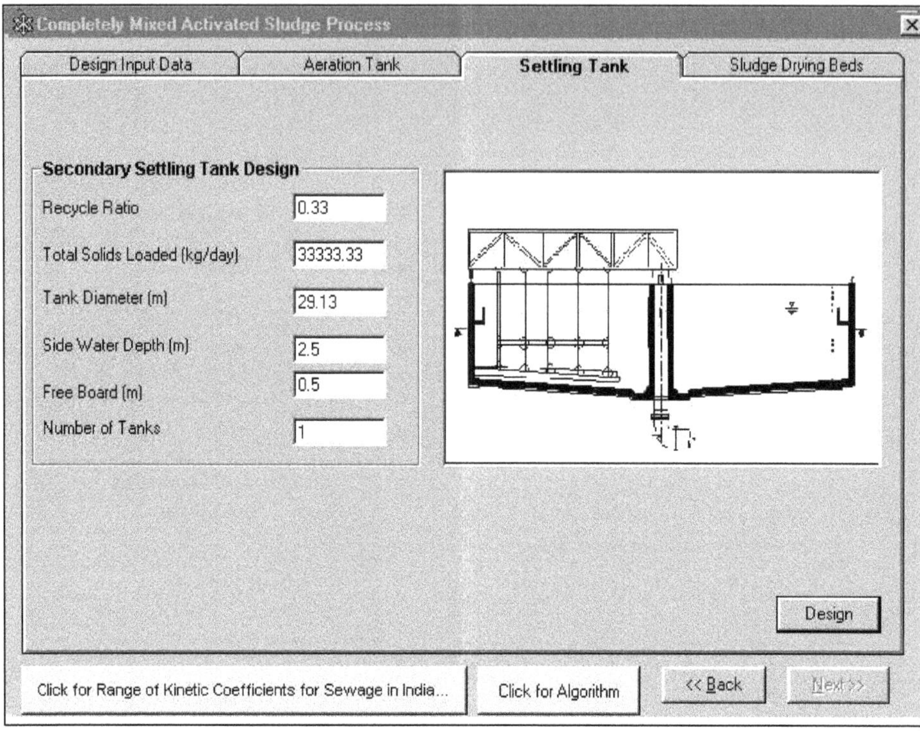
Completely Mixed Activated Sludge Process
Design Input Data
Aeration Tank
Settling Tank
Sludge Drying Beds
Secondary Settling Tank Design
Recycle Ratio 0.33
Total Solids Loaded (kg/day) 33333.33
Tank Diameter (m) 29.13
Side Water Depth (m) 2.5
Free Board (m) 0.5
Number of Tanks 1
Design
Click for Range of Kinetic Coefficients for Sewage in India...
Click for Algorithm
<< Back
Next >>

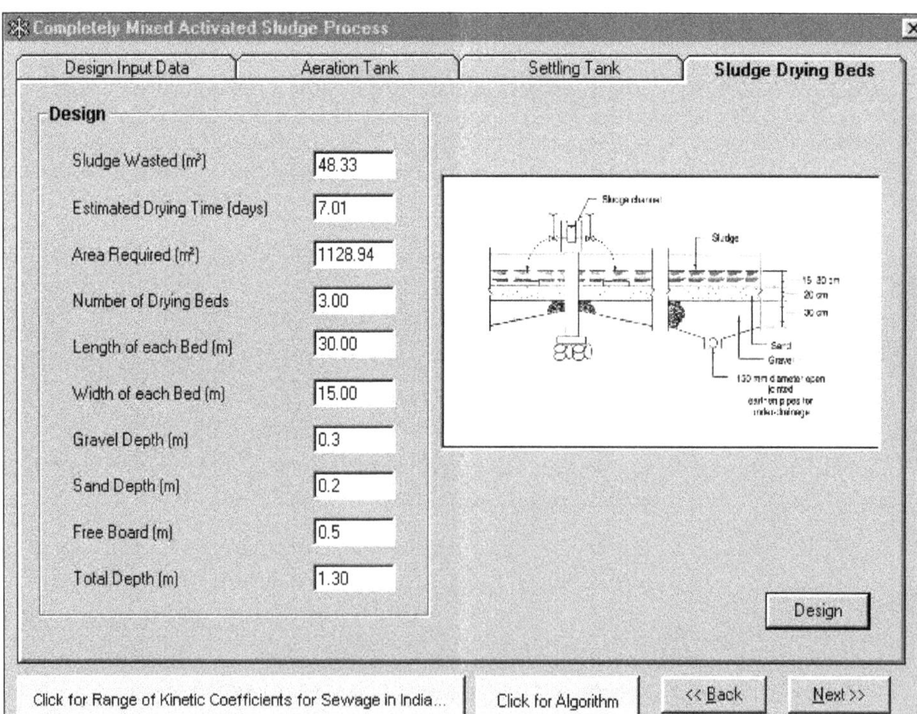
Completely Mixed Activated Sludge Process
Design Input Data
Aeration Tank
Settling Tank
Sludge Drying Beds
Design
Sludge Wasted (m³) 48.33
Estimated Drying Time (days) 7.01
Area Required (m²) 1128.94
Number of Drying Beds 3.00
Length of each Bed (m) 30.00
Width of each Bed (m) 15.00
Gravel Depth (m) 0.3
Sand Depth (m) 0.2
Free Board (m) 0.5
Total Depth (m) 1.30
Sludge
Design
Click for Range of Kinetic Coefficients for Sewage in India...
Click for Algorithm
<< Back
Next >>

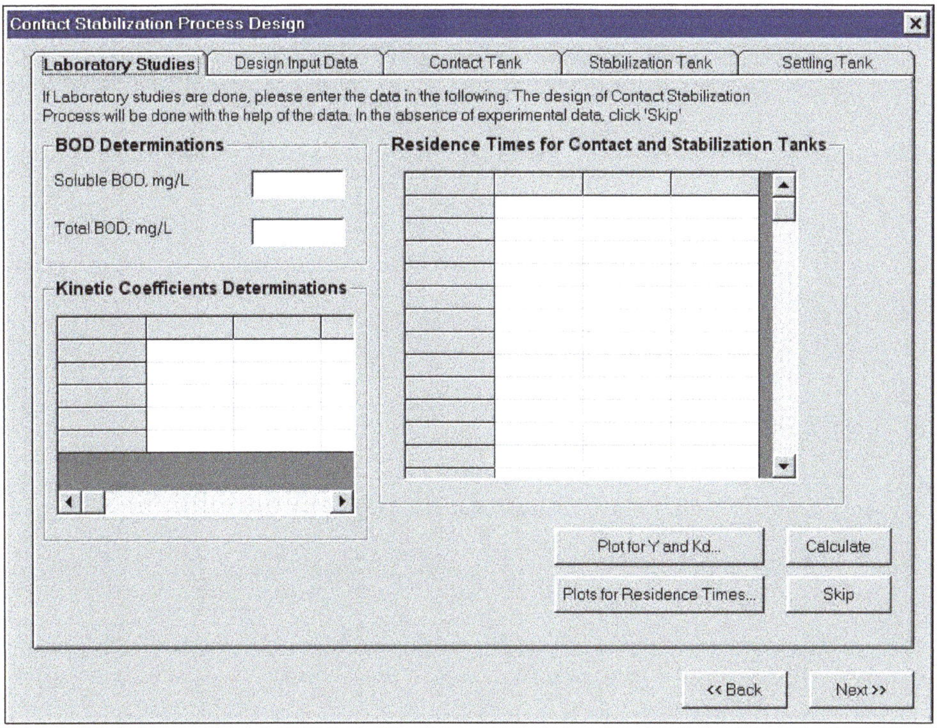
Contact Stabilization Process Design
Laboratory Studies
Design Input Data
Contact Tank
Stabilization Tank
Settling Tank
If Laboratory studies are done, please enter the data in the following. The design of Contact Stabilization Process will be done with the help of the data. In the absence of experimental data, click 'Skip'
BOD Determinations
Soluble BOD, mg/L
Total BOD, mg/L
Residence Times for Contact and Stabilization Tanks
Kinetic Coefficients Determinations
Plot for Y and Kd...
Calculate
Plots for Residence Times...
Skip
<< Back
Next >>

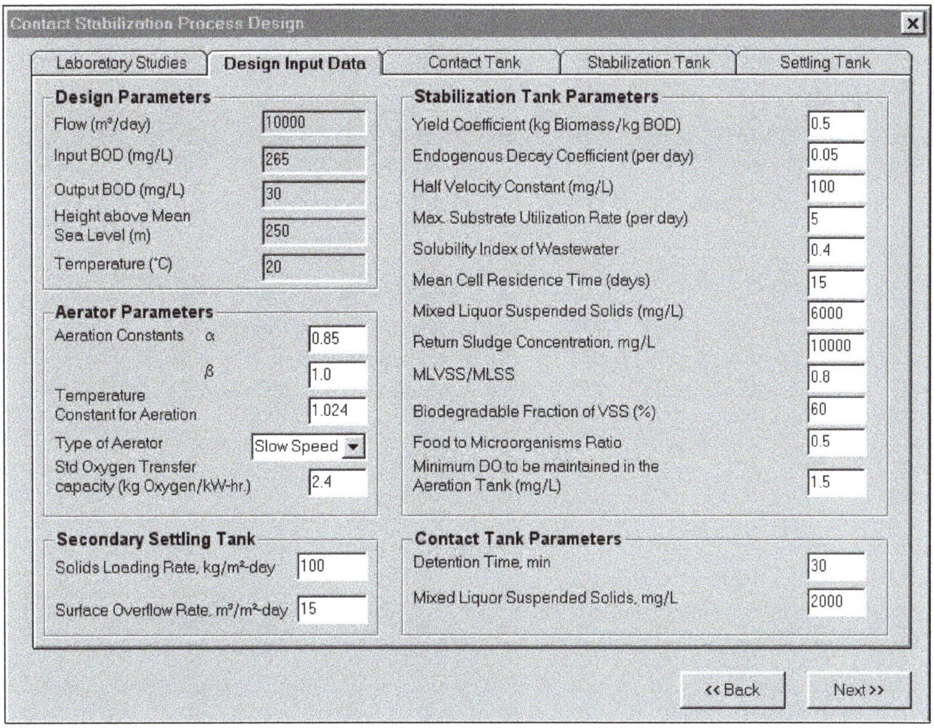
Contact Stabilization Process Design
Laboratory Studies
Design Input Data
Contact Tank
Stabilization Tank
Settling Tank
Design Parameters
Flow (m³/day) 10000
Input BOD (mg/L) 265
Output BOD (mg/L) 30
Height above Mean Sea Level (m) 250
Temperature (°C) 20
Aerator Parameters
Aeration Constants α 0.85
β 1.0
Temperature Constant for Aeration 1.024
Type of Aerator Slow Speed
Std Oxygen Transfer capacity (kg Oxygen/kW-hr.) 2.4
Stabilization Tank Parameters
Yield Coefficient (kg Biomass/kg BOD) 0.5
Endogenous Decay Coefficient (per day) 0.05
Half Velocity Constant (mg/L) 100
Max. Substrate Utilization Rate (per day) 5
Solubility Index of Wastewater 0.4
Mean Cell Residence Time (days) 15
Mixed Liquor Suspended Solids (mg/L) 6000
Return Sludge Concentration, mg/L 10000
MLVSS/MLSS 0.8
Biodegradable Fraction of VSS (%) 60
Food to Microorganisms Ratio 0.5
Minimum DO to be maintained in the Aeration Tank (mg/L) 1.5
Secondary Settling Tank
Solids Loading Rate, kg/m²-day 100
Surface Overflow Rate, m³/m²-day 15
Contact Tank Parameters
Detention Time, min 30
Mixed Liquor Suspended Solids, mg/L 2000
<< Back
Next >>

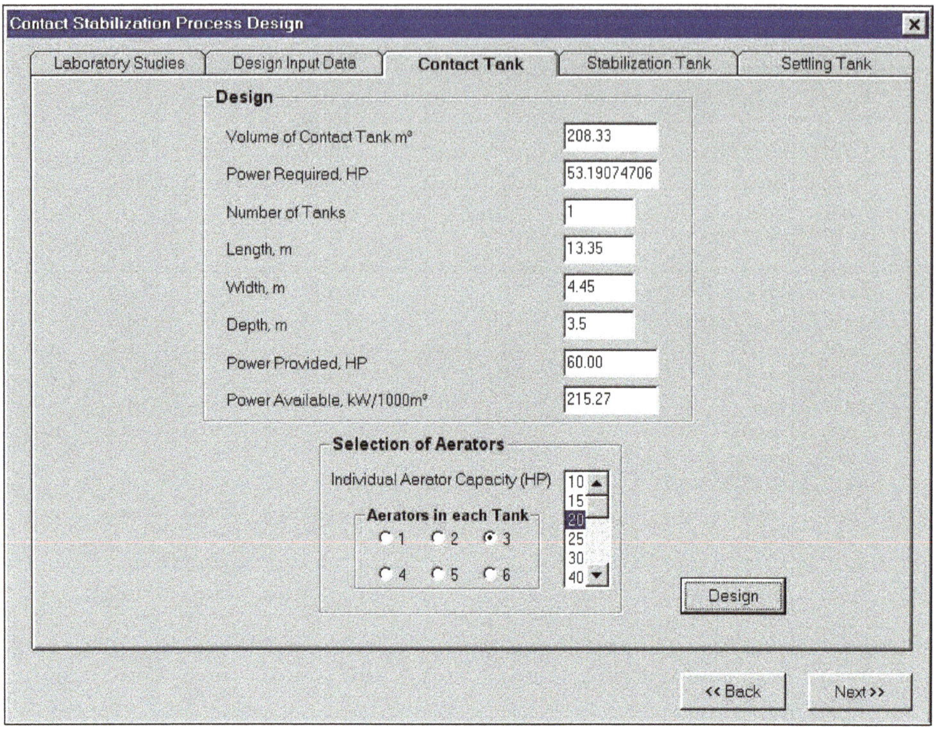
Contact Stabilization Process Design
Laboratory Studies
Design Input Data
Contact Tank
Stabilization Tank
Settling Tank
Design
Volume of Contact Tank m³
208.33
Power Required, HP
53.19074706
Number of Tanks
1
Length, m
13.35
Width, m
4.45
Depth, m
3.5
Power Provided, HP
60.00
Power Available, kW/1000m³
215.27
Selection of Aerators
Individual Aerator Capacity (HP)
10
15
20
25
30
40
Aerators in each Tank
1
2
3
4
5
6
Design
<< Back
Next >>

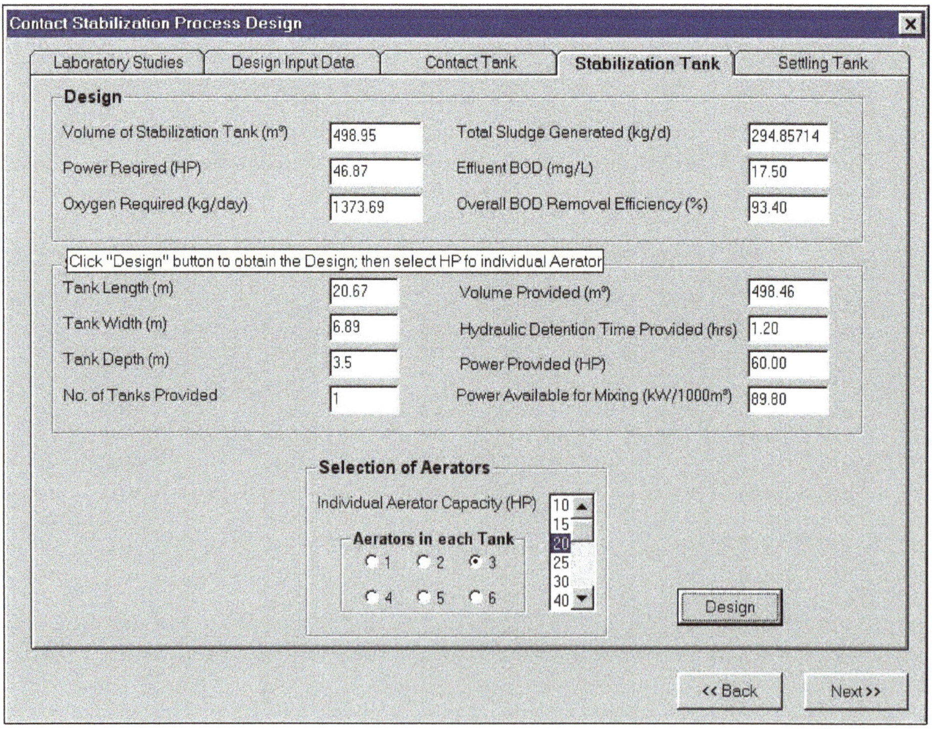
Contact Stabilization Process Design
Laboratory Studies
Design Input Data
Contact Tank
Stabilization Tank
Settling Tank
Design
Volume of Stabilization Tank (m³)
498.95
Power Reqired (HP)
46.87
Oxygen Required (kg/day)
1373.69
Total Sludge Generated (kg/d)
294.85714
Effluent BOD (mg/L)
17.50
Overall BOD Removal Efficiency (%)
93.40
Click "Design" button to obtain the Design; then select HP fo individual Aerator
Tank Length (m)
20.67
Tank Width (m)
6.89
Tank Depth (m)
3.5
No. of Tanks Provided
1
Volume Provided (m³)
498.46
Hydraulic Detention Time Provided (hrs)
1.20
Power Provided (HP)
60.00
Power Available for Mixing (kW/1000m³)
89.80
Selection of Aerators
Individual Aerator Capacity (HP)
10
15
20
25
30
40
Aerators in each Tank
1
2
3
4
5
6
Design
<< Back
Next >>

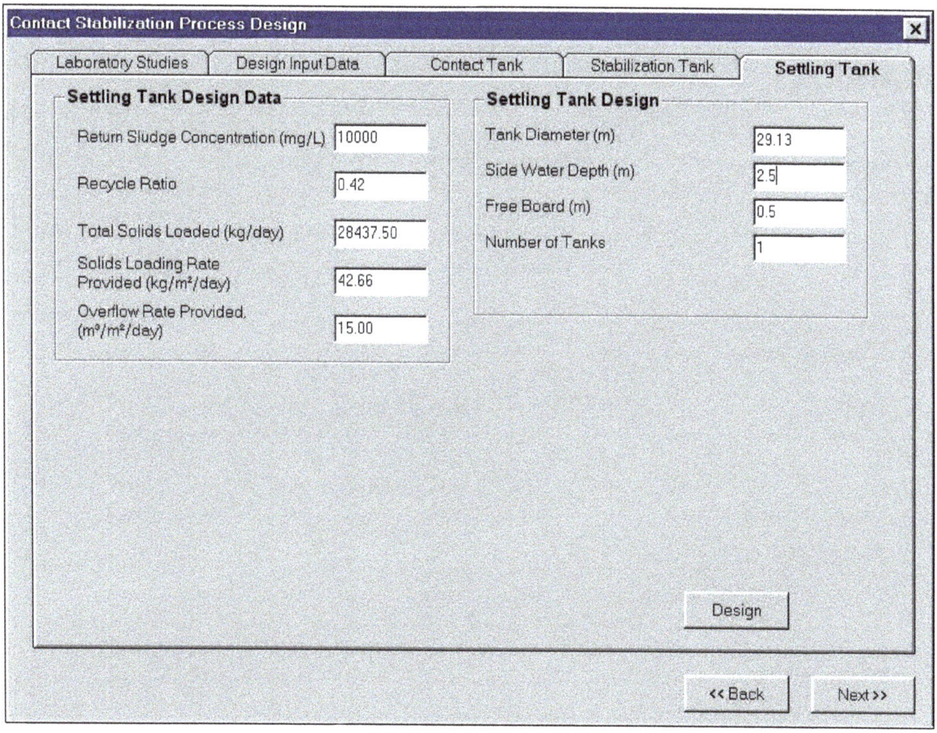
Contact Stabilization Process Design
Laboratory Studies
Design Input Data
Contact Tank
Stabilization Tank
Settling Tank
Settling Tank Design Data
Return Sludge Concentration (mg/L) 10000
Recycle Ratio 0.42
Total Solids Loaded (kg/day) 28437.50
Solids Loading Rate Provided (kg/m²/day) 42.66
Overflow Rate Provided. (m³/m²/day) 15.00
Settling Tank Design
Tank Diameter (m) 29.13
Side Water Depth (m) 2.5
Free Board (m) 0.5
Number of Tanks 1
Design
<< Back
Next >>

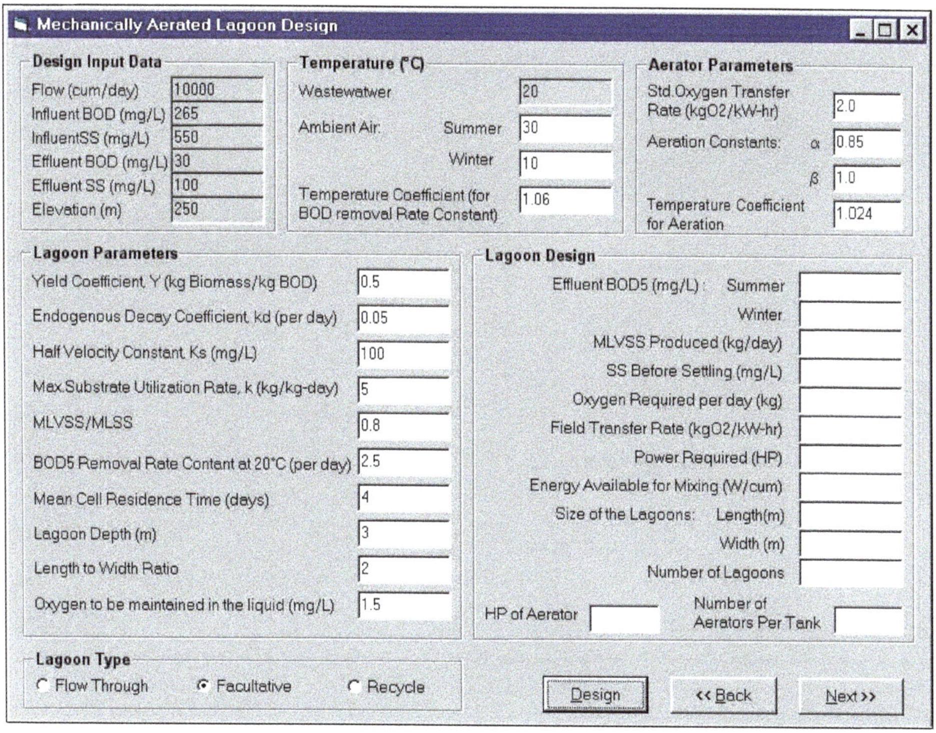
Mechanically Aerated Lagoon Design
Design Input Data
Flow (cum/day) 10000
Influent BOD (mg/L) 265
InfluentSS (mg/L) 550
Effluent BOD (mg/L) 30
Effluent SS (mg/L) 100
Elevation (m) 250
Temperature (°C)
Wastewatwer 20
Ambient Air: Summer 30
Winter 10
Temperature Coefficient (for BOD removal Rate Constant) 1.06
Aerator Parameters
Std.Oxygen Transfer Rate (kgO2/kW-hr) 2.0
Aeration Constants: α 0.85
β 1.0
Temperature Coefficient for Aeration 1.024
Lagoon Parameters
Yield Coefficient, Y (kg Biomass/kg BOD) 0.5
Endogenous Decay Coefficient, kd (per day) 0.05
Half Velocity Constant, Ks (mg/L) 100
Max.Substrate Utilization Rate, k (kg/kg-day) 5
MLVSS/MLSS 0.8
BOD5 Removal Rate Content at 20°C (per day) 2.5
Mean Cell Residence Time (days) 4
Lagoon Depth (m) 3
Length to Width Ratio 2
Oxygen to be maintained in the liquid (mg/L) 1.5
Lagoon Design
Effluent BOD5 (mg/L) : Summer
Winter
MLVSS Produced (kg/day)
SS Before Settling (mg/L)
Oxygen Required per day (kg)
Field Transfer Rate (kgO2/kW-hr)
Power Required (HP)
Energy Available for Mixing (W/cum)
Size of the Lagoons: Length(m)
Width (m)
Number of Lagoons
HP of Aerator
Number of Aerators Per Tank
Lagoon Type
Flow Through
Facultative
Recycle
Design
<< Back
Next >>

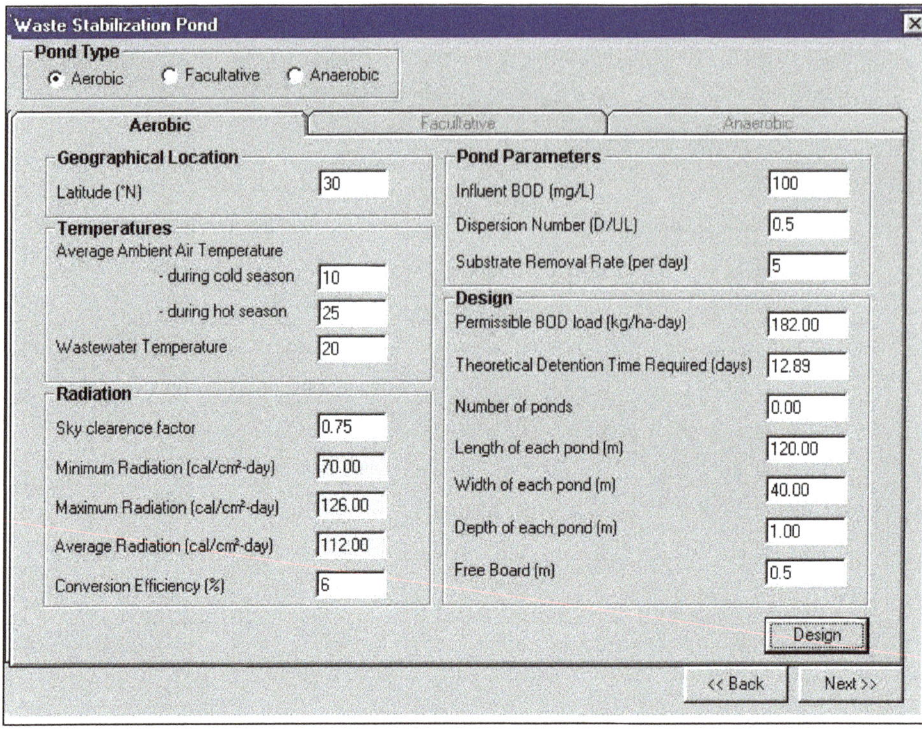
Waste Stabilization Pond
Pond Type
Aerobic
Facultative
Anaerobic
Aerobic
Facultative
Anaerobic
Geographical Location
Latitude (°N)
30
Temperatures
Average Ambient Air Temperature
- during cold season
10
- during hot season
25
Wastewater Temperature
20
Radiation
Sky clearence factor
0.75
Minimum Radiation (cal/cm²-day)
70.00
Maximum Radiation (cal/cm²-day)
126.00
Average Radiation (cal/cm²-day)
112.00
Conversion Efficiency (%)
6
Pond Parameters
Influent BOD (mg/L)
100
Dispersion Number (D/UL)
0.5
Substrate Removal Rate (per day)
5
Design
Permissible BOD load (kg/ha-day)
182.00
Theoretical Detention Time Required (days)
12.89
Number of ponds
0.00
Length of each pond (m)
120.00
Width of each pond (m)
40.00
Depth of each pond (m)
1.00
Free Board (m)
0.5
Design
<< Back
Next >>

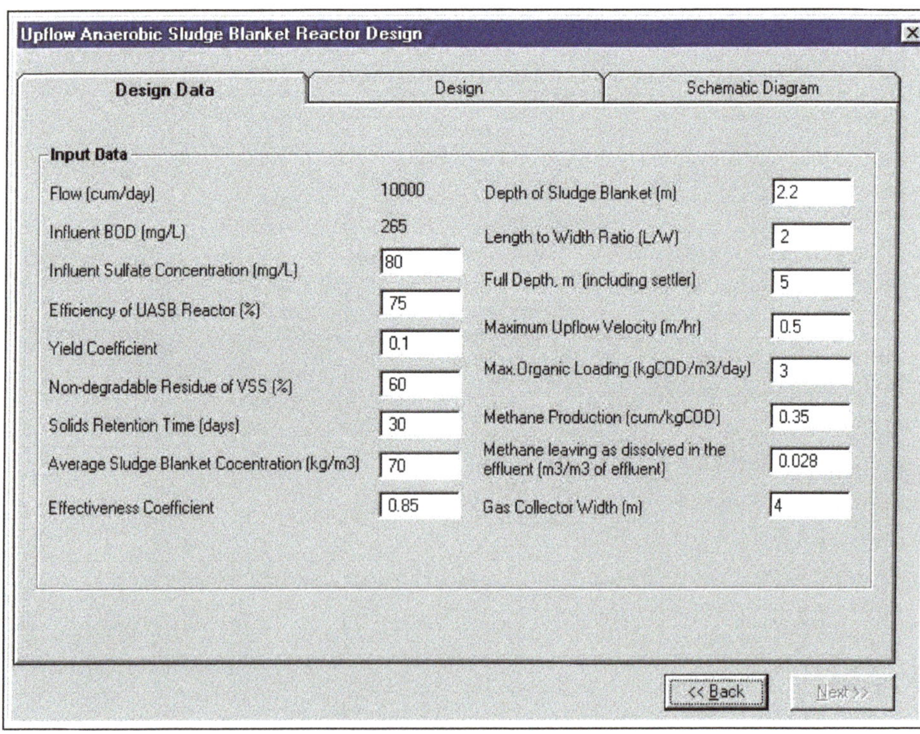
Upflow Anaerobic Sludge Blanket Reactor Design
Design Data
Design
Schematic Diagram
Input Data
Flow (cum/day)
10000
Influent BOD (mg/L)
265
Influent Sulfate Concentration (mg/L)
80
Efficiency of UASB Reactor (%)
75
Yield Coefficient
0.1
Non-degradable Residue of VSS (%)
60
Solids Retention Time (days)
30
Average Sludge Blanket Cocentration (kg/m3)
70
Effectiveness Coefficient
0.85
Depth of Sludge Blanket (m)
2.2
Length to Width Ratio (L/W)
2
Full Depth, m (including settler)
5
Maximum Upflow Velocity (m/hr)
0.5
Max.Organic Loading (kgCOD/m3/day)
3
Methane Production (cum/kgCOD)
0.35
Methane leaving as dissolved in the effluent (m3/m3 of effluent)
0.028
Gas Collector Width (m)
4
<< Back
Next >>

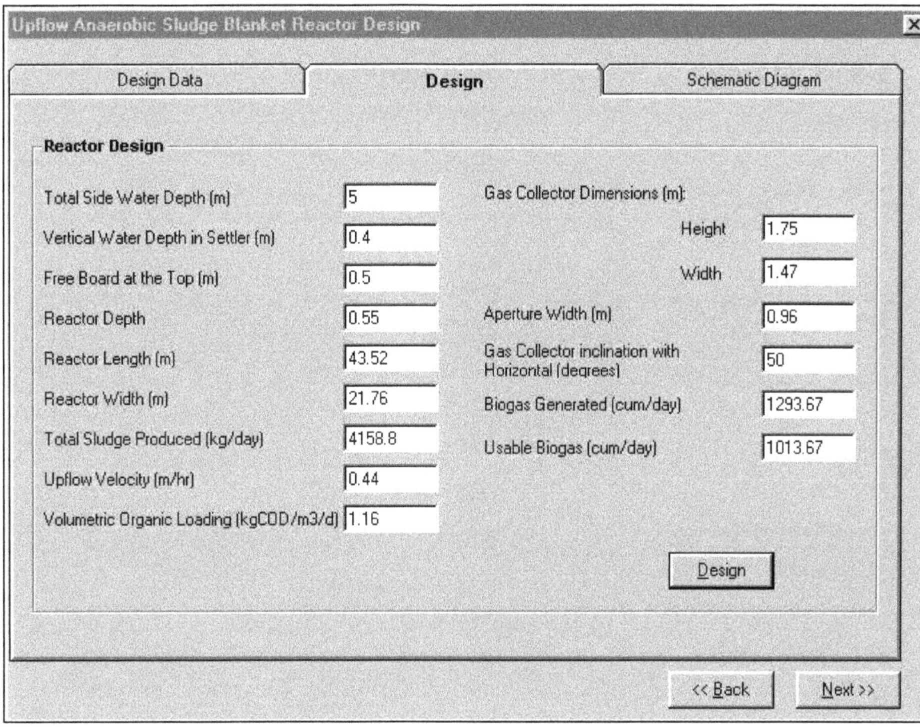
Upflow Anaerobic Sludge Blanket Reactor Design
Design Data
Design
Schematic Diagram
Reactor Design
Total Side Water Depth (m)
5
Vertical Water Depth in Settler (m)
0.4
Free Board at the Top (m)
0.5
Reactor Depth
0.55
Reactor Length (m)
43.52
Reactor Width (m)
21.76
Total Sludge Produced (kg/day)
4158.8
Upflow Velocity (m/hr)
0.44
Volumetric Organic Loading (kgCOD/m3/d)
1.16
Gas Collector Dimensions (m):
Height
1.75
Width
1.47
Aperture Width (m)
0.96
Gas Collector inclination with Horizontal (degrees)
50
Biogas Generated (cum/day)
1293.67
Usable Biogas (cum/day)
1013.67
Design
<< Back
Next >>

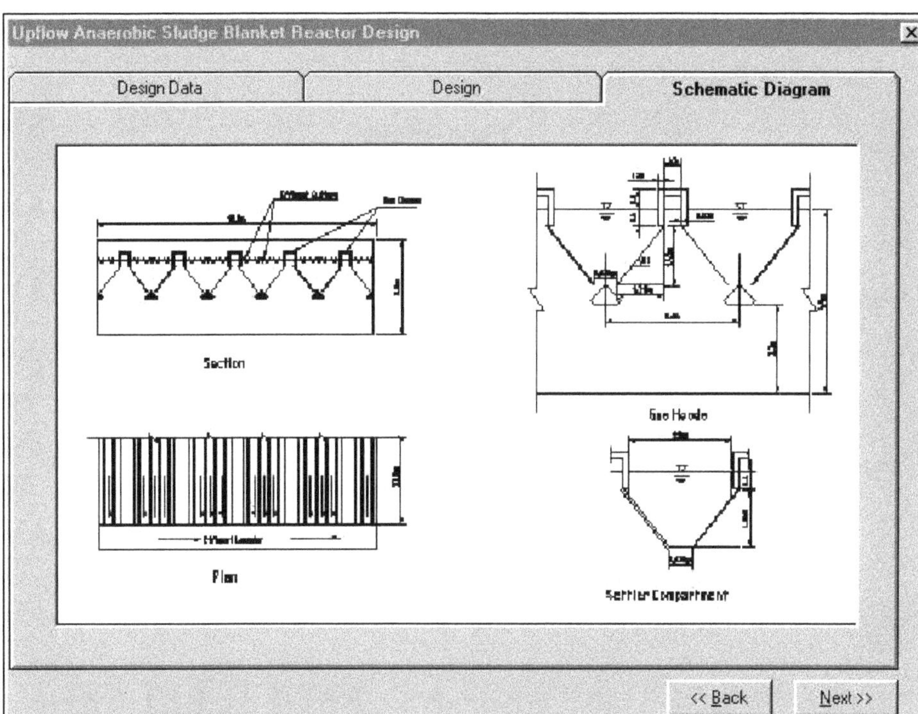
Upflow Anaerobic Sludge Blanket Reactor Design
Design Data
Design
Schematic Diagram
Section
Plan
<< Back
Next >>

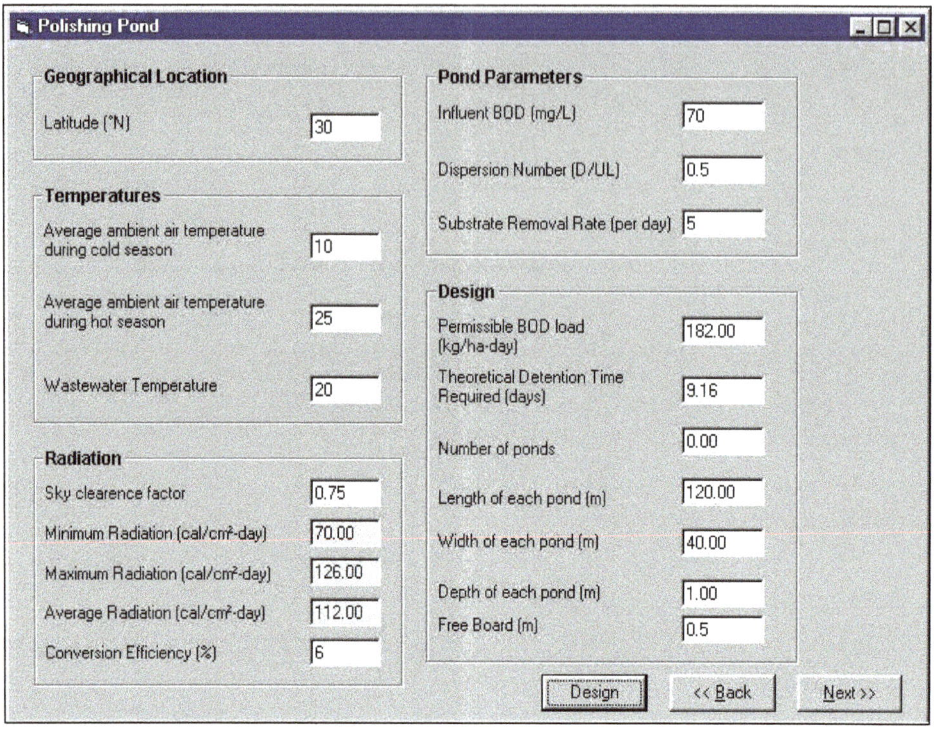
Polishing Pond
Geographical Location
Latitude (°N) 30
Temperatures
Average ambient air temperature during cold season 10
Average ambient air temperature during hot season 25
Wastewater Temperature 20
Radiation
Sky clearence factor 0.75
Minimum Radiation (cal/cm²-day) 70.00
Maximum Radiation (cal/cm²-day) 126.00
Average Radiation (cal/cm²-day) 112.00
Conversion Efficiency (%) 6
Pond Parameters
Influent BOD (mg/L) 70
Dispersion Number (D/UL) 0.5
Substrate Removal Rate (per day) 5
Design
Permissible BOD load (kg/ha-day) 182.00
Theoretical Detention Time Required (days) 9.16
Number of ponds 0.00
Length of each pond (m) 120.00
Width of each pond (m) 40.00
Depth of each pond (m) 1.00
Free Board (m) 0.5
Design
<< Back
Next >>

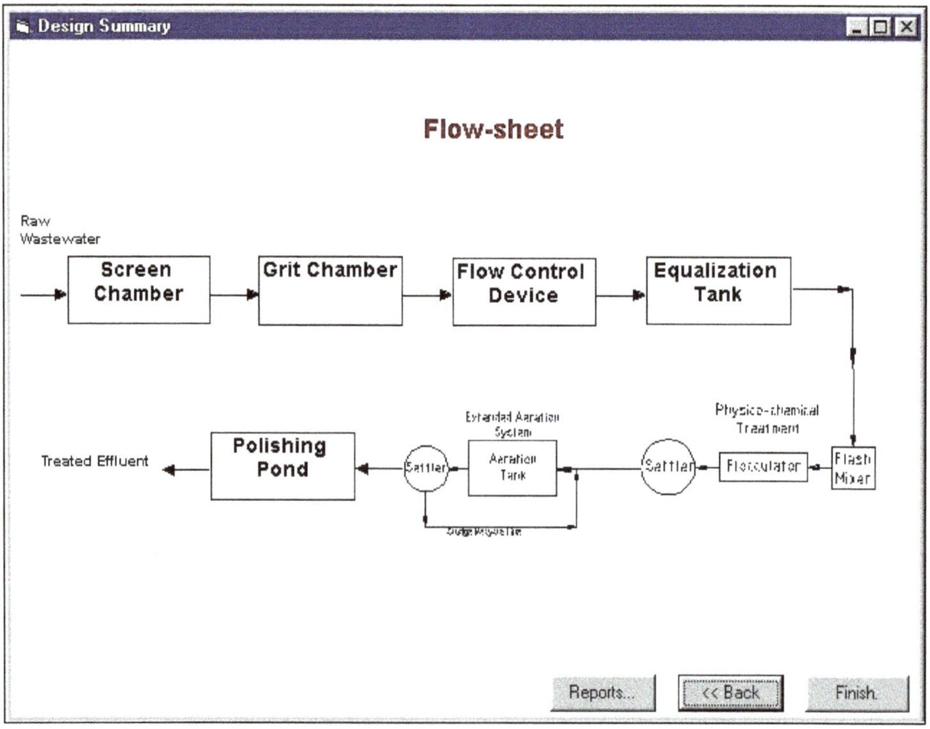
Design Summary
Flow-sheet
Raw Wastewater
Screen Chamber
Grit Chamber
Flow Control Device
Equalization Tank
Physico-chemical Treatment
Extended Aeration System
Flash Mixer
Flocculator
Settler
Aeration Tank
Settler
Polishing Pond
Treated Effluent
Reports...
<< Back
Finish

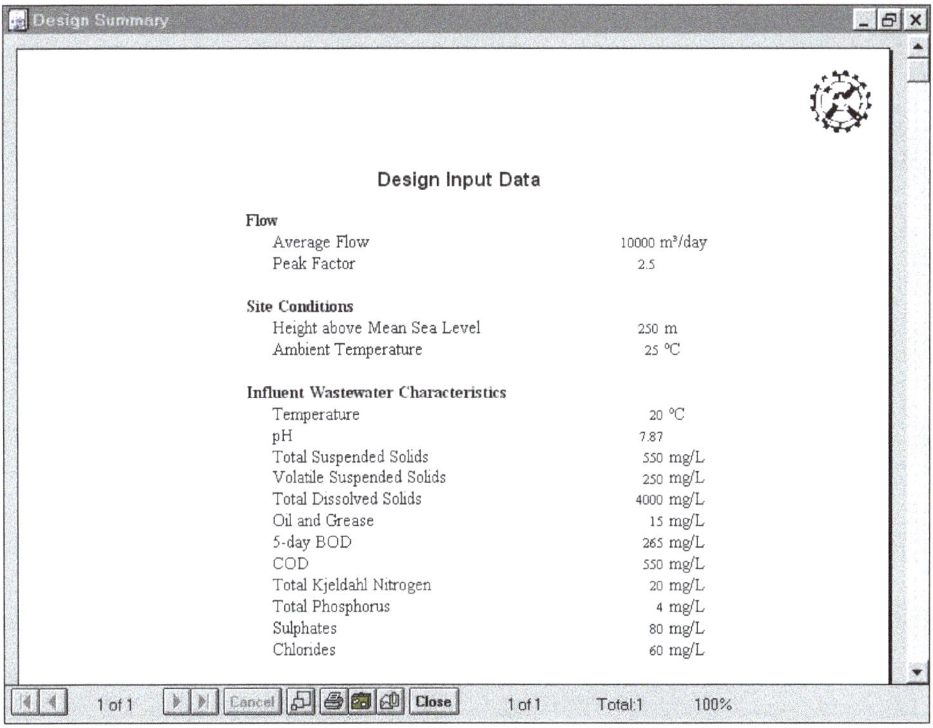

Design Input Data

Flow	
Average Flow	10000 m^3/day
Peak Factor	2.5
Site Conditions	
Height above Mean Sea Level	250 m
Ambient Temperature	25 °C
Influent Wastewater Characteristics	
Temperature	20 °C
pH	7.87
Total Suspended Solids	550 mg/L
Volatile Suspended Solids	250 mg/L
Total Dissolved Solids	4000 mg/L
Oil and Grease	15 mg/L
5-day BOD	265 mg/L
COD	550 mg/L
Total Kjeldahl Nitrogen	20 mg/L
Total Phosphorus	4 mg/L
Sulphates	80 mg/L
Chlorides	60 mg/L

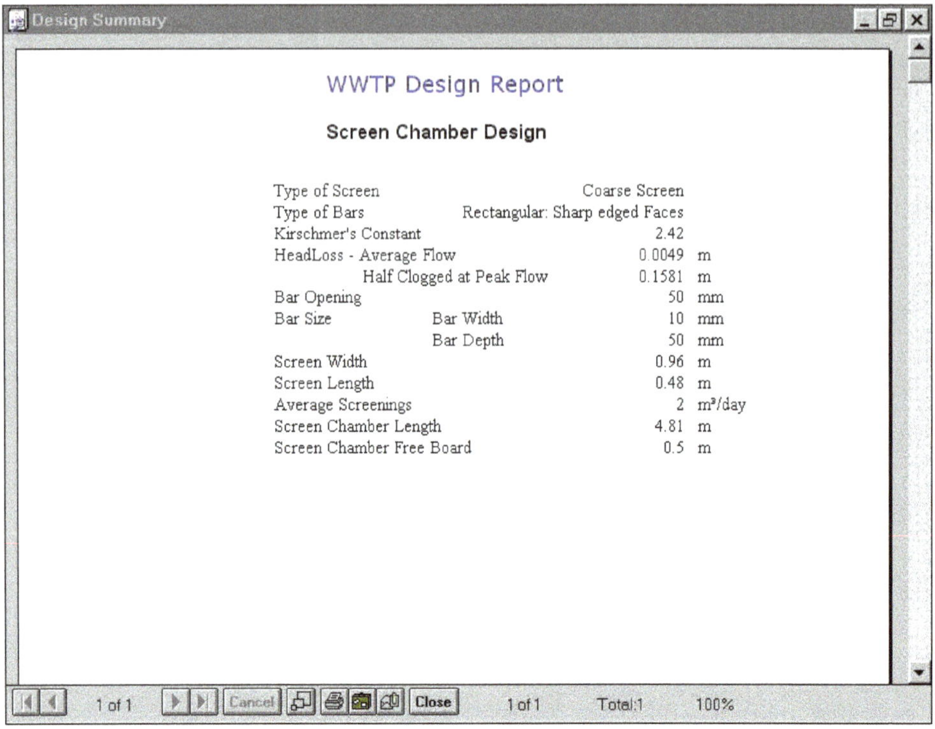

WWTP Design Report

Screen Chamber Design

Type of Screen		Coarse Screen	
Type of Bars		Rectangular: Sharp edged Faces	
Kirschmer's Constant		2.42	
HeadLoss - Average Flow		0.0049	m
Half Clogged at Peak Flow		0.1581	m
Bar Opening		50	mm
Bar Size	Bar Width	10	mm
	Bar Depth	50	mm
Screen Width		0.96	m
Screen Length		0.48	m
Average Screenings		2	m³/day
Screen Chamber Length		4.81	m
Screen Chamber Free Board		0.5	m

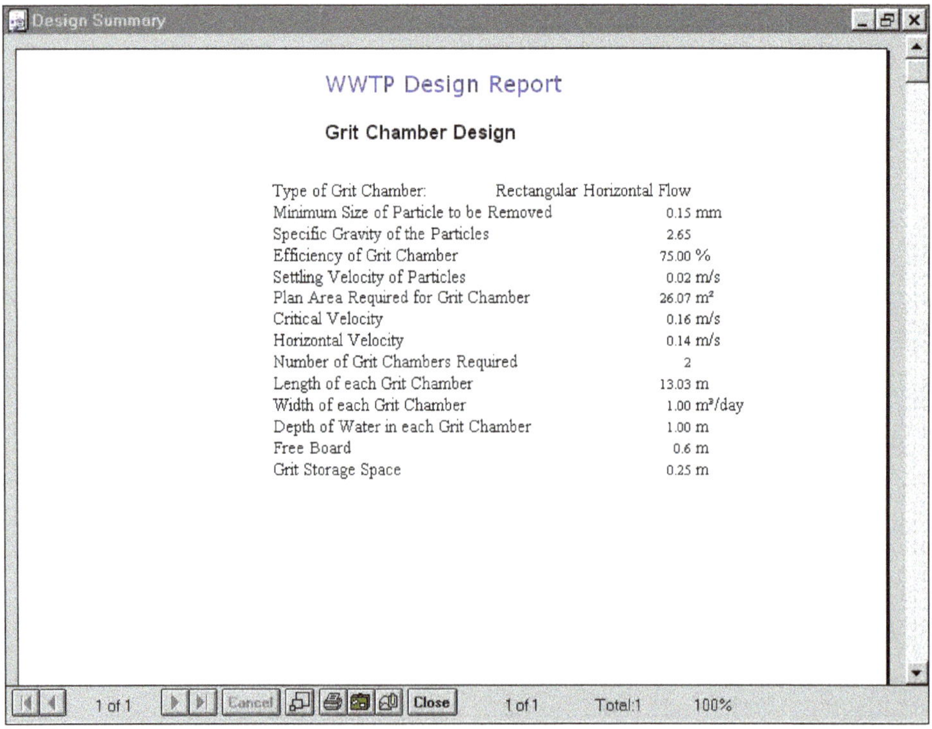

WWTP Design Report

Grit Chamber Design

Type of Grit Chamber:	Rectangular Horizontal Flow	
Minimum Size of Particle to be Removed	0.15	mm
Specific Gravity of the Particles	2.65	
Efficiency of Grit Chamber	75.00	%
Settling Velocity of Particles	0.02	m/s
Plan Area Required for Grit Chamber	26.07	m²
Critical Velocity	0.16	m/s
Horizontal Velocity	0.14	m/s
Number of Grit Chambers Required	2	
Length of each Grit Chamber	13.03	m
Width of each Grit Chamber	1.00	m³/day
Depth of Water in each Grit Chamber	1.00	m
Free Board	0.6	m
Grit Storage Space	0.25	m

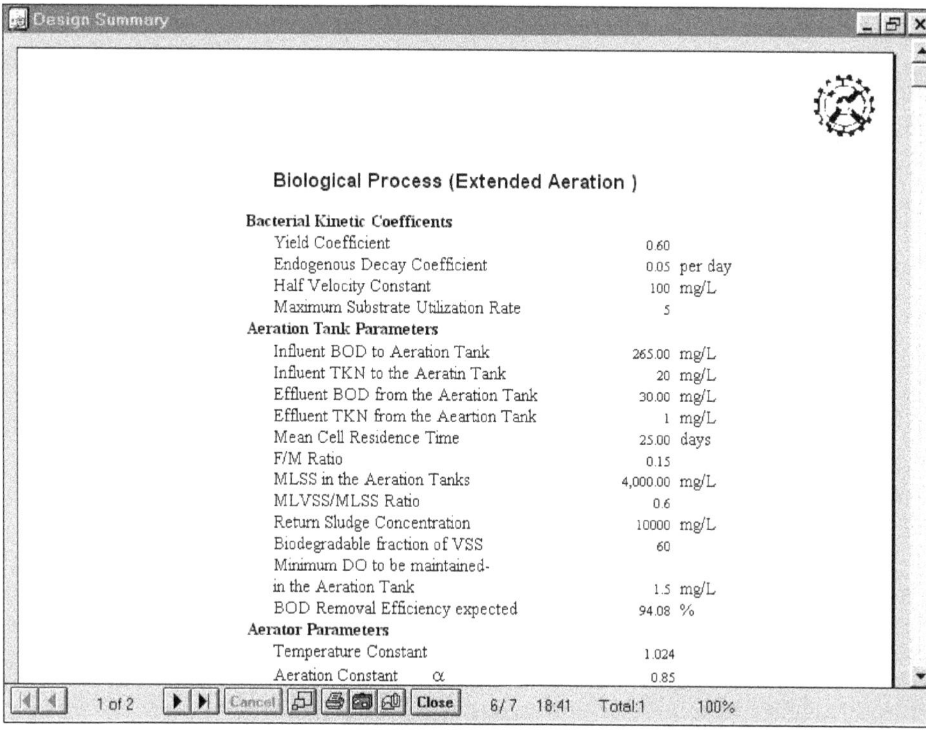

Biological Process (Extended Aeration)

Bacterial Kinetic Coefficents		
Yield Coefficient	0.60	
Endogenous Decay Coefficient	0.05	per day
Half Velocity Constant	100	mg/L
Maximum Substrate Utilization Rate	5	
Aeration Tank Parameters		
Influent BOD to Aeration Tank	265.00	mg/L
Influent TKN to the Aeratin Tank	20	mg/L
Effluent BOD from the Aeration Tank	30.00	mg/L
Effluent TKN from the Aeartion Tank	1	mg/L
Mean Cell Residence Time	25.00	days
F/M Ratio	0.15	
MLSS in the Aeration Tanks	4,000.00	mg/L
MLVSS/MLSS Ratio	0.6	
Return Sludge Concentration	10000	mg/L
Biodegradable fraction of VSS	60	
Minimum DO to be maintained- in the Aeration Tank	1.5	mg/L
BOD Removal Efficiency expected	94.08	%
Aerator Parameters		
Temperature Constant	1.024	
Aeration Constant α	0.85	

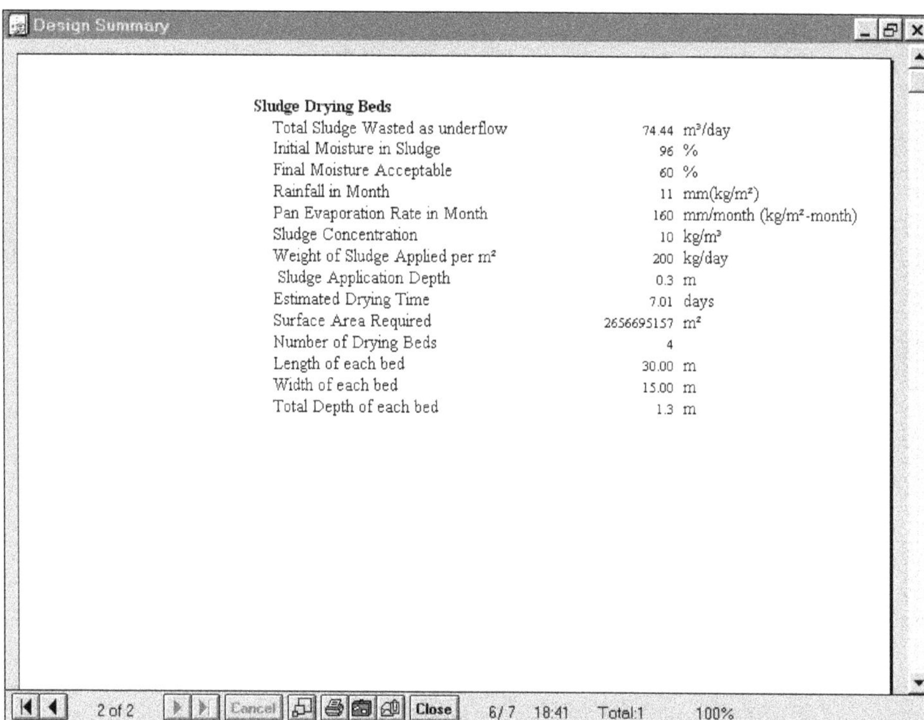

Sludge Drying Beds		
Total Sludge Wasted as underflow	74.44	m³/day
Initial Moisture in Sludge	96	%
Final Moisture Acceptable	60	%
Rainfall in Month	11	mm(kg/m²)
Pan Evaporation Rate in Month	160	mm/month (kg/m²-month)
Sludge Concentration	10	kg/m³
Weight of Sludge Applied per m²	200	kg/day
Sludge Application Depth	0.3	m
Estimated Drying Time	7.01	days
Surface Area Required	2656695157	m²
Number of Drying Beds	4	
Length of each bed	30.00	m
Width of each bed	15.00	m
Total Depth of each bed	1.3	m

Annexure-II
Sample Design Calculations for Typical Wastewater Treatment Plant

This annexure presents the basic calculations to be carried out for design of a treatment plant. In this case study, the raw wastewater characteristics are presented below:

Parameter	*Value*	*Parameter*	*Value*
Flow	20,000 m^3/d	Colour	200
pH	7.6	SS	640
TDS	2000	COD	700
BOD	300		

All results are expressed in mg/l except pH, Colour and Flow.

Colour is expressed in Pt-Co units.

The treatment flow scheme comprises the following unit operations/processes: screen chamber, grit chamber, Parshall flume, equalization tank, extended aeration system, and sludge drying beds. Design procedure/calculations for each of these units are is given below.

(i) Screen chamber

Average flow = 0.2315 m^3/s; Peak flow = 0.463 m^3/s

Velocity through screens–at Average flow = 0.3 m/s

at Peak flow = 0.8 m/s

Net area required for screen-

At average flow = (0.2315/0.3) = 0.7717 m^2

At peak flow = (0.463/0.8) = 0.5787 m^2

Provide net area = 0.772m^2. Provide bars of size 10x50 mm with clear opening 20 mm

Gross area required = [0.772 x 35)/20)] = 1.158 m^2

Providing screen width = 1.5 m, Depth of wastewater = (1.1.58/1.5)

Velocity of flow at up-stream side of screen at peak flow = [0.463/ (0.772 x 1.5)] = 0.4 m/s

Head loss at the screen = 0.0729 (0.8^2–0.4^2) = 0.035 m

For half clogged screen and velocity doubled,

Head loss = 0.0729 (1.6^2–0.4^2) = 0.113 m

For manual removal of screenings, inclination angle of screen with horizontal = 60^0

Gross area screen required = 1.158/sin 60° = 1.337

Length of screens theoretically required = (1.337/1.5)

Provide screen of size 1.0 m x 1.5 m at an inclination to horizontal 60^0 in two channels of width 1.5 m each (one standby).

Provide screen chamber of size 3.0 m x 1.5 m x 1.0 m side water depth and 0.6 m free board for screen. Provide a drop of 0.2 m after screen chamber.

Quantity of screenings = 0.2 m^3/MLD = 4 m^3

(ii) Grit Chamber

Average flow = 0.2315 m^3/s; Peak flow = 0.463 m^3/s; Minimum flow = 0.116 m^3/s

Settling velocity by Stoke's law – $Vs = \frac{g\ x\ d^2}{18\ x\ \upsilon}(Ss - 1)$

where,

g: Acceleration due to gravity (m/s^2)

d: Minimum diameter of the particle to be settled (m)

Ss: Specific gravity of the particle

υ Kinematic viscosity of water (1.003 x 10^{-6} m^2/s at 20°C)

Vs = [9.81 x (0.15 x $10^{-3)2/}$18 x (1.003 x 10^{-6})]* (2.65–1) = 0.0202 m/s

Check for Reynold's number, R = (Vs d/u) = 3.0

Reynold's number is greater than 0.5 hence Stoke's law does not apply.

Applying transition law – (for 0.5 < R < 10^3), Vs = [0.707 (Ss -1) [$(d)^{1.6}$/ $(u)^{0.6}]^{0.714}$

Vs = [0.707 (2.65 -1) [$(0.15 \times 10^{-3})^{1.6}$/ $(1.003 \times 10^{-6})^{0.6}]^{0.714}$ = 0.0178 m/s

Surface Overflow Rate (SOR)

For 100 per cent removal efficiency, SOR = 0.0178 x 24 x 60 x 60
= 1537.9 $m^3/m^2/d$

To calculate actual overflow rate, Assume removal efficiency, e = 75 per cent

e = 1 – [1 + n {Vs/ (Q/A)}]$^{1/n}$ *i.e.* Q/A = (Vs x n)/[$(1–e)^{-n}$ – 1]

Q/A = (1537.9 x 0.125)/ $(1 – 0.75)^{-0.125}$ – 1 = 1016.03 $m^3/m^2/d$

Dimensions of the Grit Chamber

Plan area = Peak flow/SOR = 40000/1016.03 = 39.4 m^2

Provide two channels of 1.5 m width and 13 m length each.

Critical displacement velocity –

V_c = (8k/f) [(Ss–1)g d]$^{1/2}$

= [(8 x 0.04)/0.03]x[(2.65–1) x 9.81 x 0.15 x $10^{-3}]^{1/2}$ = 0.161 m/s

The horizontal velocity should be kept less than critical displacement velocity, V_c.

Assume a depth of 1 m,

V_h = 0.463/ (1 x 1.5 x 2) = 0.15 m/s (< 0.161 m/s, hence okay)

HRT at peak flow is = 10 x 1.5 x 2 x 1/0.463 = 84.2 s

Total depth of grit chamber = Water depth + free board + grits storage space = 1+0.25+0.25 = 1.5m

Provide two grit chambers each of size 13 m x 1.5 m x 1.5 m

(iii) Equalization Tank

Volume of basin by mass curve = 1133 m^3.

Providing detention time = 6 hr, $\text{Volume} = \frac{20000 \times 6}{24} 5000\ m^3$

Providing a circular tank with depth = 4 m,

$$\text{Diameter} = \left[\frac{5000 \times 4}{4 \times 3.142}\right]^{1/2} = 39.89\ m$$

Provide a circular tank of 39.9 m diameter, 4 m depth and a free board of 0.6 m.

Power required for mixing at the rate of 15 KW/1000 m^3-

$$= \frac{15 \times 5000}{1000} = 75\ kW = 100\ HP$$

Provide four aerators of 25 HP each for mixing.

(iv) OPTION–A: Extended Aeration System

A) Aeration Tank

Mean Cell Residence Time, q_c = 25 days

Average Flow, Q = 20,000 m^3/d

Influent BOD, S_o = 300 mg/L

Effluent BOD, S = 20 mg/L

Mixed Liquor Suspended Solids (MLSS), X = 4000 mg/L

Mixed Liquor Volatile Suspended Solids (MLVSS), X_v = 2400 mg/L (X_v/X = 0.6)

Yield Coefficient, Y = 0.6 mg biomass/mg BOD

Endogenous Respiration Coefficient, K_d = 0.05 day^{-1}

1. Volume of Aeration Tank

$$V = \frac{\theta\, Y\, (S_o - S)\, Q}{X_v\, (1 + K_d\, \theta)}$$

$$= \frac{25 \times 0.6\,(300 - 20)\,20000}{2400\,(1 + 0.05 \times 25)} = 15555.56\ \text{m}^3$$

2. *Hydraulic Detention Time (HRT)*

$$\theta = \frac{15555.56}{20000} \times 24 = 18.6\ \text{hrs}$$

3. *Net Increase in MLVSS*

$P_x = Y_{observed} \times Q\,(S_o - S);$

$$Y_{observed} = \frac{Y}{1 + K_d\theta_c} = \frac{0.6}{1 + (0.05 \times 25)} = 0.267$$

$P_x = 0.267 \times 20000\,(300-20) = 1495.2$ kg/d

Mass of sludge wasted through the secondary settling tank effluent:

Assuming the concentration of 30 mg/L in the effluent,

Mass of sludge wasted = $30 \times 10^{-3} \times 20000 = 600$ kg/d

Sludge to be wasted as the underflow from Secondary Settling Tank,

$$= \frac{1495.2}{0.6} - 600 = 1892\ \text{kg/d}$$

Concentration of the sludge = 1 per cent

$$\text{Sludge volume} = \frac{1892}{10} = 189.2\ \text{m}^3/\text{d}.$$

4. *Recirculation Ratio*

$X_v\,(Q + Q_r) = X_{vr} \times Q_r$

where,

Q_r = return sludge flow

X_{vr} = return sludge volatile suspended solids concentration

$2400\,(Q + Q_r) = 6000\,(Q_r) \implies Q/Q_r = 0.67$

5. *Oxygen Requirement*

$$\text{kg}\ O_2/\text{d} = \frac{Q(S_o - S)}{f} - 1.42\,(P_x)$$

$$= \frac{20000(210 - 20)10^{-3}}{0.68} - (1.42 \times 1495.2) = 6112.11\ \text{kg/d}$$

6. Aeration

Provide Mechanical Surface low speed aerators.

O_2 transferred under field conditions:

$$N = N_o \times \frac{\beta C_W - C_L}{C_{S\,20}} \times 1.024^{(T\text{-}20)}\,\alpha$$

where,

N: kg O_2/kWhr transferred under field conditions,

N_o: kg O_2/kWhr transferred in water at 20°C and zero dissolved Oxygen,

β: Salinity- surface tension correction factor,

C_W: O_2 saturation concentration for tap water at given temp. and altitude, mg/L

C_{S20}: Oxygen saturation concentration in tap water 20°C, mg/L

C_L: Operating oxygen concentration, mg/L

T: Temperature, °C

α: Oxygen transfer correction factor for waste

- Temperature = 20 °C
- Salinity of the tap water = 1500 mg/L
- Altitude above MSL = 260 m

O_2 solubility correction factor, F_a = 0.73

$C_W = C_{Sea} \times F_a = 8.999 \times 0.73 = 6.57$ mg/L

$$N = N_o \frac{0.9 \times (6.57 - 2)}{8.999} \times 1.024^{(20-20)} \times 0.82$$

Standard oxygen transfer capability for a surface low-speed aerator = 2 kg O_2/kWhr

∴ O_2 transfer under field conditions is = 2 x 0.36 = 0.72 kg O_2/kWhr

$$\text{Power requirement} = \frac{6112.11}{0.72 \times 24} = 353.7 \text{ kW} = 474\text{HP}.$$

7. Size of the Aeration Tank and Placing of Aerators

Volume of aeration tank required = 15555.56 m^3; Provide a depth of 3.0 m

$$\text{Surface area required} = \frac{15555.56}{3.0} = 5185.2 \text{ m}^2$$

Provide 5 nos. of tanks, each of size 64m x 16m x 3m, with a free board of 0.8m.

Provide 4 aerators each of 25 HP in each tank.

Power available for mixing = 24.4 W/m^3

$$HRT = \frac{5 \times 64 \times 16 \times 3}{20000} \times 24 \text{ hrs} = 18.4 \text{ hrs}$$

$$F/M \text{ ratio} = \frac{(300 \times 10^{-3})\,20000}{15360 \times 2400} = 0.16 \{\text{within } 0.1 - 0.18, \text{ hence okay}\}.$$

(B) Secondary Settling Tank:

MLSS in the aeration tank = 4000 mg/L

1. Solids Loading

(Q + Qr) X = (20000 + 13400) 4000 x 10^{-3} = 133600 kg/d

2. Surface Area Required

Assuming a solids loading rate of 100 $kg/m^2/d$,

$$\text{Surface area required will be} = \frac{133600}{100} = 1336 \text{ m}^2$$

3. Size of the Sedimentation Tank

Provide 2 tanks each of 29m diameter. Provide a depth of 2.5m.

$$\text{Surface overflow rate} = \frac{20000}{2(\pi 29^2/4)} = 15 \text{ m}^3/\text{m}^2/\text{d}$$ (within the range of 8-15 $m^3/m^2/d$, hence okay)

$$\text{Solids loading rate} = \frac{133600}{2(\pi 29^2/4)} = 101 \text{ kg/m}^2/\text{d}$$ (within the range of 25-125 $kg/m^2/d$, hence okay)

$$\text{Detention Time} = \frac{1321 \times 2.5}{20000} \times 24 = 3.9 \text{ hrs.}$$

(C) Sludge Drying Beds

Solids wasted as underflow from secondary settling tank = 189.2 m^3/d

Average drying period = 10 days; Sludge application depth = 0.3m

$$\therefore \text{Area required} = \frac{189.2 \times 10}{0.3} = 6306.7 \text{ m}^2$$

Provide 14 nos. of beds each of size 30m x 16m x 0.3m and a free board of 0.5m

(v) OPTION-B: Physico-chemical Process

(A) Flash Mixer

Design flow = 20000 m^3/d = 0.2315 m^3/s

Detention time for a flash mixer varies between 30–90 seconds

Adopting a detention time of 60 sec,

Volume of the flash mixer = 0.2315 x 60 = 13.89 m^3

Provide a rectangular flash mixer of size–2.3m x 2.3m x 2.7m depth and a free board of 0.6m

B) Clari-flocculator

Flocculator

Detention time = 20 min; Velocity gradient = 40 s^{-1}

Volume = 13.89 x 20 = 277.8 m^3

Assume a depth of 2.5m, Diameter = 11.89m

Provide 11.9m dia. flocculator.

Clarifier

With surface overflow rate = 25 $m^3/m^2/d$; Area required $= \frac{20000}{25} = 800\ m^2$

Clarifier diameter (d); $800 = \frac{\pi}{4}(d^2 - 11.9^2) = 34.09m$

Provide a clarifier of 34m dia. Assume a depth of 3m, Detention time = 2.88 hrs.

(C) Overhead Tanks

Ferrous Sulphate Tank

Dose 450 mg/L; Purity of $FeSO_4$ = 80 per cent; Requirement of a day = 9000/0.8 kg = 11250 kg

For a 10 per cent solution,

Volume of the tank required for the storage of solution for 8 hrs

$$= \frac{11250 \times 10}{3} = 37.5\ m^3$$

Provide 2 tanks of size 3.5m x 3.5m x 2.5m.

Lime Tank

Dose 400 mg/L; Assuming 50 per cent hydroxide, Requirement of a day

$$= \frac{8000}{0.5} = 16000\ kg$$

For a 10 per cent solution,

Volume of the tank required for the storage of solution for 8 hrs =

$$= \frac{16000 \times 10}{3} = 53.3\ m^3$$

Provide 2 tanks of size 3.5m x 3.5m x 2.5m.

Water Tank

Total water requirement in a shift of 8 hrs is = 37.5 + 53.3 = 90.8 m^3

Provide a detention time of 3 hrs, Provide a tank of size 3.5m x 3.5m x 2.5m.

(D) Sludge Drying Beds

Total solids coming to the drying beds = (400 + 450 + 384) x 10^{-3} x 20000 = 24680 kg/d

Sludge concentration = 5 per cent; ∴ Volume of sludge = 493.6 m^3

Assuming a sludge application depth = 0.3m and 10 days drying period,

$$\text{Area required for the beds} = \frac{493.6}{0.3} \times 10 = 16453\ m^2$$

Provide 36 nos. of beds of size 30m x 16m x 0.3m

(E) Holding Tank

Detention time = 1 day; ∴ Volume of tank = 20000 m^3

Provide a depth of 1.5m; ∴ Area = 13333.33 m^2

Provide 2 units each of size 116m x 57m x 1.5m; Provide a free board of 0.5m

(vi) OPTION-C: Conventional Activated Sludge Process

(A) Primary Settling Tank

Average flow = 20000 m^3/d; Suspended solids = 640 mg/L

Assuming a surface overflow rate = 25 $m^3/m^2/d$ at average flow,

$$\text{Area required} = \frac{20000}{25} = 800\ m^2$$

Provide a 32m-diameter tank with a depth of 2.5 m

$$\text{Detention time} = \frac{(\pi/4)32^2 \times 2.5}{20000/24} = 2.4\ \text{hrs}$$

BOD removal in primary settling tank = 30 per cent

SS removal in primary settling tank = 50 per cent

B) Aeration Tank

Mean Cell Residence Time, q_c = 8 days

Average Flow, Q = 20,000 m^3/d

Influent BOD, S_o = 210 mg/L

Effluent BOD, S = 20 mg/L

Mixed Liquor Suspended Solids (MLSS), X = 2500 mg/L

Mixed Liquor Volatile Suspended Solids (MLVSS), X_v = 2000 mg/L

Yield Coefficient, Y = 0.6 mg biomass/mg BOD

Endogenous Respiration Coefficient, K_d = 0.05 day^{-1}.

1. Volume of Aeration Tank

$$V = \frac{\theta\, Y\, (S_0 - S)\, Q}{X_v\, (1 + K_d\, \theta)} = \frac{8 \times 0.6\, (210 - 20)\, 20000}{2000\, (1 + 0.05 \times 8)} = 6514\ m^3$$

2. Hydraulic Detention Time (HRT),

$$\theta = \frac{6514}{20000} \times 24 = 7.8\ hrs$$

Provide a HRT of 6 hrs; ∴ Volume of the aeration tank = 5000 m^3

3. Net Increase in MLVSS

$P_x = Y_{observed} \times Q\, (S_o\text{-}S)$

$$Y_{observed} = \frac{Y}{1 + K_d \theta_c} = \frac{0.6}{1 + (0.05 \times 8)} = 0.43$$

P_x = 0.43 x 20000 (210-20) 10^{-3} = 1634 kg/d

Mass of sludge wasted in the secondary settling tank effluent:

Assuming the concentration of 30 mg/L in the effluent,

Mass of sludge wasted in the effluent = 30 x 10^{-3} x 20000 = 600 kg/d

Sludge to be wasted as the underflow from Secondary Settling Tank,

$$= \frac{1634}{0.8} - 600 = 1442.5 \quad kg/d$$

Concentration of the sludge = 1 per cent

$$\text{Sludge volume} = \frac{1442.5}{10} = 144.25\ m^3/d$$

4. Recirculation Ratio

$X_v (Q + Q_r) = X_{vr} \times Q_r$

where, Q_r = return sludge flow

X_{vr} = return sludge volatile suspended solids concentration

$2000 (Q + Q_r) = 8000 (Q_r)$ ==> $Q_r/Q = 0.33$

5. Oxygen Requirement

$$\text{kg } O_2/d = \frac{Q(S_o - S)}{f} - 1.42\,(P_X)$$

$$= \frac{20000(210 - 20)10^{-3}}{0.68} - (1.42 \times 1634) = 3268 \text{ kg/d}$$

6. Aeration

Provide Mechanical Surface low speed aerators. O_2 transferred under field conditions: -

$$N = N_o \times \frac{\beta C_W - C_L}{C_{S\,20}} \times 1.024^{(T-20)}\,\alpha$$

where,

N: kg O_2/kWhr transferred under field conditions,

N_o: kg O_2/kWhr transferred in water at 20°C and zero dissolved Oxygen,

β: Salinity- surface tension correction factor,

C_W: O_2 saturation concentration for tap water at given temp. and altitude, mg/L

C_{S20}: Oxygen saturation concentration in tap water 20°C, mg/L

C_L: Operating oxygen concentration, mg/L

T: Temperature, °C

α: Oxygen transfer correction factor for waste

- Temperature = 20 °C
- Salinity of the tap water = 1500 mg/L,
- Altitude = 250 m,

Oxygen solubility correction factor, Fa = 0.73

$C_W = C_{Sea} \times Fa = 8.999 \times 0.73 = 6.57$ mg/L

$$N = N_o \frac{0.9 \times (6.57 - 2)}{8.999} \times 1.024^{(20-20)} \times 0.82$$

Standard oxygen transfer capability for a surface low-speed aerator = 2 kg O_2/ kWhr

$\therefore O_2$ transfer under field conditions is = 2 x 0.36 = 0.72 kg O_2/kWhr

$$\therefore \text{ Power requirement} = \frac{3268}{0.72 \times 24} = 189 \text{ kW} = 252 \text{ HP}$$

7. Size of the Aeration Tank and Placing of Aerators

Volume of aeration tank required = 5000 m^3 Providing a depth of 3.0 m

$$\text{Surface area required} = \frac{5000}{3.0} = 1666.67 \text{ m}^2$$

Provide 4 nos. of tanks, each of size 30m x 15m x 3m, with a free board of 0.8m.

Provide 2 aerators each of 30 HP in each tank.

$$\text{HRT} = \frac{4 \times 30 \times 15 \times 3}{20000} \times 24 \text{ hrs} = 6.48 \text{ hrs}$$

$$\text{F/M ratio} = \frac{(210 \times 10^{-3})\, 20000}{5400 \times (2000 \times 10^{-3})} = 0.38 \text{ \{within 0.2 - 0.4, hence okay\}}.$$

(C) Secondary Settling Tank

MLSS in the aeration tank = 2500 mg/L

1. Solids Loading

$(Q + Q_r)$ X = (20000 + 6666.6) 2500 x 10^{-3} = 66666.5 kg/d.

2. Surface Area Required

Assuming a solids loading rate of 100 kg/m^2/d,

$$\text{Surface area required will be} = \frac{66666.5}{100} = 666.67 \text{ m}^2$$

$$\text{Surface overflow rate (SOR) for this area will be} = \frac{20000}{666.67} = 30 \text{ m}^3/\text{m}^2/\text{d}$$

SOR is too large for a secondary settling tank

$$\text{Adopting a SOR} = 15 \text{ m}^3/\text{m}^2/\text{d, Area required} = \frac{20000}{15} = 1333.33\text{m}^2.$$

3. Size of the Sedimentation Tank:

Provide 2 tanks, each of 29m diameter. Provide a depth of 2.5m.

$$\text{Surface overflow rate} = \frac{20000}{2(\pi 29^2/4)} = 15\ m^3/m^2/d$$ (within the range of 8-15 $m^3/m^2/d$, hence okay)

$$\text{Solids loading rate} = \frac{66666.5}{2(\pi 29^2/4)} = 50.5\ kg/m^2/d$$ (within the range of 25-125 $kg/m^2/d$, hence okay)

$$\text{Detention Time} = \frac{1321 \times 2.5}{20000} \times 24 = 3.9\ \text{hrs.}$$

(D) Anaerobic Digester

Primary settled solids = (640–320)10^{-3} x 20000 = 6400 kg/d

Sludge concentration = 3 per cent; $\therefore$ $\text{Volume} = \dfrac{6400}{30} = 213.33\ m^3/d$

Secondary settled solids = 144.25 m^3/d

Total sludge produced per day = 213.33 + 144.25 = 357.58 m^3

Detention time of the digester = 30 days

Digester volume = 357.58 x 30 = 10727.4 m^3

Provide 4 units each of 19 m diameter and depth 10 m

Power required for mixing = 5-8 W/m^3;

$\therefore$ Provide 25 HP drive in each of the digesters.

(E) Sludge Drying Beds

Digested solids from anaerobic digesters = 357.58 m^3/d

Average drying period = 10 days; Sludge application depth = 0.3m

$$\text{Area required} = \frac{357.58 \times 10}{0.3} = 11919.33\ m^2$$

Provide 24 nos. of beds each of size 32m x 16m x 0.3m and a free board of 0.5m.

References

1. ——— (1999). *Process Advisor, Classic Edition, Version 1.6*, Optimal Control Research, http: //www.treat-wastewater.com

2. Alderman, B.J., Theis, T.L. and Collins, A.G. (1998). Optimal Design for Anaerobic Pretretament of Municipal Wastewater, *Journal of Environmental Engineering*, ASCE, Vol. 124, No. 1, pp. 4-10.

3. Al-Ghusain, I.A., Huang, J., Hao, O.J. and Lim, B.S. (1994). Using pH as a Real-time Control Parameter for Wastewater Treatment and Sludge Digestion Processes, *Wat. Sci. Tech.*, Vol. 30, No. 4, pp.159-168.

4. Andrews, J.F. (1993). Modelling and Simulation of Wastewater Treatment Processes, *Wat. Sci. Tech.*, Vol. 28, No. 11-12, pp.141-150.

5. Arceivala, S.J. (1981). *Wastewater Treatment and Disposal: Engineering and Ecology in Pollution Control*, Marcel Dekker, Inc., New York.

6. Arceivala, S.J. (1998). *Wastewater Treatment for Pollution Control*, 2nd Edition, Tata McGraw-Hill Publishing Company Limited, New Delhi.

7. Armiger, W.B., Lee, G.J., Schewler, B.R. Jr. and Mah, T.J. (1993). Control of BNR Processes by Regulation of the VFA/M Ratio, *Wat. Sci. Tech.*, Vol. 28, No. 11-12, pp. 303-314.

8. Ayesa, E., Florez, J., Larrea, L. and Garcia-Heras, J.L. (1993). Evaluation of Sensitivity and Observability of the State Vector for System Identification and Experimental Design, *Wat. Sci. Tech.*, Vol. 28, No. 11-12, pp. 209-218.

9. Balmer, P. and Mattsson, B. (1994). Wastewater treatment plant operation costs, *Wat. Sci. Tech.*, Vol. 30, No. 4, pp.7-15.

10. Baykal, B.B. (1993). Small Activated Sludge Performance and Viability Characteristics under Intermittent and Reduced Flow, *Wat. Sci. Tech.*, Vol. 28, No. 10, pp. 309-316.

11. Bozinis, N.A., Alexiou, *I.E.* and Pistikopoulos, E.N. (1996). A mathematical Model for the Optimal Design and Operation of an Anaerobic Co-digestion Plant, *Wat. Sci. Tech.*, Vol. 34, No. 5-6, pp. 383-392.

12. Brands, E., Liebeskind, M. and Dohmann, M. (1994). Parameters for Dynamic Simulation of Wastewater Treatment Plants with High-rate and Low-rate Activated Sludge Tanks, *Wat. Sci. Tech.*, Vol. 30, No. 4, pp. 211-214.

13. Brouwer, H., Klapwijk, A. and Keesman, K.J. (1994). Modelling and Control of Activated Sludge Plants on the Basis of Respirometry, *Wat. Sci. Tech.*, Vol. 30, No. 4, pp.265-274.

14. Cantet, J., Paul, E. and Clauss, F. (1996). Upgrading Performance of an Activated Sludge Process through Addition of Talqueous Powder, *Wat. Sci. Tech.*, Vol. 34, No. 5-6, pp. 75-83.

15. Cecil, L.K. (1972). Interrelationship between freshwater and wastewater plants, in *Applications of New Concepts of Physical-Chemical Wastewater Treatment*, Sept. 18-22.

16. Chen, J. and Beck, M.B. (1993). Modelling, Control and On-line Estimation of Activated Sludge Bulking, *Wat. Sci. Tech.*, Vol. 28, No. 11-12, pp. 249-256.

17. Collins, A.G., Theis, T.L., Kilambi, S., He, Lin and Pavlostathis, S.G. (1998). Anaerobic Treatment of Low-strength Domestic Wastewater using an Anaerobic Expanded Bed Reactor, *Journal of Environmental Engineering*, ASCE, Vol. 124, No. 7, pp. 652-659.

18. CPHEEO (1993). *Manual on Sewerage and Sewage Treatment* (2nd Edition), Central Public Health and Environmental Engineering Organization, Ministry of Urban Development, New Delhi.

19. Criddle, C.S. (1992). The Kinetics of Co-metabolism, *Biotechnology and Bioengineering*, Vol. 41, No. 11,pp 1048-1056.
20. Deininger, A. (1994). Influence of combined sewage influent on secondary clarifiers of activated sludge plants, *Wat. Sci. Tech.*, Vol. 30, No. 4, pp. 67-70.
21. Droste, S.J. (1997). *Theory and Practice of Water and Wastewater Treatment*, John Wiley and Sons, Inc., New York.
22. Dupont, R. and Sinkjaer, O. (1994). Optimization of Wastewater Treatment Plants by Means of Computer Models, *Wat. Sci. Tech.*, Vol. 30, No. 4, pp.181-190.
23. Eckenfelder, W.W. Jr. (1972). Interrelationships between Biological Treatment and Physical-chemical Treatment, *in Applications of New Concepts of Physical-Chamical Wastewater Treatment*, Sept. 18-22, pp. 75-84.
24. Eckenfelder, W.W. Jr. (1989). *Industrial Water Pollution Control*, 2nd Edition, McGraw-Hill Book Co., New York.
25. Enbutsu, I., Baba, K., Hara, N., Waseda, K. and Nogita, S. (1993). Integration of Multi AI Paradigms for Intelligent Operation Support Systems – Fuzzy Rule Extraction from a Neural Network, *Wat. Sci. Tech.*, Vol. 28, No. 11-12, pp. 333-340.
26. Evans, B. and Laughton, P. (1994). Emerging Trends in Electrical Energy Usage at Canadian (Ontario). Municipal Wastewater Treatment Facilities and Strategies for Improving Energy Efficiency, *Wat. Sci. Tech.*, Vol. 30, No. 4, pp.17-23.
27. Finnson, A. (1993). Simulation of a strategy to start a nitrification at Bromma sewage plant using based on the IAWPRC Model No.1, *Wat. Sci. Tech.*, Vol. 28, No. 11-12, pp. 185-195.
28. Funamizu, N. and Takakuwa, T. (1994). Simulation of the Operating Conditions of the Municipal Wastewater Treatment Plant at Low Temperature using a Model that includes the IAWPRC Activated Sludge Model, *Wat. Sci. Tech.*, Vol. 30, No. 4, pp.105-113.
29. Henze, M., Grady Jr., C.P.L., Gujer, W., Marais, G.v.R., Matsuo, T. (1987). Activated Sludge Model No. 1. IAWQ Scientific and Technical Reports No. 1, IAWQ, London, England.
30. Hjorth, B. (1993). Extended Nutrient Removal in Small One-tank Activated Sludge Plants, *Wat. Sci. Tech.*, Vol. 28, No. 10, pp. 317-327.
31. Hulsman, A. and Swartz, C.D. (1993). Development of an Improved Compact Package Plant for Small Community Wastewater Treatment, *Wat. Sci. Tech.*, Vol. 28, No. 10, pp. 283-288.
32. Jeppsson, U. and Diehl, S. (1996). An evaluation of a dynamic model of the secondary clarifier, *RaRrr*, F. (1996). Improving settling dynamics of activated sludge by adding fine talc powder, *Wat. Sci. Tech.*, Vol. 34, No. 5-6, pp. 19-26.
33. Joshi, S. (1997). *M.E. Dissertation on Study of the Sewage Treatment Plant, Jaipur*; MREC, Rajasthan University, Jaipur

34. Joshi, V. and Khanna, P. (1983). Optimal Design of Municipal Wastewater Treatment Systems using Dynamic Programming, *Asian Environment*, pp. 25-36.

35. Keller, J., Romli, M., Lee, P.L. and Greenfield, P.F. (1993). Dynamic Model Simulation and Verification of a Two-stage High-rate Anaerobic Treatment Process with Recycle, *Wat. Sci. Tech.*, Vol. 28, No. 11-12, pp.197-207.

36. Krebs, P., Stamou, A.I., Garcia-Heras, J.L. and Rodi, W. (1996). Influence of inlet and outlet configuration on the flow in secondary clarifiers, *Wat. Sci. Tech.*, Vol. 34, No. 5-6, pp. 1-9.

37. Kroiss, H. (1994). Design and Design Evaluation of Biological Wastewater Treatment Plants, *Wat. Sci. Tech.*, Vol. 30, No. 4, pp. 1-6, 1994

38. Ladiges, G. and Kayser, R. (1993). On-line and Off-line Expert Systems for the Operation of Wastewater Treatment Plants, *Wat. Sci. Tech.*, Vol. 28, No. 11-12, pp. 315-323.

39. Lefevre, F., Audic, J.M. and Hujon, B. (1993). Automatic Regulation for Activated Sludge Aeration – Single-tank Nitrification-Denitrification, *Wat. Sci. Tech.*, Vol. 28, No. 10, pp. 289-298.

40. Liebeskind, M., Schapers, D., Bornemann, C., Brands, E., Freund, M. and Rolfs, T. (1996). Parameter determination and model fitting – two approaches for modelling processes in wastewater treatment plants, *Wat. Sci. Tech.*, Vol. 34, No. 5-6, pp. 27-33.

41. Lin, S.H., Lin, T.M. and Leu, H.G. (1998). Determination of Oxygen Diffusion Coefficient in Wastewaters, *Journal of Environmental Engineering*, ASCE, Vol. 123, No. 3, pp. 265-271.

42. Mahesh, B.V., Rajesh, H.S. and Manmohan Rao, N. (1992). Simplified Design of Aeration Tanks, *Journal of Indian Association for Environmental Management*, Vol. 19, pp. 96-103.

43. Matsche, N. and Moser, D. (1993). Operation of a Two-Stage Activated Sludge Package Plant for High Efficiency Treatment, *Wat. Sci. Tech.*, Vol. 28, No. 10, pp. 299-307.

44. Metcalf and Eddy, Inc. (1997). *Wastewater Engineering: Treatment, Disposal, and Reuse*, 3rd Edition, Reviewed by Tchobanoglous, G. and Burton, F.L., Tata McGraw-Hill Publishing Co. Ltd., New Delhi.

45. Metzger, M. (1994). Fast–model real time simulator for the wastewater treatment process, *Wat. Sci. Tech.*, Vol. 30, No. 4, pp.191-197.

46. Mhaisalkar, V.A, Bassin, J.K. Paramasivam, R. and Khanna, P. (1993). Dynamic Programming Optimization of Water Treatment Plant Design, *Journal of Environmental Engineering*, ASCE, Vol. 119, No. 6, pp. 1158-1175.

47. Microsoft (1997). *Programmer's Guide: Microsoft Visual Basic Version 5*, Microsoft Corporation, USA.

48. Minami, S., Nagasawa, H., Saito, Y., Yamagishi, M., Hiraoka, M., Takeda, N. and Okajima, S. (1993). Statistical Analysis of Operation Data and Automatic Control

on Fluidized Bed Incineration Plant with Dryer, *Wat. Sci. Tech.*, Vol. 28, No. 11-12, pp. 341-345.

49. Murphy, K.L. and Timpany, P.L. (1967). Design and Analysis of Mixing for an Aeration Tank, *Journal of Sanitary Engineering Division*, ASCE, Vol. 93, No. SA5, pp. 1-15.

50. NEERI (1992). *NEERI-92: Annual Report*, National Environmental Engineering Research Institute, Nagpur

51. NEERI (1994). *Final Report: Performance Evaluation of Sewage Treatment Plants in India*, National Environmental Engineering Research Institute, Nagpur.

52. Park, Y.K. and Lee, C.H. (1996). Dyeing Wastewater Treatment by Activated Sludge Process with a Polyurethane Fluidized Bed Biofilm, *Wat. Sci. Tech.*, Vol. 34, No. 5-6, pp. 193-200.

53. Qasim, S.R. (1985). *Wastewater Treatment Plants: Planning, Design, and Operation*, CBS Publishing Japan Ltd., New York.

54. Ramalho, R.S. (1983). *Introduction to Wastewater Treatment Processes*, Second Edition, Academic Press, Inc. New York.

55. Rasmussen, M.R., Larsen, T. and Clauss, F. (1996). Improving settling dynamics of activated sludge by adding fine talc powder, *Wat. Sci. Tech.*, Vol. 34, No. 5-6, pp. 11-18.

56. Rich, L.G. (1996). Modification of Design Approach to Aerated Lagoons, *Journal of Environmental Engineering*, ASCE, Vol. 122, No. 2, pp. 149-153.

57. Selvapathy, P. and Subhash Babu, P. (1995). Heavy Metals Removal from Wastewater by Duckweed, *3rd International Conference Appropriate Waste Management Technologies for Developing Countries*, NEERI, Nagpur, Feb 25-26, pp. 647-653

58. Shiono, S., Usui, T., Iriyama, M., Suzuki, K. and Naka, Y. (1993). Stable Combustion in Sludge Melting Furnace by Fuzzy Logic Control, *Wat. Sci. Tech.*, Vol. 28, No. 11-12, pp. 347-354.

59. Sincero, A.P. Sr. and Sincero, G.A. (1996). *Environmental Engineering: A Design Approach*, Prentice-Hall, New Jersey.

60. Singh, R.C. (1987). Rational Design of Grit Removing Chambers or Channels, IE (I). Journal-EN, Vol. 67, pp. 43-49.

61. Smith, J.M., Masse, A.N. and Feige, W.A. (1972). Upgrading Existing Wastewater Treatment Plants, *Applications of New Concepts of Physical-Chemical Wastewater Treatment*, Sept. 18-22, pp. 103-117.

62. Sperling, M.v. (1993). Parameter Estimation and Sensitivity Analysis of an Activated Sludge Model using Monte-Carlo Simulation and the Analyst's Involvement, *Wat. Sci. Tech.*, Vol. 28, No. 11-12, pp. 219-229.

63. Stimson, K.R. (1993). Activated Sludge Control Adviser, *Wat. Sci. Tech.*, Vol. 28, No. 11-12, pp. 295-302.

64. Stokes, L., Takacs, I., Watson, B. and Watts, J.B. (1993). Dynamic Modelling of an ASP Sewage Works – A Case Study, *Wat. Sci. Tech.*, Vol. 28, No. 11-12, pp.151-161.

65. Thornberg, D.E. and Thomsen, H.A. (1994). Interaction Between Computer Simulation and Control using On-line Nitrogen Measurements, *Wat. Sci. Tech.*, Vol. 30, No. 4, pp.199-206.

66. Tsai, Y.P., Ouyang, C.P., Wu, M.Y. and Chiang, W.L. (1993). Fuzzy Control of A Dynamic Activated Sludge Process for the Forecast and Control of Effluent Suspended Solid Concentration, *Wat. Sci. Tech.*, Vol. 28, No. 11-12, pp. 355-367.

67. Van Vuuren, L. R. J. and Henzen, M. R. (1972). Process selection and cost of advanced wastewater treatment in relation to the quality of secondary effluents and quality requirements for various uses, in *Applications of New Concepts of Physical-Chamical Wastewater Treatment*, Sept. 18-22, pp.371-383.

68. Walton, W.C. (1995). *Designing Groundwater Models with Windows*, Lewis Publishers, Boca Raton.

69. Watanabe, S., Baba, K., Yoda, M., Wu, W.C., Enbutsu, I., Hiraoka, M. and Tsumura, K. (1993). Intelligent Operation Support System for Activated Sludge Process, *Wat. Sci. Tech.*, Vol. 28, No. 11-12, pp. 325-332.

70. Watts, J.B. and Garber, W.F. (1993). On-line respirometry: A powerful tool for activated sludge plant operation and. design, *Wat. Sci. Tech.*, Vol. 28, No. 11-12, pp. 389-399.

www.ingramcontent.com/pod-product-compliance
Ingram Content Group UK Ltd.
Pitfield, Milton Keynes, MK11 3LW, UK
UKHW022214051026
15624UKWH00001B/152

9 789351 309116